AF352848

Brain Function and Health, Sports, and Exercise

Brain Function and Health, Sports, and Exercise

Editors

Filipe Manuel Clemente
Ana Filipa Silva

MDPI • Basel • Beijing • Wuhan • Barcelona • Belgrade • Manchester • Tokyo • Cluj • Tianjin

Editors
Filipe Manuel Clemente
Escola Superior de Desporto e Lazer
Instituto Politécnico de Viana do Castelo
Viana do Castelo
Portugal

Ana Filipa Silva
Escola Superior de Desporto e Lazer
Instituto Politécnico de Viana do Castelo
Melgaço
Portugal

Editorial Office
MDPI
St. Alban-Anlage 66
4052 Basel, Switzerland

This is a reprint of articles from the Special Issue published online in the open access journal *Brain Sciences* (ISSN 2076-3425) (available at: www.mdpi.com/journal/brainsci/special_issues/ Brain_Health_Sports_Exercise).

For citation purposes, cite each article independently as indicated on the article page online and as indicated below:

LastName, A.A.; LastName, B.B.; LastName, C.C. Article Title. *Journal Name* **Year**, *Volume Number*, Page Range.

ISBN 978-3-0365-3866-2 (Hbk)
ISBN 978-3-0365-3865-5 (PDF)

Contents

About the Editors

Filipe Manuel Clemente

Filipe Manuel Batista Clemente has been a university professor since the 2012/2013 academic year and is currently an assistant professor at Escola Superior de Desporto e Lazer de Melgaço (IPVC, Portugal). Filipe holds a Ph.D. in Sports Sciences –Sports Training from the University of Coimbra; his dissertation entitled "Towards a new approach to match analysis: understanding football players'synchronization using tactical metrics"involved observation and match analysis in soccer

As scientific merit, Filipe has had 264 articles published and/or accepted by journals indexed with an impact factor (JCR), as well as over 105 scientific articles that have been peer-reviewed indexed in other indexes. In addition to scientific publications in journals and congresses, he is also the author of six international books and seven national books in the areas of sports training and football. He has also edited various special editions subordinate to sports training in football in journals with an impact factor and/or indexed in SCImago. Additionally, he is a frequent reviewer for impact factor journals in quartiles 1 and 2 of the JCR.

Although he started producing research in 2011, he was included in the restricted list of the world's most-cited researchers in the world (where only eight other Portuguese researchers in sports sciences appear), which was published in the journal Plos Biology in 2020. In 2021, the list was updated, with Filipe Manuel Clemente being again included in the top 2% of the world researchers, in which he was positioned in the second place in six Portugueses included in the area of sports sciences. Filipe M. Clemente's SCOPUS h-index is 24 (with a total of 2605 citations), and his Google h-index is 35 (5305 citations). In a list promoted by independent website Expert Escape, he was ranked 40th of 14,875 researchers of football (soccer) in 2020 and in 19th of 15949 in 2021.

Ana Filipa Silva

Ana Filipa Silva currently works as Assistant Professor in the Polytechnic Institute of Viana do Castelo (IPVC) in Melgaço, Portugal and as a researcher in Research Centre in Sports Sciences, Health Sciences and Human Development (CIDESD, Portugal). She has a European Ph.D. in Sport Sciences –Sports Training in Faculty of Sport, University of Porto, Porto, Portugal (2017, Portugal). Among others, her main publications are within the following topics: (i) motor control, (ii) youth sports performance, (iii) decision making in sports, and (iv) cognitive performance in sports. She is currently guest associate editor for Frontiers in Physiology, Frontiers in Sport and Active Living and Human Movement. She is guest editor of different Special Issues: (i) Training and Performance in Youth Sports at International Journal Environmental Research and Public Health; (ii) In Search of Individually Optimal Movement Solutions in Sport: Learning between Stability and Flexibility at Frontiers in Physiology and Frontiers in Sport and Active Living and Human Movement; (iii) Decision-Making in Youth Sport Flexibility at Frontiers in Physiology and Frontiers in Sport and Active Living and Human Movement; and (iv) Children's Exercise Physiology, Volume II at Frontiers in Physiology and Frontiers in Sport and Active Living and Human Movement.

Preface to "Brain Function and Health, Sports, and Exercise"

Sports and exercise have been related to acute and chronic changes in brain health and function. Regular exercise has been used as a non-pharmacological approach for protecting brain health while improving some brain functions. With benefits observed in young and old individuals and healthy and clinical populations, sports and exercise seem to play an important role in contributing to brain health and function. Despite some evidence regarding the contributions of sports and exercise to brain health and function, there is an increasing number of original research papers and systematic reviews with or without meta-analysis that may help professionals to identify which types of sport and exercise are suitable for specific improvements and the adequate duration of carrying out such activities. Additionally, there is space for further analysis of the contribution of sports and exercise to both the improvement of efficiency in work and to the mitigation of the effects of specific neurodegenerative diseases. Therefore, the Special Issue "Brain Function and Health, Sports, and Exercise"welcomed contributions from different areas of knowledge that may assist in improving our understand ing of the relationships between sports and exercise and brain health and function. Original studies, systematic reviews, and meta-analysis on the following main topics were received: (i) role of exercise in neurodegenerative diseases; (ii) role of sport and exercise in cognitive performance; (iii) role of sport and exercise in brain health; (iv) effects of different sport and exercise modes on brain function and health; and (v) dose–response relationships between exercise and brain health and function.

Filipe Manuel Clemente and Ana Filipa Silva
Editors

Article

Acute Effect of a Simultaneous Exercise and Cognitive Task on Executive Functions and Prefrontal Cortex Oxygenation in Healthy Older Adults

Manon Pellegrini-Laplagne [1,2], Olivier Dupuy [1,3,*], Philippe Sosner [1,4,5] and Laurent Bosquet [1,3]

1   Laboratoire MOVE (UR 20296), Faculté des Sciences du Sport, Université de Poitiers, TSA 31113, CEDEX 9, 96073 Poitiers, France; manon.pellegrini.laplagne@gmail.com (M.P.-L.); psosner@monstade.fr (P.S.); laurent.bosquet@univ-poitiers.fr (L.B.)
2   Société Rev'Lim, 87000 Limoges, France
3   Faculty of Medicine, School of Kinesiology and Physical Activity Sciences (EKSAP), University of Montreal, Montreal, QC HC3 3J7, Canada
4   Centre Médico-Sportif MON STADE, 75013 Paris, France
5   AP-HP Hôtel-Dieu, Centre de Diagnostic et de Thérapeutique, 75004 Paris, France
*   Correspondence: olivier.dupuy@univ-poitiers.fr; Tel.: +33-(0)5-49-45-33-43

Abstract: The rapid increase in population aging and associated age-related cognitive decline requires identifying innovative and effective methods to prevent it. To manage this socio-economic challenge, physical, cognitive, and combined stimulations are proposed. The superiority of simultaneous training compared to passive control and physical training alone seems to be an efficient method, but very few studies assess the acute effect on executive function. This study aimed to investigate the acute effect of simultaneous physical and cognitive exercise on executive functions in healthy older adults, in comparison with either training alone. Seventeen healthy older adults performed three experimental conditions in randomized order: physical exercise, cognitive exercise, and simultaneous physical and cognitive exercise. The protocol involved a 30 min exercise duration at 60% of theoretical maximal heart rate or 30 min of cognitive exercise or both. Executive functions measured by the Stroop task and pre-frontal cortex oxygenation were assessed before and after the intervention. We found a main effect of time on executive function and all experimental condition seems to improve inhibition and flexibility scores (<0.05). We also found a decrease in cerebral oxygenation ($\Delta[HbO_2]$) in both hemispheres after each intervention in all cognitive performance assessed ($p < 0.05$). Simultaneous physical and cognitive exercise is as effective a method as either physical or cognitive exercise alone for improving executive function. The results of this study may have important clinical repercussions by allowing to optimize the interventions designed to maintain the cognitive health of older adults since simultaneous provide a time-efficient strategy to improve cognitive performance in older adults.

Keywords: physical exercise; cognitive exercise; simultaneous training; healthy aging; executive function; cerebral oxygenation

Citation: Pellegrini-Laplagne, M.; Dupuy, O.; Sosner, P.; Bosquet, L. Acute Effect of a Simultaneous Exercise and Cognitive Task on Executive Functions and Prefrontal Cortex Oxygenation in Healthy Older Adults. *Brain Sci.* **2022**, *12*, 455. https://doi.org/10.3390/brainsci12040455

Academic Editors: Ana Filipa Silva and Filipe Manuel Clemente

Received: 3 February 2022
Accepted: 18 March 2022
Published: 28 March 2022

Publisher's Note: MDPI stays neutral with regard to jurisdictional claims in published maps and institutional affiliations.

1. Introduction

Age-related decline in cognitive performance, such as a decrease in processing speed, attention, and executive functions can be associated with cognitive disorders such as dementia [1–4]. Considering the demographic evolution in many countries, the increase in the incidence of cognitive impairment in the near future is expected to double within the next 20 years [5]. Consequently, there is a need to identify innovative and effective treatments to prevent, reduce, and delay functional impacts of normal age-related cognitive decline on the daily life of healthy older adults. In this context, one of the main challenges consists of developing efficient strategies to strengthen the "cognitive reserve" of the elderly [6]. Fortunately, drug-free interventions aimed at mitigating cognitive decline are attracting great research and societal interest. To date, three strategies have been used to prevent cognitive decline and dementia: (i) cognitive training, (ii) exercise training, and (iii) simultaneous cognitive and exercise training [7]. Due to the positive impact of separate cognitive and physical training, Kraft put forward the hypothesis that simultaneous cognitive and physical training could induce a better cognitive improvement than physical training or cognitive training alone [8]. This hypothesis has been supported by a recent meta-analysis by Gavelin et al., 2021 who reported that combined cognitive and exercise training was more efficient to improve cognition performances than each stimulation alone [9].

For several years, a growing body of literature has investigated the acute effect of a single bout of aerobic exercise on cognitive functions. Although this effect has been mostly described in healthy young adults [10–14], the interest in the specific response of older adults is more recent [15–17]. The main observation is a positive impact of acute exercise on executive functions [16], the magnitude of the effect being larger in older vs. younger adults [17]. Surprisingly, the acute effect of a task combining simultaneously a physical and a cognitive exercise has received very little attention. To the best of our knowledge, the work by Ji et al., 2019 [18] is the single one that experimental session involving a cognitive exercise, a physical exercise, a combined cognitive and physical exercise, and a reading exercise that served as a control condition. Pre-frontal cortex oxygenation was measured during a Stroop task test that was performed before and after each exercise. The authors failed to report greater effect for combined modality and showed an improvement in executive performance after the physical exercise and the combined cognitive and physical exercise, but not after the cognitive exercise and the control condition. Interestingly, they also reported greater oxygenation of the pre-frontal cortex after the combined physical and cognitive exercise in comparison with the other conditions in post-exercise condition. These results provide interesting insights into the adaptation process of the pre-frontal cortex to an acute bout of combined cognitive and physical exercise, but they still need to be confirmed [18]. The lack of superiority of simultaneous exercise on cognitive functions in the elderly reported by Ji et al., 2019 [18] can potentially be explained by the cognitive exercise used. Indeed, these authors used verbal fluency as a cognitive exercise and tested its effect on Stroop performance, particularly on flexibility. It is known to date that cognitive training has a positive effect on the stimulated cognitive function only and the transfer to another cognitive process is difficult [19]. It is now important to test the effect of multiple cognitive exercises by exercising several cognitive and executive functions.

The aim of this study was therefore to investigate the acute effect of simultaneous physical and cognitive exercise using multiple stimulated cognitive functions on executive functions in healthy older adults in comparison with either exercise alone. Based on Kraft theory [8], we hypothesized that simultaneously combined cognitive and physical exercise would be effective in improving executive performance, that the magnitude of improvement would be greater than either exercise mode alone, and that this effect would be mediated by a greater increase in pre-frontal cortex oxygenation.

2. Methods

2.1. Population

In total, 17 men ($n = 5$) and women ($n = 12$) aged between 57 and 69 years old were recruited to participate in this study. Their characteristics are presented in Table 1. None of them suffered from cognitive impairment (as defined as a MOCA score inferior or equal to 24) or from cardiovascular, metabolic, neurological, or psychiatric diseases. Participants who have been prescribed with pharmacological treatment that could modify cardiovascular or neuromuscular functions were excluded. The protocol was reviewed and approved by a national ethics committee for non-interventional research (CERSTAPS # 17 November 2017) and was conducted by recognized ethical standards and national/international laws. All participants signed a written statement of informed consent.

Table 1. Anthropometric, blood pressure, and cognitive characteristics of participants.

Sex	Women	12
	Men	5
Age		62.4 ± 3.9
Height (cm)		166 ± 8.1
Weight (Kg)		69.0 ± 12.9
Arterial pressure (Mean of 3 measures)	Systolic	124.0 ± 13.8
	Diastolic	77.3 ± 11.4
MOCA		27.9 ± 1.7
BDI		3.0 ± 3.55

Results are presented: mean ± SD. BDI: Beck's Depression Inventory. MoCA: Montreal Cognitive Assessment.

2.2. Study Design

Participants completed four different experimental sessions. The first session consisted of familiarization with the computerized modified Stroop task test and a submaximal intensity exercise test to determine the power necessary to achieve 60% of theoretical maximal heart rate (*t*HRmax). In the following sessions, participants performed the computerized modified Stroop task before and after a 30 min bout of either cognitive (CE), physical (PE), or combined exercise (SE), in random order. The period separating each of these three experimental visits was one week. Each participant was tested each time at the same time of day. Pre-frontal cortex oxygenation was measured continuously during the cognitive test. Considering the observation by Chang et al., 2012 that greater benefits of physical exercise on cognitive performance were observed 15 min after exercise cessation, the Stroop task was performed 10 min after each condition cessation [20]. The study design is presented in Figure 1.

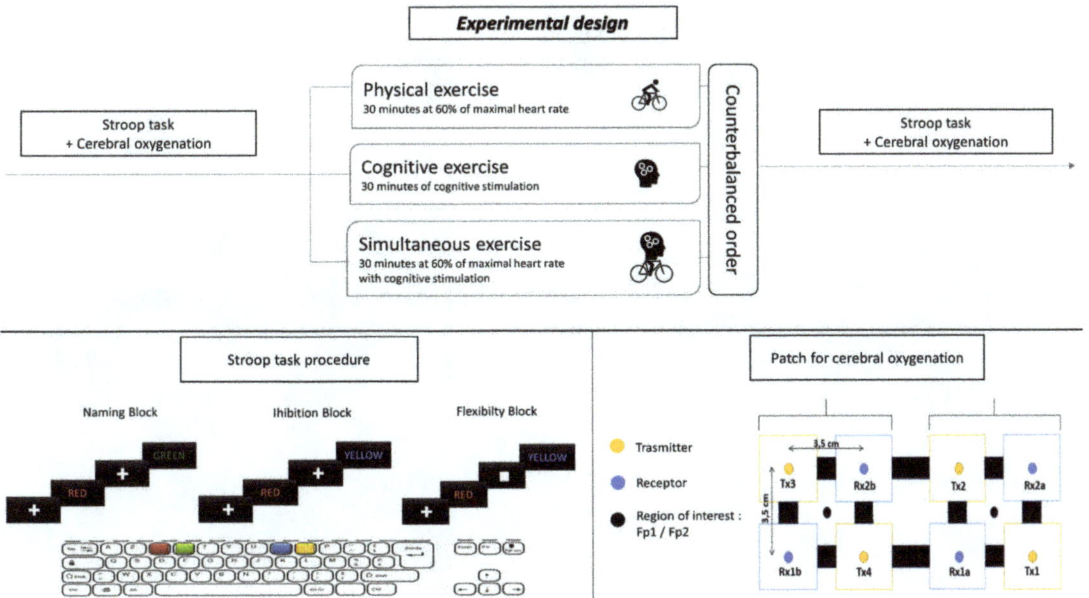

Figure 1. Illustration of experimental design, Stroop task and patch used for cerebral oxygenation.

2.2.1. Submaximal Intensity Exercise Test

This test was performed on a stationary bicycle (Monark LC6, Monark Exercise AB, Vansbro, Sweden). Theoretical maximal heart rate was determined by the formula proposed by Gellish and colleagues [21]:

$$tHR_{max} = 207 - 0.7 * age \tag{1}$$

The test began at 30 W and increased by 20 W every two minutes until the participant reached 60% of tHR_{max}. Participants had to pedal at 60 rotations per minute (rpm). Oxygen uptake (VO_2 mL·kg^{-1}·min^{-1}) was determined continuously on a 30-s basis using a portable cardiopulmonary exercise testing system (Metalyser Cortex 3B, CORTEX Biophysik GmbH, Germany). Gas analyzers were calibrated before each test using ambient air and a gas mixture of known concentrations (15% O_2 and 5% CO_2). The turbine was calibrated before each test using a 3-l syringe at several flow rates. The highest VO_2 over a 30-s period during the last stage was considered as the oxygen uptake at 60% of tHR_{max} ($VO_{2\ peak(60\%)}$, mL·kg^{-1}·min^{-1}). Heart rate was measured continuously using a heart rate monitor (Polar RS800 cx, Polar Electro, Kempele, Finland).

2.2.2. Physical Exercise

Participants completed a 30 min session of stationary bicycle at 60% of their tHR_{max}, using a specific device (Velo-Cognitif, REV'LIM, Limoges, France, Figure 2). The corresponding power was estimated during the submaximal intensity exercise test. Heart rate was recorded continuously to adjust power output during the session. The choice of exercise intensity of 60% of tHR_{max} is based on some observations suggesting that cognitive performance was improved at this intensity [22] and fulfils the recommendations by Lauenroth et al., 2016 [23].

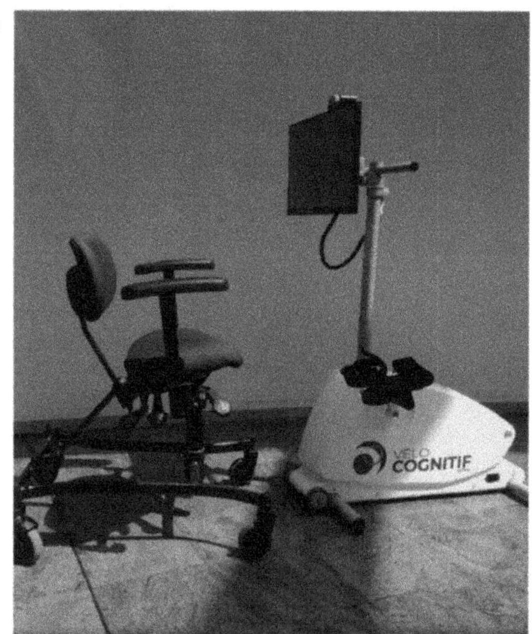
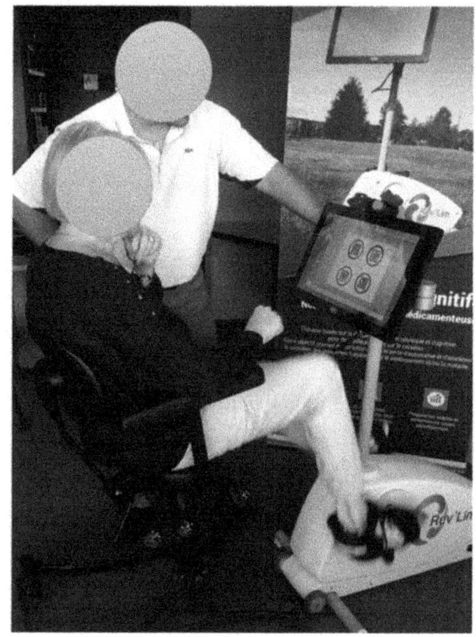

Figure 2. (**A**) Illustration of 'Velo-Cognitif', (**B**) example of participants during the simultaneous physical and cognitive exercises on 'Velo-Cognitif'.

2.2.3. Cognitive Exercise

Participants completed a 30 min session of cognitive games in PRESCO (HappyNeuron, Grenade Sur Garonne, France). This software was already used by [24]. The cognitive exercise consisted of 4 different games (among 32 possible games), which were the same for all participants but with a level of difficulty adapted to their cognitive capacities. These 4 games were used to train mainly executive functions. The testing order, the objective, and the description of each cognitive game are presented in Table 2. The procedure of the cognitive exercise session is illustrated in Figure 3. Each cognitive game had an approximate duration of 7 min, and 30 s of rest was used between each game. Cognitive games were presented on a touch screen, and subjects were asked to perform these games by responding on the touch screen (Figure 2).

Table 2. Cognitive exercise (game) used to perform cognitive training.

Name of The Cognitive Game	Testing Order	Cognitive Function	Aim of the Cognitive Exercise
Catch the ladybird	1	Processing speed	Press a ladybird as quickly as possible
'Vive l'alternance'	2	Flexibility-Working memory	Based on the principle of the Trail Making Test Sort a list of words in alphabetical order and a list of numbers in ascending order, alternating between the two lists.
'Hanoi Tower'	3	Planification	Principle of the Hanoi Tower test with different degrees of difficulty.
N-back	4	Working memory	Principle of the n-back test with different elements (numbers, shapes, colours) and different degrees of difficulty.

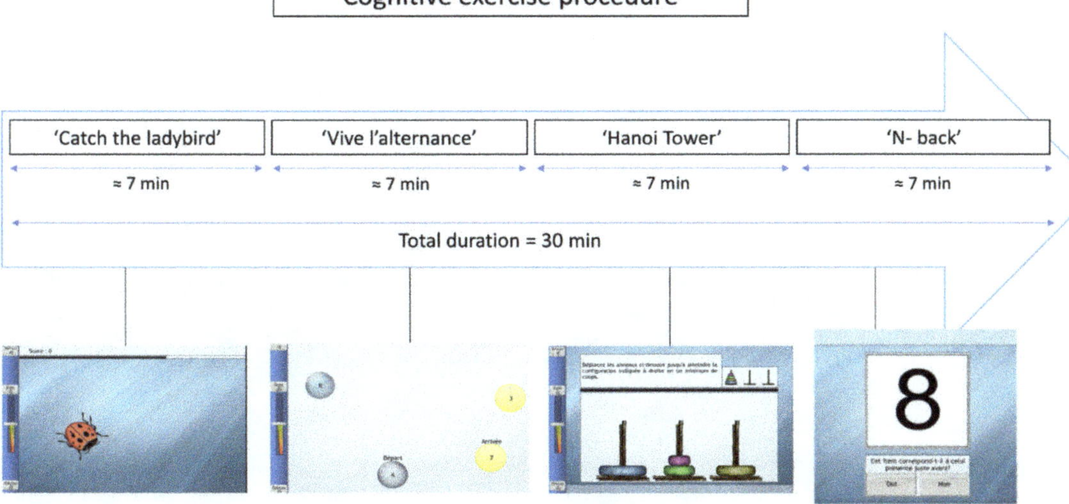

Figure 3. Illustration of cognitive exercise procedure.

2.2.4. Simultaneously Combined Physical and Cognitive Exercise

Participants performed a 30 min session of simultaneously combined physical and cognitive exercise, using a specific device (Velo-Cognitif, REV'LIM, Limoges, France, Figure 2) and the same modalities as previously described for the cognitive exercise and physical exercise. Generally, the cognitive exercise started after 30 s of the beginning of pedaling.

2.3. Neuropsychological Assessment

2.3.1. Global Cognitive Test

The Montreal Cognitive Assessment test (MOCA) was used to evaluate global intellectual efficiency. Briefly, this test consists of a 30-point test divided into 8 parts and 14 subtests. It is a very complete test because it targets most cognitive functions: visuospatial skills, executive functions, attention, working memory, short-term memory, delayed recall, and language. A score higher than 26 is a normal score, and below this, people have a mild cognitive impairment [25]. The dependent variable was the total score.

2.3.2. Executive Functions

The Computerized Modified Stroop task was used to assess executive functions. The test used in this study is based on the Modified Stroop Color Test and included three experimental conditions: naming, inhibiting, and switching. This task was already used in several articles [26–29] with young and older people. Each block lasted between 2–4 min and was interspersed with 60-s resting blocks. Overall, there were 3 experimental task blocks (1 naming, 1 inhibition, and 1 switching) and 2 resting blocks, for a total length between 8 and 14 min. In total, there were 60 Naming trials (Block 1), 60 Inhibition trials (Block 2), and 60 Switching trials (Block 3). All trials began with a fixation cross (or square for switching condition) for 1.5 s, and all visual stimuli appeared in the center of the computer screen for 2.5s. Participants responded with two fingers (index and major finger) from each of their hands on an AZERTY keyboard. In the Naming block, participants were presented with a visual stimulus of the name of colors (RED/BLUE/GREEN/YELLOW) in French presented in the color that is congruent with the word (i.e., RED presented in red ink). Participants were asked to identify the color of the ink with a button press. In the Inhibition block, each stimulus consisted of a color word (RED/BLUE/GREEN/YELLOW) printed in the incongruent ink color (e.g., the word RED was presented in blue ink). Participants were asked to identify the color of the ink (e.g., blue). In the Switching block, in 25% of the trials, a square replaced the fixation cross. When this occurred, participants were instructed to read the word instead of identifying the color of the ink (e.g., RED). As such, within the Switching block, there were both inhibition trials in which the participant had to inhibit their reading of the word and correctly identify the color of the ink, and there were switch trials in which the participant had to switch their response mode to read the word instead of identifying the color of the ink when a square appeared before the word presented. Visual feedback on performance was presented after each trial. A practice session was completed before the acquisition run to ensure the participants understood the task. The practice consisted of a shorter version of the task. Dependent variables were reaction times (ms) and the number of errors committed (%). We also calculated two scores, the first being the inhibition score, which is the result of the inhibition block's score minus naming blocks' performances on number or error, and corresponds to pure inhibition capacities. The second, flexibility cost, is equal to the results of flexibility block's score minus inhibition blocks' performances. The order of handover of executive blocks was counterbalanced between subjects. The Stroop task is presented in Figure 1.

2.4. Prefrontal Cortex Oxygenation Measurement

Cerebral oxygenation and, more precisely, the concentration changes in $[HbO_2]$ ($\Delta[HbO_2]$), HHb ($\Delta[HHb]$ and $\Delta[THb]$) were recorded during the Stroop test with the OxyMon fNIRS system (Artinis Medical Systems, Elst, Netherlands). This system utilizes near-infrared light, which penetrates the skull and brain but is absorbed by hemoglobin (Hb) chromophores in capillary, arteriolar, and venular beds (Ferrari & Quaresima, 2012) [30]. The light was transmitted with two wavelengths, 764 and 857 nm, and data were sampled with a frequency of 10 Hz. This procedure measures relative changes in $[HbO_2]$ and $[HHb]$

using the modified Beer–Lambert law. This law takes into account the differential path-length factor (DPF), which is determined using the following formula: DPF ($\lambda = 807$ nm, A) $= 4.99 + 0.067 \times$ (age 0.814). In our study, the DPF ranged from 5.69 to 6.60. The region of interest is the Fp1 and Fp2 of the prefrontal cortex (PFC). The patch used in this study is presented in Figure 1 and used eight optical channels, comprising four emitters and four receptors, covering the right and left DLPFC and ventrolateral PFC (VLPFC) (Brodmann areas, BAs 9/46 and 47/45/44), which were located using the 10/20 international system. The distance between each emitter and receptor was 3.5 cm. The sensors were shielded from ambient light with a black cloth. Oxysoft version 3.0 (Artinis Medical Systems, Elst, Nether-lands) was used for data collection. Initially resting PFC oxygenation was acquired in a seated position for 2 min before each Stroop task (before and after experimental conditions). Because continuous-wave technology does not allow quantifying absolute concentration due to the incapacity of measuring optical path lengths, the mean of [HbO$_2$], [HHb], and [THb] during the total duration of each block of the Stroop were compared to the minute preceding each block. Our participants were asked to always face forward during the test, avoid making a sudden head movement, clenching their jaw, frowning, and other facial expressions. This procedure was already used by two different teams [26–28,31].

2.5. Statistical Analysis

Standard statistical methods were used for the calculation of means and standard deviations. Normal Gaussian distribution of the data was verified by the Shapiro–Wilks's test and homoscedasticity by a modified Levene Test. Repeated measures ANOVA was used to test the interaction between time and exercise condition on cognitive performance. When the main effect was found, a Bonferroni post-hoc test was performed. All statistical analyses were made with SPSS 17.0, and all statistical analyses with a *p*-value < 0.05 were considered significant. Effects sizes (ES) were also calculated with Hedges' g formula, previously described by Dupuy et al., 2015, and interpreted with Cohen's scale, where EF ≤ 0.2 (trivial), >0.2 (small), >0.5 (moderate), and >0.8 (large) [26].

3. Results

3.1. Participants

The study included nineteen healthy adults aged between 57 and 69 years old. We removed two of them because of health considerations during the study. The final sample included 17 participants (12 women and 5 men). Their characteristics are presented in Table 1.

3.2. Cognitive Assessment

All cognitive results are presented in Table 3. We found no difference between con-dition and no interaction between time and condition in the congruent condition of the Stroop test (i.e., naming). In contrast, we found a main effect of time on the executive performance, including the inhibition reaction time (RT) ($F_{(1.45)} = 7.4$, $p < 0.01$, ES $= -0.18$) and the flexibility errors number ($F_{(1.44)} = 6.65$, $p = 0.01$, ES $= -0.31$) and RT ($F_{(1.45)} = 16.5$, $p < 0.01$, ES $= -0.24$)]. Regarding inhibition cost, we found no main effect of time [errors number ($F_{(1.44)} = 1.37$, $p = 0.24$); RT ($F_{(1.44)} = 0.1$, $p = 0.80$)]. In contrast, we found a main effect of time on the RT flexibility cost score ($F_{(1.44)} = 4.17$ $p = 0.04$, ES $= -0.15$) and no interaction between time and condition ($p = 0.12$). For all significant results, the effects sizes are calculated and presented in Figure 4. Based on the ES analysis (Figure 4), we found no effect of physical exercise (ES $= -0.01$) and cognitive exercise (ES $= 0.11$) on flexibility cost, whereas we found a moderate effect of combined cognitive and physical exercise (ES $= -0.67$).

Table 3. Cognitive performance before and after intervention.

		Overall		PE		CE		SE		ANOVA Analysis	
		Pre	Post	Pre	Post	Pre	Post	Pre	Post	Main Effect (p Value)	Interaction (p Value)
Naming	Errors (nb)	0.35 ± 0.62	0.50 ± 0.81	0.47 ± 0.71	0.53 ± 0.87	0.35 ± 0.61	0.69 ± 0.94	0.24 ± 0.56	0.30 ± 0.60	0.14	0.45
	RT (ms)	809.8 ± 121.3	796.8 ± 149.2	805.1 ± 133.7	798.1 ± 165.7	819.2 ± 119.8	811.6 ± 151.3	805.2 ± 116.7	781.5 ± 137.0	0.08	0.83
Inhibition	Errors (nb)	0.73 ± 1.8	0.36 ± 0.53	1.06 ± 2.9	0.24 ± 0.4	0.65 ± 1.1	0.44 ± 0.6	0.47 ± 0.9	0.41 ± 0.5	0.59	0.69
	RT (ms)	962.6 ± 197.5	926.8 ± 185.2	975.8 ± 231.7	928.6 ± 212.1	976.4 ± 185.6	926.6 ± 171.3	935.6 ± 180.3	925.3 ± 181.2	<0.01	0.23
Inhibition Cost	Errors (nb)	0.36 ± 1.7	-0.14 ± 0.9	0.58 ± 2.6	-0.29 ± 0.9	0.25 ± 1.1	-0.25 ± 1.2	0.23 ± 1.0	0.11 ± 0.8	0.24	0.61
	RT (ms)	156.4 ± 139.6	135.2 ± 107.5	174.5 ± 175.1	133.9 ± 112.3	160.8 ± 126.5	130.4 ± 100.4	134.0 ± 115.3	141.1 ± 115.1	0.70	0.59
Flexibility	Errors (nb)	4.0 ± 4.8	2.7 ± 3.2	4.1 ± 5.3	2.3 ± 2.8	4.1 ± 5.5	3.4 ± 4.0	3.7 ± 3.6	2.5 ± 2.6	0.01	0.83
	RT (ms)	1220.3 ± 210.5	1161.1 ± 222.5	1232.1 ± 200.7	1183.6 ± 211.4	1216.8 ± 199.3	1186.0 ± 241.8	1212.0 ± 241.2	1114.0 ± 218.9	<0.01	0.40
Flexibility Cost	Errors (nb)	3.3 ± 5.1	2.3 ± 3.1	3.1 ± 6.5	2.1 ± 2.6	3.5 ± 5.3	2.8 ± 4.0	3.2 ± 3.2	2.1 ± 2.6	0.06	0.98
	RT (ms)	285.7 ± 140.5	250.6 ± 154.8	281.5 ± 101.5	267.3 ± 104.8	276.7 ± 170.2	281.7 ± 205.8	298.9 ± 148.7	202.7 ± 133.3	0.01	0.17

Results are presented: mean $\pm$ standard deviation; RT: reaction time; PE: Physical Exercise; CE: Cognitive exercise; SE: simultaneous exercise; nb: number.

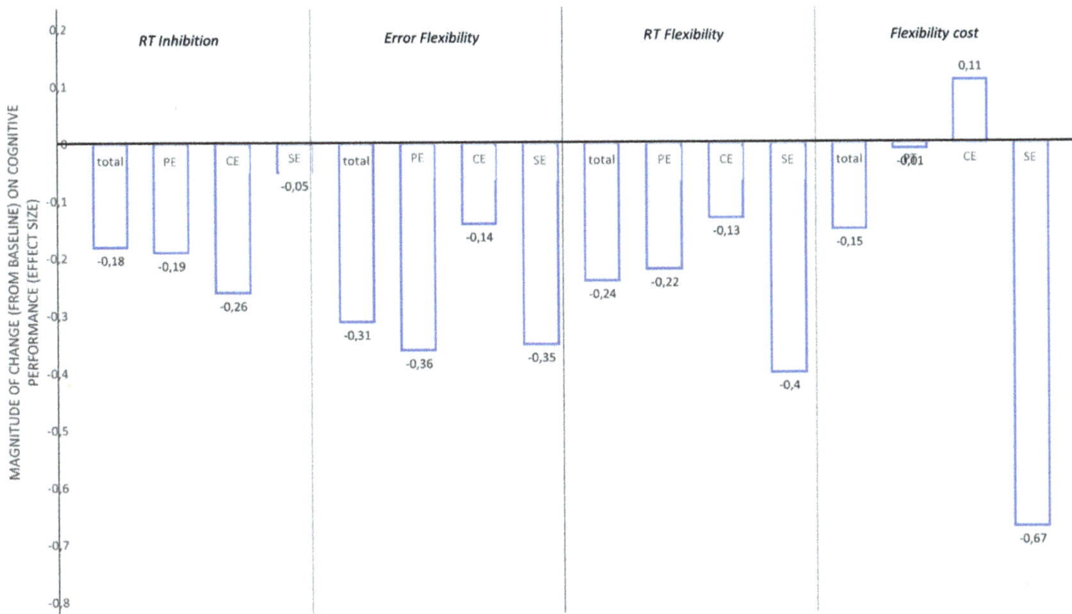

Figure 4. Magnitude of change (from baseline) on cognitive performance (Effect size). Negative effect size indicates a reduced reaction time and a lesser error produced after each intervention. PE: Physical Exercise; CE: Cognitive exercise; SE: simultaneous exercise.

3.3. PFC Oxygenation

All results for right and left hemispheres are presented in the Table 4. The results for the whole PFC are presented in the Figure 5. Repeated measured ANOVA revealed a main effect of time on $\Delta[HbO_2]$ and on $\Delta[THB]$ during naming, inhibition and flexibility conditions of the Stroop test. More precisely, we found a decrease in $[\Delta HbO_2]$ during naming ($F_{(1.44)} = 10.6$, $p < 0.01$, ES $= -0.51$), inhibition ($F_{(1.44)} = 7.22$, $p = 0.01$, ES $= -0.44$) and flexibility ($F_{(1.44)} = 8.5$, $p < 0.01$, ES $= -0.45$). Similarly, total $\Delta[THB]$ decreased during naming ($F_{(1.44)} = 9.7$, $p < 0.01$, ES $= -0.58$), inhibition ($F_{(1.44)} = 7.2$, $p < 0.01$, ES $= -0.49$) and flexibility ($F_{(1.44)} = 4.28$, $p = 0.04$, ES $= -0.30$). In contrast, total $\Delta[HHB]$ remained stable during each condition of the Stroop test [naming ($F_{(1.44)} = 0.05$, $p = 0.81$), inhibition ($F_{(1.44)} = 0.19$, $p = 0.65$) and flexibility ($F_{(1.44)} = 0.31$, $p = 0.57$)]. We found no differences between the left and right hemispheres.

Table 4. Cerebral oxygenation during Stroop test before and after intervention for the right and left hemisphere.

			Overall		PE		CE		SE		ANOVA Analysis	
			Pre	Post	Pre	Post	Pre	Post	Pre	Post	Main Effect (p Value)	Interaction (p Value)
Naming	Left	$\Delta[HbO_2]$ ($\mu mol \cdot L^{-1}$)	0.04 ± 1.2	-0.53 ± 1.1	-0.04 ± 1.5	-0.80 ± 1.2	0.19 ± 0.99	-0.27 ± 1.1	-0.002 ± 1.2	-0.54 ± 0.86	**<0.01**	0.63
		$\Delta[Hbb]$ ($\mu mol \cdot L^{-1}$)	0.20 ± 0.58	0.17 ± 0.47	0.16 ± 0.75	0.25 ± 0.50	0.29 ± 0.53	0.05 ± 0.29	0.15 ± 0.45	0.20 ± 0.58	0.92	0.58
		$\Delta[THb]$ ($\mu mol \cdot L^{-1}$)	0.25 ± 1.0	-0.36 ± 1.2	0.12 ± 0.92	-0.54 ± 1.36	0.48 ± 0.94	-0.22 ± 1.0	0.15 ± 1.1	-0.33 ± 1.1	**<0.01**	0.89
	Right	$\Delta[HbO_2]$ ($\mu mol \cdot L^{-1}$)	0.05 ± 1.0	-0.40 ± 0.94	-0.11 ± 1.1	-0.58 ± 0.98	0.11 ± 0.94	-0.16 ± 0.84	0.14 ± 1.1	-0.47 ± 0.99	**<0.01**	0.87
		$\Delta[Hbb]$ ($\mu mol \cdot L^{-1}$)	0.23 ± 0.67	0.20 ± 0.49	0.21 ± 0.87	0.35 ± 0.52	0.27 ± 0.58	0.07 ± 0.40	0.20 ± 0.58	0.18 ± 0.53	0.74	0.46
		$\Delta[THb]$ ($\mu mol \cdot L^{-1}$)	0.28 ± 0.99	-0.20 ± 1.1	0.11 ± 1.2	-0.23 ± 1.2	0.38 ± 0.78	-0.1 ± 0.85	0.34 ± 1.0	-0.28 ± 1.4	**<0.01**	0.91
Inhibition	Left	$\Delta[HbO_2]$ ($\mu mol \cdot L^{-1}$)	-0.08 ± 1.0	-0.57 ± 1.0	0.14 ± 1.2	-0.28 ± 1.0	-0.13 ± 0.99	-0.59 ± 1.1	-0.24 ± 0.96	-0.82 ± 0.93	**0.02**	0.90
		$\Delta[Hbb]$ ($\mu mol \cdot L^{-1}$)	-0.05 ± 0.36	-0.05 ± 0.49	-0.01 ± 0.37	-0.15 ± 0.42	0.05 ± 0.40	0.05 ± 0.53	-0.18 ± 0.27	-0.06 ± 0.53	0.86	0.53
		$\Delta[THb]$ ($\mu mol \cdot L^{-1}$)	-0.13 ± 0.99	-0.62 ± 1.1	0.13 ± 1.0	-0.43 ± 1.01	-0.08 ± 0.96	-0.54 ± 1.4	-0.42 ± 0.95	-0.88 ± 0.71	**0.03**	0.88
	Right	$\Delta[HbO_2]$ ($\mu mol \cdot L^{-1}$)	0.02 ± 1.0	-0.35 ± 0.87	0.03 ± 0.96	-0.04 ± 0.99	-0.03 ± 1.0	-0.43 ± 0.77	0.07 ± 1.2	-0.57 ± 0.81	**0.02**	0.50
		$\Delta[Hbb]$ ($\mu mol \cdot L^{-1}$)	0.03 ± 0.36	-0.001 ± 0.50	-0.01 ± 0.40	-0.09 ± 0.43	0.11 ± 0.31	-0.04 ± 0.26	-0.01 ± 0.39	0.12 ± 0.72	0.55	0.64
		$\Delta[THb]$ ($\mu mol \cdot L^{-1}$)	0.06 ± 1.0	-0.35 ± 0.87	0.02 ± 0.76	-0.13 ± 1.1	0.08 ± 0.99	-0.47 ± 0.75	0.06 ± 1.4	-0.45 ± 0.80	**0.02**	0.82
Flexibility	Left	$\Delta[HbO_2]$ ($\mu mol \cdot L^{-1}$)	0.55 ± 1.0	0.12 ± 0.94	0.98 ± 1.1	0.18 ± 0.96	0.50 ± 0.73	0.11 ± 1.0	0.21 ± 1.1	0.07 ± 0.93	**<0.01**	0.06
		$\Delta[Hbb]$ ($\mu mol \cdot L^{-1}$)	-0.34 ± 0.46	-0.22 ± 0.69	-0.43 ± 0.49	-0.18 ± 0.62	-0.38 ± 0.41	-0.15 ± 0.92	-0.22 ± 0.49	-0.35 ± 0.47	0.54	0.65
		$\Delta[THb]$ ($\mu mol \cdot L^{-1}$)	0.21 ± 1.0	-0.1 ± 1.3	0.55 ± 1.0	-0.003 ± 1.2	0.12 ± 0.79	-0.04 ± 1.6	-0.01 ± 1.2	-0.28 ± 0.98	0.09	0.36
	Right	$\Delta[HbO_2]$ ($\mu mol \cdot L^{-1}$)	0.66 ± 0.91	0.31 ± 0.91	0.89 ± 0.84	0.37 ± 1.0	0.65 ± 0.75	0.25 ± 0.90	0.47 ± 1.1	0.31 ± 0.89	**<0.01**	0.21
		$\Delta[Hbb]$ ($\mu mol \cdot L^{-1}$)	-0.31 ± 0.44	-0.20 ± 0.49	-0.37 ± 0.55	-0.19 ± 0.58	-0.37 ± 0.35	-0.25 ± 0.46	-0.18 ± 0.40	-0.18 ± 0.47	0.70	0.96
		$\Delta[THb]$ ($\mu mol \cdot L^{-1}$)	0.35 ± 1.0	0.10 ± 0.98	0.52 ± 1.1	0.18 ± 1.1	0.28 ± 0.87	0.01 ± 0.80	0.28 ± 1.1	0.12 ± 1.1	**0.03**	0.34

PE: Physical Exercise; CE: Cognitive exercise; SE: simultaneous exercise. Bold indicates significant results.

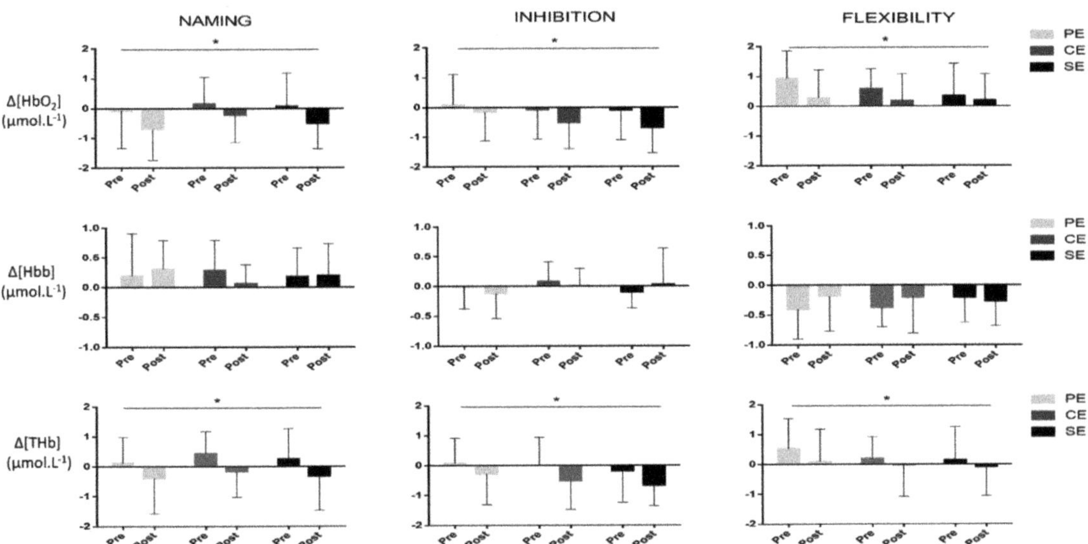

Figure 5. Cerebral oxygenation during Stroop test before and after intervention for the whole PFC (Prefrontal Cortex); * significant ($p < 0.05$). PE: Physical Exercise; CE: Cognitive Exercise; SE: simultaneous Exercise.

4. Discussion

This study aimed to investigate the acute effect of simultaneous physical and cognitive exercise on executive functions in healthy older adults, in comparison with either exercise alone. Based on Kraft theory [8], we hypothesized that simultaneously combined cognitive and physical exercise would be effective in improving executive performance and that the magnitude of improvement would be greater than either exercise mode alone. We also hypothesized that this effect would be mediated by a greater increase in pre-frontal cortex oxygenation. Contrary to our hypotheses, we do not find either in our cognitive results, or in our cerebral oxygenation results, a superior effect of simultaneous exercise compared to physical or cognitive exercises alone. Our main findings were (1) an improvement in executive performance (i.e., inhibition and flexibility conditions of the Stroop task) after each exercise condition; (2), a larger decrease in flexibility cost (based on the ES) after simultaneous cognitive and physical exercise; and (3) a decrease in $\Delta[HbO_2]$ and $\Delta[THb]$ after each exercise condition, both in the left and right PFC.

The effect of acute physical exercise on the cognitive performance of older adults has been summarized in systematic reviews and meta-analyses [15–17]. The Stroop task is probably the test that is the most widely used to assess the acute effect of physical exercise on cognitive performance. The results of the literature seem unclear with (i) specific positive effects on executive functions such as inhibition, (ii) effects only on processing speed, or (iii) even a general effect. It would seem that these contradictory effects are due to the multitude of exercise modalities used. Of the six studies identified by McSween's systematic review (2019), exercise duration ranged from 10 min to 30 min using either cycling, walking, g, or stepping exercise modalities [15]. The Stroop task that we use contained three blocks that allow us to evaluate naming, inhibition, and flexibility functions. Our results showed that the exercise condition alone induces a beneficial effect on the reaction time of the inhibition and flexibility block, as well as the flexibility cost. However, we do not observe any effect on the processing speed on the inhibition cost. Our results validated the hypothesis that executive functions are more sensitive to acute physical exercise as shown by several authors [32–34]. Indeed, results from Abe et al., 2018, Hyodo et al., 2012, and

Johnson et al., 2016 suggested an enhancement of the higher levels of executive functions assessed by the Stroop interference in the post-exercise condition, which is consistent with previous findings in young adults [35,36]. In addition, these authors reported no enhancement of information processing speed. Furthermore, our results validate the hypothesis that the most 'demanding' functions are more sensitive to the effect of acute exercise. Indeed, several researchers [37–39] suggest that more demanding tasks are likely to be more sensitive to the effects of physical exercise in comparison with automatic effortless tasks, which supports the results of Abe et al., 2018, Hyodo et al., 2012 and Johnson et al., 2016 [32–34].

The acute effects of cognitive exercise or simultaneous cognitive and physical exercise on cognitive function in the elderly have been little studied. To our knowledge, only Ji et al., 2019 [18] have studied this effect and reported a positive effect on inhibition and flexibility processes. Our results also confirm this effect. However, like Ji et al., 2019 [18] our results do not corroborate Kraft's hypothesis [8] that double stimulation could have a greater effect than other conditions alone. It should be remembered that this hypothesis was formulated on long-term chronic effects and that we tested this hypothesis on acute effects after a single session. However, although our statistical approach did not report interaction x time on executive function, we can observe the largest effect (based on the ES) of the simultaneous stimulation on the flexibility cost compared to other stimulations (cognitive and physical alone). This result seems encouraging and confirms that this stimulation induces a beneficial effect and can be considered as a cognitive enhancement time efficiency strategy for elderly people.

The effects observed in this study on cerebral oxygenation showed a decrease in cerebral oxygenation during the three blocks of the Stroop task after three types of exercise. Indeed, compared to baseline, we observe a smaller increase in $\Delta[HbO_2]$ and $\Delta[THB]$ during the naming, inhibition, and flexibility tasks. The effect of acute exercise on brain oxygenation during a cognitive task has been reviewed recently by Herold et al., 2018 [40]. Usually, in the vast majority of studies, greater brain oxygenation is observed during a cognitive task after exercise. However, this finding was observed in young subjects in most studies, which potentially explains the difference with the results obtained in our study. In addition, several studies that report a greater cerebral oxygenation during a cognitive task after a physical exercise uses a baseline before exercise and not just before each cognitive task. This procedure unfortunately does not allow us to assess only the cortical activity. However, Murata et al., 2015 [41] observed the same result after exercise of comparable intensity (i.e., 50% of VO_{2max}). The oxygenated hemoglobin concentration quantified across the whole brain was lower after exercise, and this was the case for go trials and no-go trials. More precisely, the oxygenated hemoglobin concentration in the dorsolateral prefrontal cortex and the supplementary motor area was significantly lower after exercise. These authors hypothesized that brain activity is less important than before exercise for the go and the no-go tasks [41]. This hypothesis can be explained by the hypothesis provided by Audiffren et al., 2008 [42] who explains that acute exercise improves 'arousal' explaining the enhancement of cognitive performance in post-exercise conditions. The facilitating effect of exercise on cognitive performance may explain the lower cortical activity observed in our study [42]. This mechanism was observed after a period of chronic physical training where less brain activity was observed during a cognitive task. Coetsee et al., 2017 observed the same results with fNIRS techniques, reporting less HBO2 after a physical training program [43]. These results are in accordance with fMRI results from Volcker-Rehage et al., 2011 [44], who found less cortical activity after physical training. We could hypothesis that the brain is more efficient after acute exercise, and therefore requires less cortical and therefore metabolic and vascular activity. The reduced cerebral oxygenation after cognitive exercise may also reflect reduced cortical activity. This phenomenon could be explained by the fact that during cognitive exercise, neural circuits are stimulated and may benefit from a facilitating effect when stimulated later. This theory is only speculative and requires further work to validate.

Though this study had several strengths, it was not without limitations, and the interpretation of our results requires some caution. One limit of this article is the small sample of this study, which does not allow us to generalize our results on a large scale. Nevertheless, the calculation of the size of the effect using the Hedge formula allows us to appreciate our results from a clinical point of view. The second limitation is the lack of a control group. The presence of a control group would allow us to fully ensure that our cognitive results and cerebral oxygenation are not due to a test–retest effect. Nevertheless, the familiarization at the first visit and the practice trials before each block of the Stroop task enables us to control this effect. In addition, the study by Ji et al., using a Stroop task and a similar design to ours, did not report any change in the reaction time or error in the older control group.

5. Conclusions

Based on Kraft theory [8], we hypothesized that simultaneously combined cognitive and physical exercise would be more effective in improving executive performance than either exercise mode alone and that this effect would be mediated by a greater increase in pre-frontal cortex oxygenation. Our main findings were (1) an improvement in executive performance (i.e., inhibition and flexibility conditions of the Stroop task) after each exercise condition; (2) a larger decrease in flexibility cost (based on the ES) after combined cognitive and physical exercise; (3) a decrease in $\Delta[HbO_2]$ and $\Delta[THB]$ after each exercise condition, both in the left and right PFC. Simultaneous physical and cognitive exercise is as effective a method as either physical or cognitive exercise alone for improving executive function and present no superiority in acute setting. The results of this study may have important clinical repercussions by allowing us to optimize the interventions designed to maintain the cognitive health of older adults since simultaneous provide a time-efficient strategy to improve cognitive performance in older adults.

Author Contributions: M.P.-L., P.S., L.B. and O.D. conceived and designed the research protocol; M.P.-L., P.S., L.B. and O.D. performed experiments; analyzed data; interpreted results of experiments; M.P.-L. and O.D. prepared the figures; M.P.-L., P.S., L.B. and O.D. drafted the manuscript; M.P.-L., P.S., L.B. and O.D. edited and revised the manuscript; All authors have read and agreed to the published version of the manuscript.

Funding: This study was supported in part by the European Union and the New Aquitaine region through the Habisan program (CPER-FEDER). This study was also supported by the Association nationale de la recherche et de la technologie (ANRT) and Rev'Lim society.

Institutional Review Board Statement: The protocol was reviewed and approved by a national ethics committee for non-interventional research (CERSTAPS # 17 November 2017).

Informed Consent Statement: All participants signed a written statement of informed consent.

Acknowledgments: Thanks are due to the Society Rev'Lim and more specifically Nicolas Troubat, for financial and technical support of Manon Pelligrini-Laplagne.

Conflicts of Interest: The authors declare no conflict of interest.

References

1. Fjell, A.M.; Walhovd, K.B. Structural Brain Changes in Aging: Courses, Causes and Cognitive Consequences. *Rev. Neurosci.* **2010**, *21*, 187–222. [CrossRef] [PubMed]
2. Borson, S. Cognition, Aging, and Disabilities: Conceptual Issues. *Phys. Med. Rehabil. Clin. N. Am.* **2010**, *21*, 375–382. [CrossRef] [PubMed]
3. Harada, C.N.; Natelson Love, M.C.; Triebel, K.L. Normal Cognitive Aging. *Clin. Geriatr. Med.* **2013**, *29*, 737–752. [CrossRef] [PubMed]
4. Peters, R. Ageing and the Brain. *Postgrad. Med. J.* **2006**, *82*, 84–88. [CrossRef] [PubMed]
5. Prince, M.; Bryce, R.; Albanese, E.; Wimo, A.; Ribeiro, W.; Ferri, C.P. The Global Prevalence of Dementia: A Systematic Review and Metaanalysis. *Alzheimer's Dement.* **2013**, *9*, 63–75.e2. [CrossRef]
6. Colzato, L.S. Editorial: One Year of Journal of Cognitive Enhancement. *J. Cogn. Enhanc.* **2018**, *2*, 1–2. [CrossRef]

7. Herold, F.; Hamacher, D.; Schega, L.; Müller, N.G. Thinking While Moving or Moving While Thinking-Concepts of Motor-Cognitive Training for Cognitive Performance Enhancement. *Front. Aging Neurosci.* **2018**, *10*, 228. [CrossRef]

8. Kraft, E. Cognitive Function, Physical Activity, and Aging: Possible Biological Links and Implications for Multimodal Interventions. *Aging Neuropsychol. Cogn.* **2012**, *19*, 248–263. [CrossRef]

9. Gavelin, H.M.; Dong, C.; Minkov, R.; Bahar-Fuchs, A.; Ellis, K.A.; Lautenschlager, N.T.; Mellow, M.L.; Wade, A.T.; Smith, A.E.; Finke, C.; et al. Combined Physical and Cognitive Training for Older Adults with and without Cognitive Impairment: A Systematic Review and Network Meta-Analysis of Randomized Controlled Trials. *Ageing Res. Rev.* **2021**, *66*, 101232. [CrossRef]

10. Ferris, L.E.E.T.; Williams, J.S.; Shen, C.-L. The Effect of Acute Exercise on Serum Brain-Derived Neurotrophic Factor Levels and Cognitive Function. *Med. Sci. Sports Exerc.* **2007**, *39*, 728–734. [CrossRef]

11. Harveson, A.T.; Hannon, J.C.; Brusseau, T.A.; Podlog, L.; Papadopoulos, C.; Durrant, L.H.; Hall, M.S.; Kang, K.D. Acute Effects of 30 Minutes Resistance and Aerobic Exercise on Cognition in a High School Sample. *Res. Q. Exerc. Sport* **2016**, *87*, 214–220. [CrossRef] [PubMed]

12. Hötting, K.; Schickert, N.; Kaiser, J.; Röder, B.; Schmidt-Kassow, M. The Effects of Acute Physical Exercise on Memory, Peripheral BDNF, and Cortisol in Young Adults. *Neural Plast.* **2016**, *2016*, 6860573. [CrossRef] [PubMed]

13. Hwang, J.; Brothers, R.M.; Castelli, D.M.; Glowacki, E.M.; Chen, Y.T.; Salinas, M.M.; Kim, J.; Jung, Y.; Calvert, H.G. Acute High-Intensity Exercise-Induced Cognitive Enhancement and Brain-Derived Neurotrophic Factor in Young, Healthy Adults. *Neurosci. Lett.* **2016**, *630*, 247–253. [CrossRef] [PubMed]

14. Winter, B.; Breitenstein, C.; Mooren, F.C.; Voelker, K.; Fobker, M.; Lechtermann, A.; Krueger, K.; Fromme, A.; Korsukewitz, C.; Floel, A.; et al. High Impact Running Improves Learning. *Neurobiol. Learn. Mem.* **2007**, *87*, 597–609. [CrossRef] [PubMed]

15. McSween, M.-P.; Coombes, J.S.; MacKay, C.P.; Rodriguez, A.D.; Erickson, K.I.; Copland, D.A.; McMahon, K.L. The Immediate Effects of Acute Aerobic Exercise on Cognition in Healthy Older Adults: A Systematic Review. *Sports Med.* **2019**, *49*, 67–82. [CrossRef] [PubMed]

16. Ludyga, S.; Gerber, M.; Brand, S.; Holsboer-Trachsler, E.; Pühse, U. Acute Effects of Moderate Aerobic Exercise on Specific Aspects of Executive Function in Different Age and Fitness Groups: A Meta-Analysis. *Psychophysiology* **2016**, *53*, 1611–1626. [CrossRef] [PubMed]

17. Chang, Y.K.; Alderman, B.L.; Chu, C.H.; Wang, C.C.; Song, T.F.; Chen, F.T. Acute Exercise Has a General Facilitative Effect on Cognitive Function: A Combined ERP Temporal Dynamics and BDNF Study. *Psychophysiology* **2017**, *54*, 289–300. [CrossRef]

18. Ji, Z.; Feng, T.; Mei, L.; Li, A.; Zhang, C. Influence of Acute Combined Physical and Cognitive Exercise on Cognitive Function: An NIRS Study. *PeerJ* **2019**, *7*, e7418. [CrossRef]

19. Basak, C.; Qin, S.; O'Connell, M.A. Differential Effects of Cognitive Training Modules in Healthy Aging and Mild Cognitive Impairment: A Comprehensive Meta-Analysis of Randomized Controlled Trials. *Psychol. Aging* **2020**, *35*, 220–249. [CrossRef]

20. Chang, Y.K.; Labban, J.D.; Gapin, J.I.; Etnier, J.L. The Effects of Acute Exercise on Cognitive Performance: A Meta-Analysis. *Brain Res.* **2012**, *1453*, 87–101. [CrossRef]

21. Gellish, R.L.; Goslin, B.R.; Olson, R.E.; McDonald, A.; Russi, G.D.; Moudgil, V.K. Longitudinal Modeling of the Relationship between Age and Maximal Heart Rate. *Med. Sci. Sports Exerc.* **2007**, *39*, 822–829. [CrossRef] [PubMed]

22. Lambourne, K.; Tomporowski, P. The Effect of Exercise-Induced Arousal on Cognitive Task Performance: A Meta-Regression Analysis. *Brain Res.* **2010**, *1341*, 12–24. [CrossRef] [PubMed]

23. Lauenroth, A.; Ioannidis, A.E.; Teichmann, B. Influence of Combined Physical and Cognitive Training on Cognition: A Systematic Review. *BMC Geriatr.* **2016**, *16*, 141. [CrossRef] [PubMed]

24. Combourieu Donnezan, L.; Perrot, A.; Belleville, S.; Bloch, F.; Kemoun, G. Effects of Simultaneous Aerobic and Cognitive Training on Executive Functions, Cardiovascular Fitness and Functional Abilities in Older Adults with Mild Cognitive Impairment. *Ment. Health Phys. Act.* **2018**, *15*, 78–87. [CrossRef]

25. Nasreddine, Z.S.; Phillips, N.A.; Bédirian, V.; Charbonneau, S.; Whitehead, V.; Collin, I.; Cummings, J.L.; Chertkow, H. The Montreal Cognitive Assessment, MoCA: A Brief Screening Tool For Mild Cognitive Impairment. *J. Am. Geriatr. Soc.* **2005**, *53*, 695–699. [CrossRef] [PubMed]

26. Dupuy, O.; Gauthier, C.J.; Fraser, S.A.; Desjardins-Crèpeau, L.; Desjardins, M.; Mekary, S.; Lesage, F.; Hoge, R.D.; Pouliot, P.; Bherer, L. Higher Levels of Cardiovascular Fitness Are Associated with Better Executive Function and Prefrontal Oxygenation in Younger and Older Women. *Front. Hum. Neurosci.* **2015**, *9*, 66. [CrossRef]

27. Goenarjo, R.; Bosquet, L.; Berryman, N.; Metier, V.; Perrochon, A.; Fraser, S.A.; Dupuy, O. Cerebral Oxygenation Reserve: The Relationship between Physical Activity Level and the Cognitive Load during a Stroop Task in Healthy Young Males. *Int. J. Environ. Res. Public Health* **2020**, *17*, 1406. [CrossRef]

28. Goenarjo, R.; Dupuy, O.; Fraser, S.; Berryman, N.; Perrochon, A.; Bosquet, L. Cardiorespiratory Fitness and Prefrontal Cortex Oxygenation during Stroop Task in Older Males. *Physiol. Behav.* **2021**, *242*, 113621. [CrossRef]

29. Mekari, S.; Earle, M.; Martins, R.; Drisdelle, S.; Killen, M.; Bouffard-Levasseur, V.; Dupuy, O. Effect of High Intensity Interval Training Compared to Continuous Training on Cognitive Performance in Young Healthy Adults: A Pilot Study. *Brain Sci.* **2020**, *10*, 81. [CrossRef]

30. Ferrari, M.; Quaresima, V. A Brief Review on the History of Human Functional Near-Infrared Spectroscopy (FNIRS) Development and Fields of Application. *NeuroImage* **2012**, *63*, 921–935. [CrossRef]

31. Agbangla, N.F.; Audiffren, M.; Pylouster, J.; Albinet, C.T. Working Memory, Cognitive Load and Cardiorespiratory Fitness: Testing the CRUNCH Model with near-Infrared Spectroscopy. *Brain Sci.* **2019**, *9*, 38. [CrossRef] [PubMed]
32. Abe, T.; Fujii, K.; Hyodo, K.; Kitano, N.; Okura, T. Effects of Acute Exercise in the Sitting Position on Executive Function Evaluated by the Stroop Task in Healthy Older Adults. *J. Phys. Ther. Sci.* **2018**, *30*, 609–613. [CrossRef] [PubMed]
33. Hyodo, K.; Dan, I.; Suwabe, K.; Kyutoku, Y.; Yamada, Y.; Akahori, M.; Byun, K.; Kato, M.; Soya, H. Acute Moderate Exercise Enhances Compensatory Brain Activation in Older Adults. *Neurobiol. Aging* **2012**, *33*, 2621–2632. [CrossRef] [PubMed]
34. Johnson, L.; Addamo, P.K.; Selva Raj, I.; Borkoles, E.; Wyckelsma, V.; Cyarto, E.; Polman, R.C. An Acute Bout of Exercise Improves the Cognitive Performance of Older Adults. *J. Aging Phys. Act.* **2016**, *24*, 591–598. [CrossRef] [PubMed]
35. Yanagisawa, H.; Dan, I.; Tsuzuki, D.; Kato, M.; Okamoto, M.; Kyutoku, Y.; Soya, H. Acute Moderate Exercise Elicits Increased Dorsolateral Prefrontal Activation and Improves Cognitive Performance with Stroop Test. *NeuroImage* **2010**, *50*, 1702–1710. [CrossRef] [PubMed]
36. Hillman, C.; Pontifex, M.; Raine, L.; Castelli, D.; Hall, E.; Kramer, A. The Effect of Acute Treadmill Walking on Cognitive Control and Academic Achievement in Preadolescent Children. *Neuroscience* **2009**, *159*, 1044–1054. [CrossRef] [PubMed]
37. Hillman, C.H.; Snook, E.M.; Jerome, G.J. Acute Cardiovascular Exercise and Executive Control Function. *Int. J. Psychophysiol.* **2003**, *48*, 307–314. [CrossRef]
38. Chodzko-Zajko, W.J.; Proctor, D.N.; Fiatarone Singh, M.A.; Minson, C.T.; Nigg, C.R.; Salem, G.J.; Skinner, J.S. Exercise and Physical Activity for Older Adults. *Med. Sci. Sports Exerc.* **2009**, *41*, 1510–1530. [CrossRef]
39. McMorris, T.; Sproule, J.; Turner, A.; Hale, B.J. Acute, Intermediate Intensity Exercise, and Speed and Accuracy in Working Memory Tasks: A Meta-Analytical Comparison of Effects. *Physiol. Behav.* **2011**, *102*, 421–428. [CrossRef]
40. Herold, F.; Wiegel, P.; Scholkmann, F.; Müller, N.G. Applications of Functional Near-Infrared Spectroscopy (FNIRS) Neuroimaging in Exercise–Cognition Science: A Systematic, Methodology-Focused Review. *J. Clin. Med.* **2018**, *7*, 466. [CrossRef]
41. Murata, Y.; Watanabe, T.; Terasawa, S.; Nakajima, K.; Kobayashi, T.; Nakade, K.; Terasawa, K.; Maruo, S.K. Moderate Exercise Improves Cognitive Performance and Decreases Cortical Activation in Go/No-Go Task. *BAOJ Med. Nurs.* **2015**, *1*, 1–7.
42. Audiffren, M.; Tomporowski, P.D.; Zagrodnik, J. Acute Aerobic Exercise and Information Processing: Energizing Motor Processes during a Choice Reaction Time Task. *Acta Psychol.* **2008**, *129*, 410–419. [CrossRef] [PubMed]
43. Coetsee, C.; Terblanche, E. Cerebral Oxygenation during Cortical Activation: The Differential Influence of Three Exercise Training Modalities. A Randomized Controlled Trial. *Eur. J. Appl. Physiol.* **2017**, *117*, 1617–1627. [CrossRef] [PubMed]
44. Voelcker-Rehage, C.; Godde, B.; Staudinger, U.M. Cardiovascular and Coordination Training Differentially Improve Cognitive Performance and Neural Processing in Older Adults. *Front. Hum. Neurosci.* **2011**, *5*, 26. [CrossRef] [PubMed]

Article

Changes in Hippocampus and Amygdala Volume with Hypoxic Stress Related to Cardiorespiratory Fitness under a High-Altitude Environment

Zhi-Xin Wang [1], Rui Su [1], Hao Li [1], Peng Dang [1], Tong-Ao Zeng [1], Dong-Mei Chen [1], Jian-Guo Wu [2], De-Long Zhang [1,3,*] and Hai-Lin Ma [1,*]

[1] Plateau Brain Science Research Center, Tibet University/South China Normal University, Lhasa 850012, China; wangzhixin666@outlook.com (Z.-X.W.); srsisu2011@163.com (R.S.); futanghu888@126.com (H.L.); 18932689504@163.com (P.D.); tsangtongao@gmail.com (T.-A.Z.); chendongmei0228@163.com (D.-M.C.)

[2] Management Department, Tibet Police College, Lhasa 850012, China; wjg198228@163.com

[3] Key Laboratory of Brain, Cognition and Education Sciences, Center for Studies of Psychological Application, Guangdong Key Laboratory of Mental Health and Cognitive Science, Ministry of Education, School of Psychology, South China Normal University, Guangzhou 510631, China

* Correspondence: delong.zhang@m.scnu.edu.cn (D.-L.Z.); David_ma79@163.com (H.-L.M.)

Citation: Wang, Z.-X.; Su, R.; Li, H.; Dang, P.; Zeng, T.-A.; Chen, D.-M.; Wu, J.-G.; Zhang, D.-L.; Ma, H.-L. Changes in Hippocampus and Amygdala Volume with Hypoxic Stress Related to Cardiorespiratory Fitness under a High-Altitude Environment. *Brain Sci.* **2022**, *12*, 359. https://doi.org/10.3390/brainsci12030359

Academic Editor: Filipe Manuel Clemente

Received: 11 February 2022
Accepted: 26 February 2022
Published: 8 March 2022

Publisher's Note: MDPI stays neutral with regard to jurisdictional claims in published maps and institutional affiliations.

Abstract: The morphology of the hippocampus and amygdala can be significantly affected by a long-term hypoxia-induced inflammatory response. Cardiorespiratory fitness (CRF) has a significant effect on the neuroplasticity of the hippocampus and amygdala by countering inflammation. However, the role of CRF is still largely unclear at high altitudes. Here, we investigated brain limbic volumes in participants who had experienced long-term hypoxia exposure in Tibet (3680 m), utilizing high-resolution structural images to allow the segmentation of the hippocampus and amygdala into their constituent substructures. We recruited a total of 48 participants (48 males; aged = 20.92 ± 1.03 years) to undergo a structural 3T MRI, and the levels of maximal oxygen uptake (VO_{2max}) were measured using a cardiorespiratory function test. Inflammatory biomarkers were also collected. The participants were divided into two groups according to the levels of median VO_{2max}, and the analysis showed that the morphological indexes of subfields of the hippocampus and amygdala of the lower CRF group were decreased when compared with the higher CRF group. Furthermore, the multiple linear regression analysis showed that there was a higher association with inflammatory factors in the lower CRF group than that in the higher CRF group. This study suggested a significant association of CRF with hippocampus and amygdala volume, which may be related to hypoxic stress in high-altitude environments. A better CRF reduced physiological stress and a decrease in the inflammatory response was observed, which may be related to the increased oxygen transport capacity of the body.

Keywords: high altitude; cardiorespiratory fitness; stress; amygdala; hippocampus

1. Introduction

Under a high-altitude environment, the decrease in atmospheric pressure and the consequent drop in the partial pressure of oxygen (PO_2) can result in hypobaric hypoxia. The human brain is the most oxygen-consuming organ, and is very susceptible to hypoxic stress [1]. Hypoxic stress has serious effects on brain structures, such as the hippocampus and amygdala [2–4]. Behavior and brain structure changes arising from hypoxia can be observed in a real high-altitude environment, and examined in simulated hypoxia situations. The hypoxia impact on the brain exhibits a significant altitude-dependent effect [5,6].

Activation of the hypothalamic–pituitary–adrenal axis (HPA) is a hallmark of the stress response [7]. At the level of the organism, a hypoxic challenge is perceived as a non-specific stress, and hypoxia could upregulate the setpoint of the HPA axis and augment adrenal steroidogenic production, resulting in neuroinflammation and neuronal cell death [8]. High-altitude hypoxia stress affects a wide range of brain areas [4,9]. Major brain regions with structural and functional abnormalities are particularly vulnerable to hypoxic stress, including the amygdala and hippocampus [8]. Changes in the amygdala and hippocampus under high-altitude hypoxia stress may be due to the influence of inflammation [10,11]. The inflammatory response to hypoxia leads to the death of hippocampal neurons [12,13]. Similar studies have found that after hypoxia stress, the inflammatory response of the hippocampus and amygdala increase, leading to the death of neurons [14,15]. Studies of hypoxia stress in the brain caused by human disease have also found that inflammation can trigger neuron damage in the hippocampus and amygdala [14]. The number of corticotropin-releasing factor- (CRF-) and neuropeptide-Y- (NPY-) positive neurons were found to be decreased in the amygdala after hypoxia-ischemia [16]. In conclusion, inflammation caused by altitude hypoxia stress seriously affects the hippocampus and amygdala.

Cardiorespiratory fitness (CRF) is an objective measure of habitual physical activity that reflects the overall capacity of the cardiorespiratory system, and has been used to assess the relationship between physical activity and health status [17]. Increasing evidence has shown that higher levels of CRF are related to better brain health [18]. Improving CRF can effectively maintain the axial homeostasis of the HPA and reduce the inflammatory response [19,20]. CRF reduces stress-induced inflammation and increases neuroplasticity in the hippocampus and amygdala [21]. For example, higher CRF was found to be associated with greater GM volumes in several AD-relevant brain regions, including the hippocampus and amygdala [22]. The CRF measured using VO_{2max} was reported to be associated with the volumetric enlargement of the hippocampal head, specifically the head region of CA1 [23]. Studies in adolescents and older adults have shown that CRF levels are associated with a greater hippocampal volume [24,25]. CRF protects neurons in the amygdala and hippocampus against Alzheimer's disease-related degeneration, probably via enhancements of brain-derived neurotrophic factor (BDNF) signaling pathways and Aβ clearance [26].

Cardiorespiratory function is an effective indicator for evaluating oxygen transport capacity [27]. Cardiorespiratory function is a term that can be used interchangeably with CRF, and indicates the VO_{2max} or ability to undertake aerobic exercises [28]. Recent studies have shown that the acclimatization and adaptive processes at high altitude in healthy individuals, with tissue hypoxia, often lead to compromised arterial oxygenation [29]. However, CRF can improve oxygen transport at high altitudes [30]. This may be associated with a lower mean RBC age, thereby improving oxygen release and increasing tissue oxygen supply [31]. At the same time, the increase in CRF improves the affinity between hemoglobin and oxygen [32]. An increase in CRF is accompanied by an increase in brain blood flow and brain metabolism [33]. In summary, CRF is an important indicator of the cardiovascular system's ability to deliver oxygen to peripheral tissues, and the tissue's ability to use that oxygen [34]. Not surprisingly, CRF is involved in adaptation to high altitudes [35].

However, it is still largely unclear whether and how the CRF regulates the impact of hypoxia on the brain under high-altitude hypoxia environments. The present study aimed to explore the relationship between the volume of the hippocampus and amygdala with CRF across participants under a high-altitude environment in Tibet (3680 m). The study recruited participants who had been exposed to high altitudes for more than 2 years. VO_{2max} was used to measure CRF via a specialized cardiorespiratory function test system, and participants were divided into high and the low-CRF groups based on levels of median VO_{2max}. MRI data were collected to segment the hippocampus and amygdala volumes, and the relationship between the hippocampal and amygdala subregions and the VO_{2max} was evaluated. To further explore the effect of CRF on the hippocampus and amygdala, we also collected biochemical indicators related to inflammation and immunity to identify the physical essence of the linkage of the hippocampus and amygdala with CRF in these immigrant participants.

2. Materials and Methods

2.1. Participants

This study recruited a total of 48 right-handed male participants who were born in and grew up in low altitude areas, and had migrated to high-altitude areas (Lhasa, 3680 m) for more than 2 years. The participants were divided into the high-CRF group ($n = 23$) and the low-CRF group ($n = 25$) according to their median VO_{2max}. In addition, the sample size was reasonable, which was measured using the G*Power ($t = 2.01$, effect size was 0.83).

All the participants had normal vision or corrected vision, and none of them had a history of mental illness or major diseases such as traumatic brain injury, hypertension, or heart disease. The two groups were matched on age (20.92 ± 1.03 years) and years of education (14.13 ± 0.33 years) (Table 1).

Table 1. Independent sample T test of demographic characteristics in the low- and high-CRF groups (mean $\pm$ SD).

	Low	High	*t*	*p*
Age	21.00 ± 1.00 (years)	20.83 ± 1.07 (years)	0.58	0.56
BMI	20.64 ± 3.34	21.65 ± 1.95	-1.26	0.20
Education	14.08 ± 0.28 (years)	14.17 ± 0.39 (years)	-0.97	0.34
Multimedia	6.04 ± 2.62 (hours)	5.57 ± 1.59 (hours)	0.75	0.46

SD: standard deviation; BMI: body-mass index; CRF: cardiopulmonary fitness; $p < 0.05$: statistical significance.

This study was approved by the local ethics committee of Tibet University, and conducted in accordance with relevant guidelines and regulations. All the participants voluntarily participated in the experiment, signed the informed consent before the experiment, and received a payment after the end of the experiment.

2.2. Experimental Design

Forty-eight male participants who had lived in Lhasa for more than two years were randomly selected, and their VO_{2max} was measured using a cardiopulmonary exercise test (CPET). Participants were divided into the low- and high-CRF groups based on their median VO_{2max}. MRI data and biochemical indicators related to inflammation were collected. Then, we tested the difference between the two groups on the subfield volume and the biochemical indicators. An across-subject regressive analysis between the biochemical parameters and the regions of the hippocampal and amygdala subfields was calculated within each group (Figure 1).

2.3. Maximum Oxygen Uptake

CRF was assessed with CPET. Oxygen, carbon dioxide, and respiratory flow data were collected in two conditions: during rest state and during exercise state (increased load pedal powered bicycle). During the test condition, an incremental protocol with a 30 W per two minutes stepwise was used to test exhaustion on a cycle ergometer. The power bike load was carried out in frequency independent mode at 60 RPM, with an accuracy of 5 W. In the preparation stage, the participants were asked to sit quietly for five minutes, and in the recovery phase, the original load was terminated and changed to 30 W. Heart rate and oxygen uptake were measured continuously during the CPET (MetaLyzer 3B, Cortex Medical GmbH, Leipzig, Germany), and the relevant indicators were calculated according to the standard Wasserman formula [36]. The VO_{2max} ($mL \cdot kg \cdot min^{-1}$) values were applied to completely characterize the aerobic predispositions of the participants. Prior to the measurement of each participant, the device was recalibrated [37]. Maximal efforts were assumed when the participant felt exhausted, had a heart rate greater than 180, or had a respiratory exchange rate (RER) equal to or exceeding 1.1.

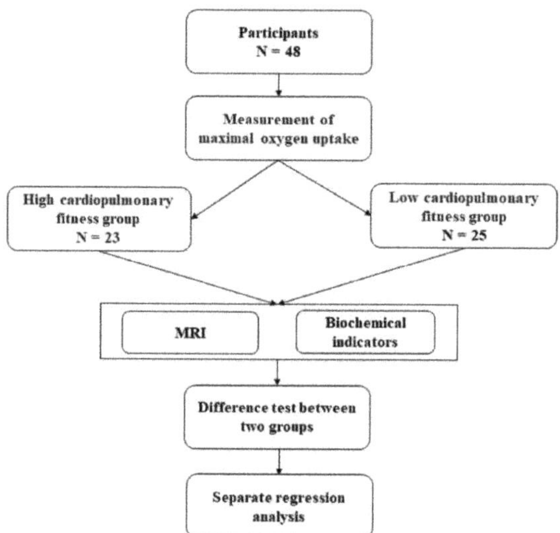

Figure 1. Study design.

2.4. MRI Data Acquisition

T1 weighted images were collected by a Siemens 3 Tesla Allegra MRI scanner at the Tibet Armed Police Corps Hospital using a magnetization-prepared rapid gradient-echo (MP-RAGE) sequence with the following parameters: slice thickness = 1 mm, TR = 1900 ms, TE = 2.41 ms, FA = 9°, FOV = 256 mm, matrix = 256 × 256, slices = 192, voxel size = 1 mm^3.

T1 weighted data were processed using Freesurfer 7.1.1 (http://surfer.nmr.mgh.harvard.edu, accessed on 25 February 2022) and MATLAB 2014 Runtime (https://surfer.nmr.mgh.harvard.edu/fswiki/MatlabRuntime, accessed on 25 February 2022). The standard volumetric pipeline was used to generate several files, including the Talairach transformation matrix for hippocampal and amygdaloid subfields segmentation. After the volumetric pipeline, quality control was performed to manually check the results of the brain extraction, Talairach transformation, and brain segmentation.

The automated segmentation of the hippocampal and amygdaloid subfields was driven by a probabilistic atlas and a Bayesian inference model, which maximized the probability of the segmentation [38,39]. A total of 64 structural subfields were extracted, including 20 amygdaloid subfields and 44 hippocampal subfields, as shown in Figure 2. The subfields of the amygdala included the mean volume of the lateral nucleus, basal nucleus, accessory basal nucleus, anterior amygdaloid area, central nucleus, medial nucleus, cortical nucleus, corticoamygdaloid transition, and paralaminar nucleus, and the whole amygdala was located at the bilateral hemispheres. The total of 44 hippocampal subfields included the mean volume of the hippocampal tail, subiculum body, cornuammonis (CA) 1 body, CA1 head, subiculum head, hippocampal fissure, presubiculum head, presubiculum body, parasubiculum, molecular layer head, molecular layer body, granule cell layers of the dentate gyrus (GC-DG) head, GC-DG body, CA3 body, CA3 head, CA4 head, CA4 body, fimbria, hippocampal amygdala transition area, hippocampal body, and hippocampal head, and the whole hippocampus was located in the bilateral hemispheres. Then, the volumes of the amygdala subfields and hippocampus subfields were extracted for statistical analysis.

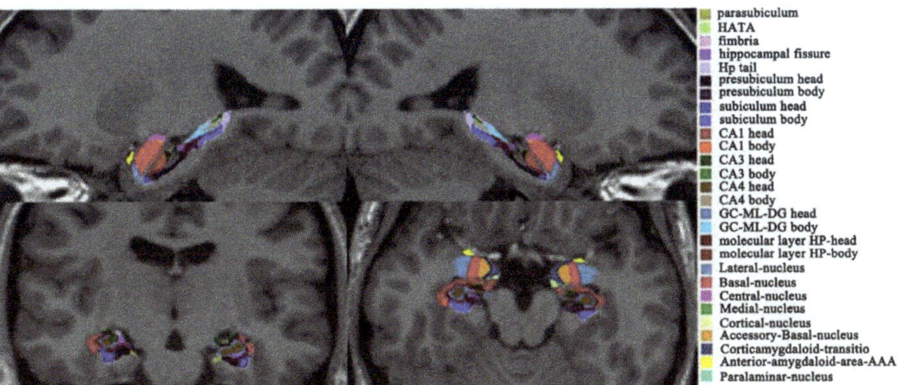

Figure 2. Subfields of the amygdala and hippocampus. CA, cornuammonis; HATA, hippocampal amygdala transition area; GC_ML_DG: granule cell layers of the dentate gyrus.

2.5. Biochemical Indicators

The biochemical indicators related to inflammation were collected from the participants by venous blood sampling in Fukang Hospital, affiliated with Tibet University (Lhasa, Tibet), corresponding to a previous study in our lab [40].

2.6. Statistical Analysis

Statistical analyses of the hippocampal subfields' volume, amygdala subfields' volume, and biochemical indicators were performed with SPSS (SPSS 20, inc./IBM, Armonk, NY, USA). We carried out the Shapiro-Wilk test first and found that the data was normally distributed ($p > 0.05$). The hippocampal subfield volume and amygdala subfield volume were tested using the analysis of covariance (ANCOVA). The ratios of the hippocampal and amygdala volume to intracranial volume were used as covariables [41]. An independent sample T test was used for the biochemical indicators, and the alpha level was set at $p < 0.05$.

To test our hypothesis, the standard multiple linear regression analysis was used to analyze the relationship between the biochemical parameters and the volume of the hippocampus and amygdala subfields in each group, in which the biochemical parameters were employed as the independent factors and the volume of the target regions as the dependent factor. Notably, only the related biochemical parameters of the existing phase were input into the stepwise regression model, and the regression analyses were controlled for age and ratio of hippocampal and amygdala volume to intracranial volume by the addition of variables into the linear model as covariates [42]. Durbin-Watson tests were performed to ensure independence of errors (residuals). We checked the tolerances and correlation coefficients to make sure that there were no collinearity problems in our data set. The assumptions of linearity, error independence, homoscedasticity, outliers, and residual normality had to be satisfied before the results which could be interpreted. The multiple linear regression hypothesis was satisfied. The significance level was assumed at $p < 0.05$, and p-values were corrected for multiple comparisons using the FDR correction [43].

3. Results

3.1. Hippocampal and Amygdala Subfields

High CRF (2.18 ± 0.20) was significantly higher than that of the low CRF (1.62 ± 0.15) ($t = -10.84$, $p < 0.001$; Figure 3A). Analysis of covariance (ANCOVA) revealed that the left GC_ML_DG head volume ($F = 4.71$, $p = 0.035$), the left CA4 head volume ($F = 4.75$, $p = 0.035$), the right subiculum body volume ($F = 4.34$, $p = 0.043$), and the right presubiculum body volume ($F = 9.06$, $p = 0.004$) in the hippocampus of the high-CRF group were significantly larger than those of the low-CRF group. As regards the amygdala subfields, we found that the left corticoamygdaloid transition volumes ($F = 0.465$, $p = 0.036$) in the high-CRF group were significantly larger than in the low-CRF group (Figure 3B).

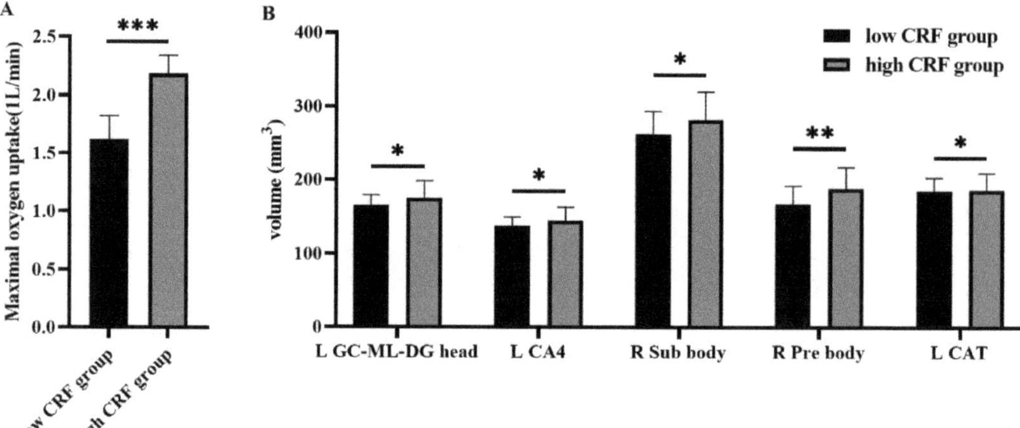

Figure 3. (**A**) The difference in maximal oxygen uptake between the two groups; (**B**) Differences between the hippocampal and amygdala subfields. L, left hemisphere; R, right hemisphere; CRF, cardiorespiratory fitness; GC-ML-DG head: granule cell layers of the dentate gyrus; CA4: cornu-ammonis 4; Sub body: subiculum body; Pre body: presubiculum body; CAT: corticoamygdaloid transition; * $p \leq 0.05$; ** $p \leq 0.01$; *** $p \leq 0.001$.

3.2. Biochemical Indicators

An independent sample T test revealed that the direct bilirubin (DBIL), total bilirubin (TBIL), and red blood cells (RBC) in the high-CRF group were significantly greater than those in the low-CRF group. The standard deviations in the red cell distribution width (RDW-SD) in the high-CRF group were significantly lower than those of the low-CRF group. There were no significant differences in the other indicators (Table 2).

Table 2. Statistical values of the independent sample T test for biochemical indexes in the low- and high-CRF groups (mean ± SD).

	Low	High	t	p
DBIL	4.9 ± 2.53 (umol/L)	6.43 ± 2.3 (umol/L)	−2.18 *	0.035
TBIL	16.56 ± 8.14 (umol/L)	21.64 ± 8.73 (umol/L)	−2.09 *	0.042
NEUT	3.19 ± 1.34 (10^9/L)	3.55 ± 1.16 (10^9/L)	−1.00	0.322
LYMPH	2.46 ± 0.56 (10^9/L)	2.56 ± 0.51 (10^9/L)	−0.64	0.528
EO	0.10 ± 0.07 (10^9/L)	0.08 ± 0.07 (10^9/L)	0.95	0.346
RBC	5.31 ± 0.53 (10^9/L)	5.87 ± 0.42 (10^9/L)	−3.97 ***	<0.001
HGB	163.20 ± 19.53 (g/L)	180.17 ± 13.41 (g/L)	−3.48 ***	<0.001
RDW-SD	42.70 ± 3.11 (%)	40.22 ± 2.56 (%)	3.00 **	0.004

TBIL: total bilirubin; DBIL: direct bilirubin; NEUT: neutrophil count; LYMPH: lymphocyte; EO: eosinophil count; RBC: red blood cells; HGB: hemoglobin; HCT: hematocrit; RDW-SD: standard deviation in red cell distribution width; * $p \leq 0.05$; ** $p \leq 0.01$; *** $p \leq 0.001$.

3.3. Multiple Regression Analysis

3.3.1. Hippocampus Subfields

Multiple linear regression analyses in the two CRF groups, adjusted for hippocampal total volume, were used to investigate possible associations between amygdala subfields volumetrics and biochemical parameters (Tables 3 and 4). The regression analysis within the two groups showed that HGB (hemoglobin) effectively positively predicted the volumes of the right subiculum body (Beta = 0.70, r^2 = 0.11, $p < 0.05$, FDR corrected) and right presubiculum body (Beta = 0.55, r^2 = 0.11, $p < 0.05$, FDR corrected). In the low-CRF group, the eosinophil (EO) predicted the left GC_ML_DG head volume (Beta = -0.53, r^2 = 0.37, $p < 0.05$, FDR corrected). The EO (Beta = -0.49, r^2 = 0.38, $p < 0.05$, FDR corrected) predicted the left CA4 head volume. The NEUT predicted the right subiculum body (NEUT, Beta = 0.48, r^2 = 0.23, $p < 0.05$, FDR corrected) and the right presubiculum body volume (Beta = 0.73, r^2 = 0.53, $p < 0.05$, FDR corrected). In the high-CRF group, the direct bilirubin (DBIL) predicted the left GC_ML_DG head (Beta = -0.45, r^2 = 0.25, $p < 0.05$, FDR corrected) volume and the left CA4 head (Beta = -0.44, r^2 = 0.19, $p < 0.05$, FDR corrected) volume.

Table 3. Multiple linear regression analysis of the hippocampal subregion, amygdala subregion, and biochemical indexes within the two groups.

Dependent Variable	Predictors	B	Ser	Beta	t
R Sub body	Constant	152.33	44.04		3.459
	HGB	0.70	2.56	0.38	2.75
R Pre body	Constant	83.47	35.86		2.33
	HGB	0.55	0.21	0.36	2.65

HGB: hemoglobin; L: left hemisphere; R: right hemisphere; Sub body: subiculum body; Pre body: presubiculum body; $p < 0.05$, FDR correct.

Table 4. Multiple linear regression analysis of the hippocampal subregion, amygdala subregion, and biochemical indexes for both groups.

	Dependent Variable	Predictors	B	Ser	Beta	t
Low CRF	L GC_ML_DG head	Constant	80.18	22.97		3.49
		EO	-101.75	26.48	-0.49	-3.84
	L CA4 head	Constant	65.92	19.07		3.43
		EO	-91.35	26.54	-0.53	-3.44
	R Sub body	Constant	228.79	14.37		15.92
		NEUT	10.86	4.17	0.48	2.61
	R Pre body	Constant	125.16	9.10		13.75
		NEUT	13.32	2.64	0.73	5.05
	L CAT	Constant	93.703	46.48		2.029
		EO	-744.84	302.94	-0.63	-3.35
High CRF	L GC_ML_DG head	Constant	208.01	13.17		15.80
		DBIL	-4.61	-1.74	-0.45	-2.65
	L CA4 head	Constant	60.99	41.39		1.47
		TBIL	-3.40	1.31	-0.44	-2.60

EO: eosinophil; NEUT: neutrophil; RBC: red blood cells; DBIL: direct bilirubin; TBIL: total bilirubin; L: left hemisphere; R: right hemisphere; GC-ML-DG head: granule cell layers of the dentate gyrus; CA4 head: cornuammonis 4 head; Sub body: subiculum body; Pre body: presubiculum body; CAT: corticoamygdaloid transition; $p < 0.05$, FDR corrected.

3.3.2. Amygdala Subfields

Multiple linear regression analysis (adjusted amygdala total volume) showed that the amygdala and biochemical parameters were also related in the two CRF groups (Table 4). In the low-CRF group, the EO predicted the left corticoamygdaloid transition volume (Beta = -0.626, r^2 = 0.21 $p < 0.05$, FDR corrected).

4. Discussion

To our knowledge, this is the first study to explore the effects of CRF on hippocampus and amygdala volumes under a real high-altitude environment. This study found that there were significant differences in the hippocampal and amygdala subregions related to

CRF levels. When compared with the low-CRF group, the volumes of the hippocampus and amygdala subfields were improved in the high-CRF group, and these changes were associated with lower inflammatory responses. This suggests the existence of a relationship between CRF levels and volume changes in the hippocampus and amygdala in high-altitude environments, which may be associated with hypoxia stress.

Previous studies have found that hypoxia severely affects gray matter volumes in the hippocampus and amygdala [44]. In this study, we found that the gray matter volume in the hippocampal subfields and amygdala subfields were increased in the high-CRF group when high-altitude participants were divided into two groups according to median VO_{2max}. This indicates that resilient cardiorespiratory function in high-altitude environments effectively protects against the negative effects of high altitude and a low oxygen environment on the hippocampus and amygdala. Previous literature has focused on analysis of the relationship between the volume of subcortical brain structures and CRF, and have found that higher levels of CRF are associated with greater volumes in the hippocampus and basal ganglia [45,46]. Correspondingly, increased CRF is associated with an increase in general cortical thickness [47], and higher CRF levels are associated with higher brain-derived neurotrophic factor (BDNF) levels, especially in the hippocampus. Increased hippocampal volume is positively correlated with BDNF levels [22]. Additionally, higher CRF is associated with increased volume of the hippocampus and amygdala in Alzheimer's patients [22]. Better CRF improves the brain and behavior, as well as neurogenesis, in both healthy and dementia models, reduces toxicity and cerebral amyloids, and reduces inflammation and oxidative stress [48]. Consistent with these prior investigations, our findings further suggested that high CRF is significantly related to increased gray matter volume in the hippocampus and amygdala at high altitude among immigrant participants.

We found that the low-CRF group was more affected by inflammation than the high-CRF group. Eosinophil effectively predicted the volume of the amygdala and hippocampal subfields in the low-CRF group. The release of cortisol relates the secretion of cytokines, especially interleukin-5, which stimulates the production and differentiation of granulocytes, such as eosinophils [49]. The inflammatory response induced by HPA axis disorder affects the volume of the hippocampus and amygdala under the stress state [50]. Inflammation is often accompanied by the apoptosis of a large number of cells and changes in the nervous system [51]. Eosinophils are important markers of inflammation and are associated with damage to neurons in the hippocampus and amygdala [52,53]. Eosinophils were observed in the hippocampal and amygdala neuron damage induced by state epilepsy in mice [54]. Previous studies have shown that improved lung function is associated with decreased eosinophils [55]. We also found that NEUT were effective predictors of hippocampal subfields volume in the low-CRF group. NEUT, as phagocytes, play an important role in inflammatory immune regulation [56]. Tissue damage caused by inflammation leads to the excessive activation of NEUT, which leads to an aggravated inflammatory response [57]. The accumulation of neutrophils in the brain has been associated with increased secondary brain damage and poor neurological outcomes [58]. Traumatic brain injury results in an inflammatory response in the brain, accompanied by an influx of neutrophils into the cerebral cortex and especially the hippocampus [59]. A lower CRF was also related to greater WBC, as well as neutrophil, lymphocyte, and monocyte, counts [60]. Higher CRF reduces NEUT content and improves the immune response [61,62]. Other studies have found that higher CRF produces oxidative damage in neutrophils and induces antioxidant defenses in lymphocytes [63]. Our results showed that the hippocampus and amygdala in the low-CRF group were more susceptible to inflammation at high altitudes, resulting in smaller amygdala and hippocampal volumes. However, the relationship between these parameters and the amygdala and hippocampus was not found in the high-CRF group, suggesting that high CRF can reduce the inflammatory response and improve the plasticity of the hippocampus and amygdala under high-altitude stress, which is similar to the findings in previous studies [64,65].

Oxygen is transported primarily by hemoglobin in red blood cells [66,67]. We found RBC and HGB were higher, and RDW-SD smaller, in the high-CRF group. RDW is a measurement of the size variation, as well as an index of the heterogeneity, of the erythrocytes (i.e., anisocytosis). Higher RDW values reflect greater variation in RBC volumes and were found to be related to many inflammation diseases in previous studies [68]. Because the erythrocyte represents the body's oxygen carrier, its redox and metabolic status is extremely important for the functioning and regulation of oxygen affinity to hemoglobin, which is determined by a number of metabolites within the erythrocyte. All tissues are dependent on RBC function, especially neurons, which use 20% of the total oxygen consumed [69]. Reduced hemoglobin levels in the hippocampus and other neurons have been found in studies of neurodegenerative diseases [70]. Increases in erythrocytes and hemoglobin are accompanied by increased hippocampal oxygenation under hypoxia [71]. On the other hand, RBCs mediate the immune system's ability to reduce inflammation and stress [72]. Another study also found that the red blood cells of physically active rats were more resistant to oxidative stress after they were deprived of oxygen [73]. Our previous study found that healthy RBC at high altitude can support the immune system [40]. Our results here showed that individuals with high CRF levels had larger hippocampal and amygdala volumes than those with low CRF levels, possibly because individuals with high CRF levels have higher oxygen transport capacity and higher immune levels.

In addition, we found that the DBIL and TBIL of the high-CRF group negatively predicted the hippocampal subfield volume, and the DBIL and TBIL in the high-CRF group was significantly increased. The increase in DBIL and TBIL may be caused by the increase in RBC and HGB. Bilirubin has been commonly considered to be simply the "final product" of heme catabolism [74]. The rise in bilirubin in rats exposed to high altitudes may be due to an increase in red blood cell counts [75]. Bilirubin is an endogenous antioxidant that plays an important role in the anti-oxidative stress and anti-inflammation of neurons [76,77]. Bilirubin enhances the bactericidal ability of neutrophils [78]. There is evidence that it protects the cardiovascular system, neuronal systems, the hepatobiliary system, the pulmonary system, and the immune system [74]. Mildly elevated serum bilirubin is generally associated with the attenuation of oxidative stress and with a decreased inflammatory status [79]. On the other hand, a decrease in bilirubin during hypoxia is associated with increased inflammation [80]. The accumulation of bilirubin acts as an effective defense mechanism against stress and increased inflammation [81]. In conclusion, our results suggest that higher CRF at high altitudes can reduce the hypoxic stress response, thereby inducing immunity and reducing inflammation, which may be related to increased oxygen transport capacity of the body.

There were several issues which should be considered in future works. First, a longitudinal trace study should be included to validate the present findings, in which the behavioral data including the gender difference, life style, personality, and so on should be considered. Second, the physiological characteristics of hypoxia tolerance are closely related to the genome, and this could provide a potential way to better understand the molecular mechanisms of human adaptation to high altitudes [82]. Third, the neuroinflammatory markers and the role of epigenetics should be explored related to the effects of exercise and oxygen interventions on the high-altitude participants for altitude adaptation.

5. Conclusions

This study indicated that higher CRF can significantly protect against the decrease in hippocampal and amygdala volume induced by high-altitude hypoxia. A better CRF might be related to a lower physiological stress response. These findings provide new insights into the stress response to hypoxia in human adaptation.

Author Contributions: Conceptualization, Z.-X.W. and T.-A.Z.; methodology, Z.-X.W. and D.-M.C.; software, Z.-X.W., P.D. and J.-G.W.; validation, H.L., R.S. and H.-L.M.; formal analysis, D.-L.Z.; investigation, D.-L.Z.; resources, H.-L.M.; data curation, Z.-X.W.; writing—original draft preparation, Z.-X.W.; writing—review and editing, Z.-X.W.; visualization, Z.-X.W.; supervision, D.-L.Z. and H.-L.M.; project administration, H.-L.M.; funding acquisition, H.-L.M. and D.-L.Z. All authors have read and agreed to the published version of the manuscript.

Funding: This work was supported by the National Natural Science Foundation of China (31660274, 31771247, and 31600907), the Reformation and Development Funds for Local Region Universities from the Chinese Government in 2020 (00060607, ZCJK 2020-11), and the High-Level Postgradu-ate Talent Training Program of Tibet University (2019-GSP-S108).

Institutional Review Board Statement: This study was conducted in accordance with the Declaration of Helsinki, and approved by the Institutional Review Board (or Ethics Committee) of the Ethics Committee of Tibet University.

Informed Consent Statement: Informed consent was obtained from all subjects involved in the study. Written informed consent was obtained from the patient(s) to publish this paper.

Data Availability Statement: Data materials can be obtained by contacting the corresponding author.

Conflicts of Interest: The authors declare no conflict of interest.

References

1. Hossmann, K.-A. The Hypoxic Brain. In *Hypoxia: Into the Next Millennium*; Roach, R.C., Wagner, P.D., Hackett, P.H., Eds.; Springer: Boston, MA, USA, 1999; pp. 155–169.
2. Kumari, P.; Kauser, H.; Wadhwa, M.; Roy, K.; Alam, S.; Sahu, S.; Kishore, K.; Ray, K.; Panjwani, U. Hypobaric hypoxia impairs cued and contextual fear memory in rats. *Brain Res.* **2018**, *1692*, 118–133. [CrossRef] [PubMed]
3. Maiti, P.; Singh, S.B.; Mallick, B.; Muthuraju, S.; Ilavazhagan, G. High altitude memory impairment is due to neuronal apoptosis in hippocampus, cortex and striatum. *J. Chem. Neuroanat.* **2008**, *36*, 227–238. [CrossRef] [PubMed]
4. Zhang, J.; Zhang, H.; Li, J.; Chen, J.; Han, Q.; Lin, J.; Yang, T.; Fan, M. Adaptive Modulation of Adult Brain Gray and White Matter to High Altitude: Structural MRI Studies. *PLoS ONE* **2013**, *8*, e68621. [CrossRef]
5. Ma, H.; Huang, X.; Liu, M.; Ma, H.; Zhang, D. Aging of stimulus-driven and goal-directed attentional processes in young immigrants with long-term high altitude exposure in Tibet: An ERP study. *Sci. Rep.* **2018**, *8*, 17417. [CrossRef]
6. Zhang, D.; Zhang, X.; Ma, H.; Wang, Y.; Ma, H.; Liu, M. Competition among the attentional networks due to resource reduction in Tibetan indigenous residents: Evidence from event-related potentials. *Sci. Rep.* **2018**, *8*, 610. [CrossRef] [PubMed]
7. McEwen, B.S. Physiology and Neurobiology of Stress and Adaptation: Central Role of the Brain. *Physiol. Rev.* **2007**, *87*, 873–904. [CrossRef]
8. Rybnikova, E.; Nalivaeva, N. Glucocorticoid-Dependent Mechanisms of Brain Tolerance to Hypoxia. *Int. J. Mol. Sci.* **2021**, *22*, 7982. [CrossRef]
9. Zhao, F.; Yang, J.; Cui, R. Effect of Hypoxic Injury in Mood Disorder. *Neural Plast.* **2017**, *2017*, 6986983. [CrossRef]
10. Gainey, S.J. Hypoxic immunomodulation results in increased disease risk and altered behavior via non-canonical pathways. Ph.D. Thesis, University of Illinois at Urbana-Champaign, Champaign, IL, USA, 28 February 2017.
11. Yang, T.; Zhou, D.; Stefan, H. Why mesial temporal lobe epilepsy with hippocampal sclerosis is progressive: Uncontrolled inflammation drives disease progression? *J. Neurol. Sci.* **2010**, *296*, 1–6. [CrossRef]
12. Zhang, J.; Malik, A.; Choi, H.B.; Ko, R.W.; Dissing-Olesen, L.; MacVicar, B.A. Microglial CR3 Activation Triggers Long-Term Synaptic Depression in the Hippocampus via NADPH Oxidase. *Neuron* **2014**, *82*, 195–207. [CrossRef]
13. Feng, J.-F.; Zhao, X.; Gurkoff, G.G.; Van, K.C.; Shahlaie, K.; Lyeth, B.G. Post-Traumatic Hypoxia Exacerbates Neuronal Cell Death in the Hippocampus. *J. Neurotrauma* **2012**, *29*, 1167–1179. [CrossRef] [PubMed]
14. Michalak, Z.; Obari, D.; Ellis, M.; Thom, M.; Sisodiya, S.M. Neuropathology of SUDEP. *Neurology* **2017**, *88*, 551–561. [CrossRef] [PubMed]
15. Xu, L.-H.; Xie, H.; Shi, Z.-H.; Du, L.-D.; Wing, Y.-K.; Li, A.M.; Ke, Y.; Yung, W.-H. Critical Role of Endoplasmic Reticulum Stress in Chronic Intermittent Hypoxia-Induced Deficits in Synaptic Plasticity and Long-Term Memory. *Antioxid. Redox Signal.* **2015**, *23*, 695–710. [CrossRef] [PubMed]
16. Carty, M.L.; Wixey, J.A.; Kesby, J.; Reinebrant, H.E.; Colditz, P.B.; Gobe, G.; Buller, K.M. Long-term losses of amygdala corticotropin-releasing factor neurons are associated with behavioural outcomes following neonatal hypoxia-ischemia. *Behav. Brain Res.* **2010**, *208*, 609–618. [CrossRef]
17. Suliga, E. Chapter 4—Lifestyle Factors Affecting Abdominal Obesity in Children and Adolescents: Risks and Benefits. In *Nutrition in the Prevention and Treatment of Abdominal Obesity*; Watson, R.R., Ed.; Academic Press: San Diego, CA, USA, 2014; pp. 39–56.
18. Johnson, N.F.; Kim, C.; Clasey, J.L.; Bailey, A.; Gold, B.T. Cardiorespiratory fitness is positively correlated with cerebral white matter integrity in healthy seniors. *NeuroImage* **2012**, *59*, 1514–1523. [CrossRef]

19. Lee, J.T.; Chaloner, E.J.; Hollingsworth, S.J. The role of cardiopulmonary fitness and its genetic influences on surgical outcomes. *Br. J. Surg.* **2005**, *93*, 147–157. [CrossRef]
20. Sapolsky, R.M. Glucocorticoids and Hippocampal Atrophy in Neuropsychiatric Disorders. *Arch. Gen. Psychiatry* **2000**, *57*, 925–935. [CrossRef]
21. Lucassen, P.J.; Meerlo, P.; Naylor, A.S.; van Dam, A.M.; Dayer, A.G.; Fuchs, E.; Oomen, C.A.; Czéh, B. Regulation of adult neurogenesis by stress, sleep disruption, exercise and inflammation: Implications for depression and antidepressant action. *Eur. Neuropsychopharmacol.* **2010**, *20*, 1–17. [CrossRef]
22. Boots, E.; Schultz, S.; Oh, J.M.; Larson, J.; Edwards, D.; Cook, D.; Koscik, R.L.; Dowling, M.N.; Gallagher, C.L.; Carlsson, C.M.; et al. Cardiorespiratory fitness is associated with brain structure, cognition, and mood in a middle-aged cohort at risk for Alzheimer's disease. *Brain Imaging Behav.* **2014**, *9*, 639–649. [CrossRef]
23. Prathap, S.; Nagel, B.J.; Herting, M.M. Understanding the role of aerobic fitness, spatial learning, and hippocampal subfields in adolescent males. *Sci. Rep.* **2021**, *11*, 9311. [CrossRef]
24. Erickson, K.I.; Prakash, R.S.; Voss, M.W.; Chaddock, L.; Hu, L.; Morris, K.S.; White, S.M.; Wójcicki, T.R.; McAuley, E.; Kramer, A.F. Aerobic fitness is associated with hippocampal volume in elderly humans. *Hippocampus* **2009**, *19*, 1030–1039. [CrossRef] [PubMed]
25. Herting, M.M.; Nagel, B.J. Aerobic fitness relates to learning on a virtual Morris Water Task and hippocampal volume in adolescents. *Behav. Brain Res.* **2012**, *233*, 517–525. [CrossRef] [PubMed]
26. Lin, T.-W.; Shih, Y.-H.; Chen, S.-J.; Lien, C.-H.; Chang, C.-Y.; Huang, T.-Y.; Chen, S.-H.; Jen, C.J.; Kuo, Y.-M. Running exercise delays neurodegeneration in amygdala and hippocampus of Alzheimer's disease (APP/PS1) transgenic mice. *Neurobiol. Learn. Mem.* **2015**, *118*, 189–197. [CrossRef]
27. Seo, B.; Kim, D.; Choi, D.; Kwon, C.; Shin, H. The Effect of Electrical Stimulation on Blood Lactate after Anaerobic Muscle Fatigue Induced in Taekwondo Athletes. *J. Phys. Ther. Sci.* **2011**, *23*, 271–275. [CrossRef]
28. Heath, E.H. *ACSM's Guidelines for Exercise Testing and Prescription*, 7th ed.; Lippincott Williams & Wilkins: Philadelphia, PA, USA, 2005; p. 37.
29. Grocott, M.; Montgomery, H.; Vercueil, A. High-altitude physiology and pathophysiology: Implications and relevance for intensive care medicine. *Crit. Care* **2007**, *11*, 203. [CrossRef]
30. Ward, S.A.; Grocott, M.P.; Levett, D.Z. Exercise testing, supplemental oxygen, and hypoxia. *Ann. Am. Thorac. Soc.* **2017**, *14*, S140–S148. [CrossRef] [PubMed]
31. Mairbäurl, H. Red blood cells in sports: Effects of exercise and training on oxygen supply by red blood cells. *Front. Physiol.* **2013**, *4*, 332. [CrossRef]
32. Kleppe, S.; Bernhardt, C.; Wölfle, J.; Breuer, J. Red Blood Cell Function in Hypoxia at Altitude and Exercise. *Int. J. Sports Med.* **1994**, *15*, 51–63.
33. Ogoh, S.; Ainslie, P.N. Regulatory Mechanisms of Cerebral Blood Flow during Exercise: New Concepts. *Exerc. Sport Sci. Rev.* **2009**, *37*, 123–129. [CrossRef]
34. Smith, T.B.; Stonell, C.; Purkayastha, S.; Paraskevas, P. Cardiopulmonary exercise testing as a risk assessment method in non cardio-pulmonary surgery: A systematic review. *Anaesthesia* **2009**, *64*, 883–893. [CrossRef]
35. Grover, R.F.; Weil, J.V.; Reeves, J.T. Cardiovascular adaptation to exercise at high altitude. *Exerc. Sport Sci. Rev.* **1986**, *14*, 269–302. [CrossRef] [PubMed]
36. Haugen, A.H.; Melanson, E.L.; Tran, Z.V.; Kearney, J.T.; Hill, J.O. Variability of measured resting metabolic rate. *Am. J. Clin. Nutr.* **2003**, *78*, 1141–1144. [CrossRef] [PubMed]
37. Mala, L.; Maly, T.; Zahalka, F.; Heller, J.; Hrasky, P.; Vodicka, P. Differences in the morphological and physiological charac-teristics of senior and junior elite Czech judo athletes. *Arch. Budo* **2015**, *11*, 217–226.
38. Iglesias, J.E.; Augustinack, J.C.; Nguyen, K.; Player, C.M.; Player, A.; Wright, M.; Roy, N.; Frosch, M.P.; McKee, A.C.; Wald, L.; et al. A computational atlas of the hippocampal formation using ex vivo, ultra-high resolution MRI: Application to adaptive segmentation of in vivo MRI. *NeuroImage* **2015**, *115*, 117–137. [CrossRef] [PubMed]
39. Saygin, Z.; Kliemann, D.; Iglesias, J.E.; van der Kouwe, A.; Boyd, E.; Reuter, M.; Stevens, A.; Van Leemput, K.; McKee, A.; Frosch, M.; et al. High-resolution magnetic resonance imaging reveals nuclei of the human amygdala: Manual segmentation to automatic atlas. *NeuroImage* **2017**, *155*, 370–382. [CrossRef]
40. Xue, X.-J.; Su, R.; Li, Z.-F.; Bu, X.-O.; Dang, P.; Yu, S.-F.; Wang, Z.-X.; Chen, D.-M.; Zeng, T.-A.; Liu, M.; et al. Oxygen Metabolism-induced Stress Response Underlies Heart–brain Interaction Governing Human Consciousness-breaking and Attention. *Neurosci. Bull.* **2021**, *38*, 166–180. [CrossRef]
41. Liu, S.; Zhao, Y.; Ren, Q.; Gong, G.; Zhang, D.; Shao, K.; Lin, P.; Yuan, Y.; Dai, T.; Zhang, Y.; et al. Research Square 2021. Available online: https://assets.researchsquare.com/files/rs-900442/v1/7c67881d-baba-4d40-b494-bd49d882bd54.pdf?c=1632165377 (accessed on 25 February 2022).
42. Brown, S.S.G.; Rutland, J.W.; Verma, G.; Feldman, R.E.; Alper, J.; Schneider, M.; Delman, B.; Murrough, J.M.; Balchandani, P. Structural MRI at 7T reveals amygdala nuclei and hippocampal subfield volumetric association with Major Depressive Disorder symptom severity. *Sci. Rep.* **2019**, *9*, 10166. [CrossRef]
43. Benjamini, Y.; Hochberg, Y. Controlling the False Discovery Rate: A Practical and Powerful Approach to Multiple Testing. *J. R. Stat. Soc. Ser. B* **1995**, *57*, 289–300. [CrossRef]

44. Haukvik, U.K.; McNeil, T.; Lange, E.H.; Melle, I.; Dale, A.M.; Andreassen, O.A.; Agartz, I. Pre- and perinatal hypoxia associated with hippocampus/amygdala volume in bipolar disorder. *Psychol. Med.* **2013**, *44*, 975–985. [CrossRef]
45. Chaddock, L.; Erickson, K.I.; Prakash, R.S.; Kim, J.S.; Voss, M.W.; Vanpatter, M.; Pontifex, M.B.; Raine, L.B.; Konkel, A.; Hillman, C.H.; et al. A neuroimaging investigation of the association between aerobic fitness, hippocampal volume, and memory performance in preadolescent children. *Brain Res.* **2010**, *1358*, 172–183. [CrossRef]
46. Chaddock, L.; Hillman, C.; Pontifex, M.; Johnson, C.R.; Raine, L.B.; Kramer, A. Childhood aerobic fitness predicts cognitive performance one year later. *J. Sports Sci.* **2012**, *30*, 421–430. [CrossRef] [PubMed]
47. Engeroff, T.; Füzéki, E.; Vogt, L.; Fleckenstein, J.; Schwarz, S.; Matura, S.; Pilatus, U.; Deichmann, R.; Hellweg, R.; Pantel, J.; et al. Is Objectively Assessed Sedentary Behavior, Physical Activity and Cardiorespiratory Fitness Linked to Brain Plasticity Outcomes in Old Age? *Neuroscience* **2018**, *388*, 384–392. [CrossRef] [PubMed]
48. Daniele, T.M.D.C.; de Bruin, P.F.C.; de Matos, R.S.; de Bruin, G.S.; Chaves, C.M.; de Bruin, V.M.S. Exercise effects on brain and behavior in healthy mice, Alzheimer's disease and Parkinson's disease model—A systematic review and meta-analysis. *Behav. Brain Res.* **2020**, *383*, 112488. [CrossRef]
49. Segerstrom, S.C.; Miller, G.E. Psychological Stress and the Human Immune System: A Meta-Analytic Study of 30 Years of Inquiry. *Psychol. Bull.* **2004**, *130*, 601–630. [CrossRef] [PubMed]
50. Majidi, J.; Kosari-Nasab, M.; Salari, A.-A. Developmental minocycline treatment reverses the effects of neonatal immune activation on anxiety- and depression-like behaviors, hippocampal inflammation, and HPA axis activity in adult mice. *Brain Res. Bull.* **2016**, *120*, 1–13. [CrossRef]
51. Dickens, A.M.; Tovar-Y-Romo, L.B.; Yoo, S.-W.; Trout, A.L.; Bae, M.; Kanmogne, M.; Megra, B.; Williams, D.W.; Witwer, K.W.; Gacias, M.; et al. Astrocyte-shed extracellular vesicles regulate the peripheral leukocyte response to inflammatory brain lesions. *Sci. Signal.* **2017**, *10*, eaai7696. [CrossRef]
52. Collombet, J.-M.; Piérard, C.; Béracochéa, D.; Coubard, S.; Burckhart, M.-F.; Four, E.; Masqueliez, C.; Baubichon, D.; Lallement, G. Long-term consequences of soman poisoning in mice: Part 1. Neuropathology and neuronal regeneration in the amygdala. *Behav. Brain Res.* **2008**, *191*, 88–94. [CrossRef]
53. Men, S.; Lee, D.H.; Barron, J.R.; Muñoz, D.G. Selective Neuronal Necrosis Associated with Status Epilepticus: MR Findings. *Am. J. Neuroradiol.* **2000**, *21*, 1837–1840.
54. Loss, C.M.; Córdova, S.D.; de Oliveira, D.L. Ketamine reduces neuronal degeneration and anxiety levels when administered during early life-induced status epilepticus in rats. *Brain Res.* **2012**, *1474*, 110–117. [CrossRef]
55. Handayani, R.N.; Yunus, F.; Ibrahim, E.I.; Rengganis, I. Correlation between Improve Lung Function with Decrease of Eosinophil Levels in Atopic Asthma Persistent After Asthma Exercise. *Ann. Trop. Med. Public Heal.* **2019**, *22*, 338–346. [CrossRef]
56. Bae, G.H.; Lee, H.Y.; Jung, Y.S.; Shim, J.W.; Kim, S.D.; Baek, S.-H.; Kwon, J.Y.; Park, J.S.; Bae, Y.-S. Identification of novel peptides that stimulate human neutrophils. *Exp. Mol. Med.* **2012**, *44*, 130–137. [CrossRef] [PubMed]
57. Pillay, J.; Hietbrink, F.; Koenderman, L.; Leenen, L. The systemic inflammatory response induced by trauma is reflected by multiple phenotypes of blood neutrophils. *Injury* **2007**, *38*, 1365–1372. [CrossRef] [PubMed]
58. Stahel, P.F.; Morganti-Kossmann, C.; Kossmann, T. The role of the complement system in traumatic brain injury. *Brain Res. Rev.* **1998**, *27*, 243–256. [CrossRef]
59. Keeling, K.; Hicks, R.; Mahesh, J.; Billings, B.; Kotwal, G. Local neutrophil influx following lateral fluid-percussion brain injury in rats is associated with accumulation of complement activation fragments of the third component (C3) of the complement system. *J. Neuroimmunol.* **2000**, *105*, 20–30. [CrossRef]
60. Craige, S.M.; Kant, S.; Jr, J.F.K. Reactive Oxygen Species in Endothelial Function: From Disease to Adaptation. *Circ. J.* **2015**, *79*, 1145–1155. [CrossRef] [PubMed]
61. Mooren, F.C.; Blöming, D.; Lechtermann, A.; Lerch, M.M.; Völker, K. Lymphocyte apoptosis after exhaustive and moderate exercise. *J. Appl. Physiol.* **2002**, *93*, 147–153. [CrossRef] [PubMed]
62. Suzuki, K.; Sato, H.; Kikuchi, T.; Abe, T.; Nakaji, S.; Sugawara, K.; Totsuka, M.; Sato, K.; Yamaya, K. Capacity of circulating neutrophils to produce reactive oxygen species after exhaustive exercise. *J. Appl. Physiol.* **1996**, *81*, 1213–1222. [CrossRef]
63. Ferrer, M.D.; Tauler, P.; Sureda, A.; Tur, J.A.; Pons, A. Antioxidant regulatory mechanisms in neutrophils and lymphocytes after intense exercise. *J. Sports Sci.* **2009**, *27*, 49–58. [CrossRef]
64. Stults-Kolehmainen, M.A.; Sinha, R. The Effects of Stress on Physical Activity and Exercise. *Sports Med.* **2013**, *44*, 81–121. [CrossRef]
65. Wärnberg, J.; Cunningham, K.; Romeo, J.; Marcos, A. Physical activity, exercise and low-grade systemic inflammation. *Proc. Nutr. Soc.* **2010**, *69*, 400–406. [CrossRef]
66. D'Alessandro, A.; Nemkov, T.; Sun, K.; Liu, H.; Song, A.; Monte, A.A.; Subudhi, A.W.; Lovering, A.T.; Dvorkin, D.; Julian, C.G.; et al. AltitudeOmics: Red Blood Cell Metabolic Adaptation to High Altitude Hypoxia. *J. Proteome Res.* **2016**, *15*, 3883–3895. [CrossRef] [PubMed]
67. Risso, A.; Turello, M.; Biffoni, F.; Antonutto, G. Red blood cell senescence and neocytolysis in humans after high altitude acclimatization. *Blood Cells Mol. Dis.* **2007**, *38*, 83–92. [CrossRef] [PubMed]
68. Celik, A.; Aydin, N.; Ozcirpici, B.; Saricicek, E.; Sezen, H.; Okumus, M.; Bozkurt, S.; Kilinc, M. Elevated red blood cell distri-bution width and inflammation in printing workers. *Med. Sci. Monit.* **2013**, *19*, 1001–1005. [PubMed]

69. Wojsiat, J.; Laskowska-Kaszub, K.; Mietelska-Porowska, A.; Wojda, U. Search for Alzheimer's disease biomarkers in blood cells: Hypotheses-driven approach. *Biomarkers Med.* **2017**, *11*, 917–931. [CrossRef]
70. Ferrer, I.; Gómez, A.; Carmona, M.; Huesa, G.; Porta, S.; Riera-Codina, M.; Biagioli, M.; Gustincich, S.; Aso, E. Neuronal Hemoglobin is Reduced in Alzheimer's Disease, Argyrophilic Grain Disease, Parkinson's Disease, and Dementia with Lewy Bodies. *J. Alzheimers Dis.* **2011**, *23*, 537–550. [CrossRef]
71. Schelshorn, D.W.; Schneider, A.; Kuschinsky, W.; Weber, D.; Krüger, C.; Dittgen, T.; Bürgers, H.F.; Sabouri, F.; Gassler, N.; Bach, A.; et al. Expression of Hemoglobin in Rodent Neurons. *J. Cereb. Blood Flow Metab.* **2008**, *29*, 585–595. [CrossRef]
72. Buttari, B.; Profumo, E.; Riganò, R. Crosstalk between Red Blood Cells and the Immune System and Its Impact on Atherosclerosis. *BioMed. Res. Int.* **2015**, *2015*, 616834. [CrossRef]
73. Devi, S.A.; Subramanyam, M.; Vani, R.; Jeevaratnam, K. Adaptations of the antioxidant system in erythrocytes of trained adult rats: Impact of intermittent hypobaric-hypoxia at two altitudes. *Comp. Biochem. Physiol. Part C Toxicol. Pharmacol.* **2005**, *140*, 59–67. [CrossRef]
74. Otero Regino, W.; Velasco, H.; Sandoval, H. Papel protector de la bilirrubina en el ser humano. *Rev. Colomb. Gastroenterología* **2009**, *24*, 293–301. (In Spanish)
75. Altland, P.D.; Parker, M.G. Bilirubinemia and intravascular hemolysis during acclimatization to high altitude. *Int. J. Biometeorol.* **1977**, *21*, 165–170. [CrossRef]
76. Dani, C.; Poggi, C.; Fancelli, C.; Pratesi, S. Changes in bilirubin in infants with hypoxic–ischemic encephalopathy. *Eur. J. Pediatr.* **2018**, *177*, 1795–1801. [CrossRef] [PubMed]
77. Vaz, A.R.; Silva, S.L.; Barateiro, A.; Falcão, A.S.; Fernandes, A.; Brito, M.A.; Brites, D. Selective vulnerability of rat brain regions to unconjugated bilirubin. *Mol. Cell. Neurosci.* **2011**, *48*, 82–93. [CrossRef] [PubMed]
78. Thong, Y.; Ness, D.; Ferrante, A. Effect of bilirubin on the fungicidal capacity of human neutrophils. *Med. Mycol.* **1979**, *17*, 125–129. [CrossRef] [PubMed]
79. Zelenka, J.; Dvořák, A.; Alán, L.; Zadinová, M.; Haluzik, M.; Vítek, L. Hyperbilirubinemia Protects against Aging-Associated Inflammation and Metabolic Deterioration. *Oxidative Med. Cell. Longev.* **2016**, *2016*, 6190609. [CrossRef] [PubMed]
80. Friedlander, A.H.; Boström, K.I.; Tran, H.-A.; Chang, T.I.; Polanco, J.C.; Lee, U.K. Severe Sleep Apnea Associated with Increased Systemic Inflammation and Decreased Serum Bilirubin. *J. Oral Maxillofac. Surg.* **2019**, *77*, 2318–2323. [CrossRef]
81. Zelenka, J.; Muchova, L.; Zelenkova, M.; Vanova, K.; Vreman, H.J.; Wong, R.J.; Vitek, L. Intracellular accumulation of bilirubin as a defense mechanism against increased oxidative stress. *Biochimie* **2012**, *94*, 1821–1827. [CrossRef]
82. Julian, C.G. Epigenomics and human adaptation to high altitude. *J. Appl. Physiol.* **2017**, *123*, 1362–1370. [CrossRef]

brain sciences

Article

Balance Expertise Is Associated with Superior Spatial Perspective-Taking Skills

Kirsten Hötting *[iD], Ann-Kathrin Rogge, Laura A. Kuhne and Brigitte Röder

Biological Psychology and Neuropsychology, Universität Hamburg, Von-Melle-Park 11, 20146 Hamburg, Germany; ann-kathrin.rogge@uni-hamburg.de (A.-K.R.); laura.andrea.kuhne@uni-hamburg.de (L.A.K.); brigitte.roeder@uni-hamburg.de (B.R.)
* Correspondence: kirsten.hoetting@uni-hamburg.de; Tel.: +49-40-42838-2621

Abstract: Balance training interventions over several months have been shown to improve spatial cognitive functions and to induce structural plasticity in brain regions associated with visual-vestibular self-motion processing. In the present cross-sectional study, we tested whether long-term balance practice is associated with better spatial cognition. To this end, spatial perspective-taking abilities were compared between balance experts ($n = 40$) practicing sports such as gymnastics, acrobatics or slacklining for at least four hours a week for the last two years, endurance athletes ($n = 38$) and sedentary healthy individuals ($n = 58$). The balance group showed better performance in a dynamic balance task compared to both the endurance group and the sedentary group. Furthermore, the balance group outperformed the sedentary group in a spatial perspective-taking task. A regression analysis across all participants revealed a positive association between individual balance performance and spatial perspective-taking abilities. Groups did not differ in executive functions, and individual balance performance did not correlate with executive functions, suggesting a specific association between balance skills and spatial cognition. The results are in line with theories of embodied cognition, assuming that sensorimotor experience shapes cognitive functions.

Keywords: motor expertise; balance; spatial cognition; physical activity

Citation: Hötting, K.; Rogge, A.-K.; Kuhne, L.A.; Röder, B. Balance Expertise Is Associated with Superior Spatial Perspective-Taking Skills. *Brain Sci.* **2021**, *11*, 1401. https://doi.org/10.3390/brainsci11111401

Academic Editors: Filipe Manuel Clemente and Ana Filipa Silva

Received: 22 September 2021
Accepted: 21 October 2021
Published: 24 October 2021

Publisher's Note: MDPI stays neutral with regard to jurisdictional claims in published maps and institutional affiliations.

1. Introduction

The study of experts who achieved very high levels of perceptual and sensorimotor performance in their field after years of extensive training has a long tradition in psychology and neuroscience, revealing fundamental principles of skill acquisition and their underlying neuronal mechanisms [1,2]. Athletes, in particular, have been extensively studied in this context, as they acquire very specific sensorimotor skills over years of regular and structured training [3,4]. For instance, Land and McLeod [5] showed that professional cricket batsmen were superior in judging when and where the ball will hit the ground compared to amateur players. The batsmen's superior performance was specifically linked to reduced latencies of their initial saccade. Elite basketball players were reported to better anticipate other basketball players' shots, but not soccer kicks, compared to non-experts, and the enhanced anticipatory skills correlated with enhanced excitability of their motor cortices [6].

It is a matter of debate whether enhanced sports-specific skills transfer to the more general cognitive functions assessed with psychometric tests. Visuo-spatial tasks, and mental rotation in particular, are the most studied tasks in this context. Several studies showed enhanced mental rotation skills in experts in combat sports, gymnastics and dancing compared to athletes of other disciplines or non-athletes, with the largest effect sizes for combat sports (for a meta-analysis, see [7]). The findings of expert studies in sports have been discussed in the context of embodied cognition. According to the embodied cognition framework, sensorimotor interactions with the environment play an important role in the development and maintenance of higher cognitive skills [8]. Exercise-related cognitive benefits may be due to a stimulation of overlapping brain networks involved in both the sensorimotor practice and specific cognitive processes, which in turn might transfer to improved performance in psychometric tests addressing these cognitive functions, such as mental rotation. Athletes performing combat sports or gymnastics are highly trained in performing and imaging their own body transformations, even in very unfamiliar body positions [9]. In addition, in combat sports, it is crucial to represent an opponent's body in space relative to one's own body to anticipate the opponent's movements. Thus, enhanced mental rotation skills, measured with psychometric tests in martial artists (combat sports) and gymnasts, might indicate a transfer of skills acquired during motor training, which is in line with predictions of embodied cognition theories.

Gymnasts have been shown to outperform athletes of many other disciplines in balance skills [10]. Balancing requires a rapid and continuous integration of vestibular, somatosensory and visual signals. The integration of these signals is not only important for postural control, but seems to be essential for cognitive functions such as self-motion perception, body self-consciousness, spatial navigation and spatial memory [11]. For instance, spatial memory deficits have been reported in patients with peripheral vestibular lesions [12]. Experimental vestibular stimulation has been shown to impair performance in perspective-taking and mental rotation tasks [13,14]. The vestibular system has widespread cortical connections to brain regions known to be involved in spatial cognition, like the posterior parietal cortex, the temporo-parietal junction, the retrosplenial cortex and the hippocampus [11]. Thus, one might hypothesize that physical exercise, such as balancing, which stimulates particularly vestibular pathways, should have an impact on visuo-spatial skills. In line with this assumption, Dordevic et al. [15] reported improvements in a spatial orientation task after one month of balance training on a slackline. Rogge et al. [16] showed that 12 weeks of a balance training compared to relaxation training improved not only dynamic balance performance, but also memory and spatial cognition in healthy adults. These cognitive effects were rather specific for spatial skills, while no group differences were found in executive functions or response speed. Moreover, cortical thickness was increased in the balance group in brain regions associated with visual and vestibular self-motion processing, such as the superior temporal cortex, visual association cortices, the posterior cingulate cortex, the superior frontal sulcus and the precentral gyri [17]. In a cross-sectional study, professional dancers and slackliners were found to have larger grey matter volumes in the posterior hippocampus compared to non-balance experts [18]. On a behavioral level, balance experts outperformed non-experts in a hippocampus-dependent configurational learning task, but not in spatial memory and navigation tasks [18].

Taken together, athletes with a history of extensive training in combat sports, gymnastics and dancing have been found to outperform athletes of many other disciplines and non-athletes in visuo-spatial tasks requiring mental transformation of their own body or objects in space. These enhanced skills might be mediated by practicing mental rotation during training. This hypothesis is further supported by the observation that users of sign language, which provides extensive practice in visuo-spatial processing, typically have superior mental rotation skills [19,20]. In addition, training involving the stimulation of vestibular networks, which are involved in visuo-spatial processes, might contribute to better mental rotation skills in athletes performing sports activities with high balance demands.

Most of the published studies have found higher mental rotation abilities in athletes practicing gymnastics and combat sports compared to sedentary groups, but not when contrasting their performance with athletes of other disciplines [21–23]. Only a few studies have shown better mental rotation skills in gymnasts compared to athletes of other disciplines, after controlling for overall physical activity [9,24]. It remains an open question whether the reported superior spatial cognitive skills in gymnasts indicate an association between practicing a specific type of exercise and cognitive functions or are due to an overall high level of physical activity.

During the last two decades, numerous epidemiological and cross-sectional studies have reported better cognitive and academic performance in physically active people compared to sedentary individuals (e.g., [25–30]). Furthermore, randomized intervention studies showed beneficial effects of regular exercise training on a wide range of cognitive functions, including memory, executive functions, visuo-spatial skills and attention [31–33]. Most published intervention studies implemented aerobic exercise training. Consequently, some authors suggested that aerobic exercise selectively improves executive functions, especially in older populations [34,35]. However, executive functions have been shown to be enhanced by complex motor training, which did not improve cardiovascular fitness [36]. A recent meta-analysis integrated 80 intervention studies on the effects of physical exercise on cognition and found the largest cognitive benefit from exercise programs that improved fine and gross-motor body coordination and balance skills over any other form of physical exercise. Overall, effect sizes did not differ significantly between cognitive domains, suggesting a rather overarching effect of physical exercise on cognition [37]. Taken together, both epidemiological and training studies in previously sedentary healthy adults have established a reliable link between regular physical exercise and better cognitive functioning.

However, it is unknown whether enhanced spatial skills in athletes highly trained in combat sports or gymnastics compared to sedentary participants are predominantly due to a high amount of physical exercise, practicing mental rotation during training by anticipating others' movements, mental imagery of one's own body movements, a stimulation of vestibular networks by balance training or a combination of these factors. The goal of the present study was to further unravel the specific contribution of balance skills on spatial cognitive performance. Therefore, we recruited balance experts who were regularly engaged in balance activities on their own, that is, without an opponent or partner. Moreover, in order to control for the balance experts' typical high overall fitness, we included endurance athletes as an active control group. We hypothesized that balance experts have specific advantages in visuo-spatial skills, both compared to endurance athletes and compared to sedentary individuals. All participants were tested on dynamic balance, spatial perspective-taking and executive functions. We additionally hypothesized that better individual balance skills are associated with better performance in the perspective-taking task. Both this correlation, as well as the overall higher performance of the balance group, were expected to be specific for the perspective-taking skills; that is, they were not expected for executive functions.

2. Materials and Methods

2.1. Participants

An a priori sample size calculation was performed with G*Power 3.1.9.2 [38]. Based on the meta-analysis of Voyer and Jansen [7], we expected a medium effect size for enhanced visuo-spatial skills in balance experts. Such an effect size can be statistically detected in a one-way Anova (three groups) with a total sample size of 159 participants (power = 0.80, alpha = 0.05). As described in detail in the next paragraphs, the data of some participants had to be discarded for the group analysis because the participants could not be unambiguously categorized as balance experts and endurance athletes, did not follow task instructions in the perspective-taking task or their data were classified as outliers. Thus, the achieved power to detect a medium effect size in the group analysis ($n = 133$) for the main outcome measure (deviation error in the perspective-taking task) was 0.72.

Individuals between 18 and 50 years of age were eligible for the study if they reported practicing either balance activities or endurance sports for at least four hours a week during the last two years. Athletes were recruited in sports clubs, at running events, through word-of-mouth, using announcements in sports-specific social media groups and via a university recruitment platform for psychological studies. Activities in the balance group included acrobatics, ballet dancing, skateboarding, bouldering, unicycling, slacklining, freestyle taekwondo (without opponent), dancing (solo dancing only), gymnastics, trampoline, tricking and yoga. Activities in the endurance group included cycling, running and aerobic fitness training. Individuals practicing team sports, ball games, combat sports with an opponent or dancing with a partner were not eligible for this study. Furthermore, participants were not considered for the study if they reported practicing both balance activities and endurance activities regularly. No current or past engagement in competitions was required for taking part in the study. Participants were screened in a telephone interview, and 97 participants were invited for testing. During the assessment session, all participants filled in a questionnaire about their physical activities, including their sports activities during the last week (Freiburg Questionnaire of Physical Activity, FQPA, [39]). Moreover, the interviewer assessed their sports activities in detail. Based on data of the FQPA and the second interview, participants were classified as balance experts or endurance athletes. Three of the invited participants did not fulfil the criteria of practicing 4 h a week on a regular basis, and 16 participants could not be unambiguously classified as balance experts or endurance athletes, and their data was thus disregarded for the group analysis. The final sample comprised 78 athletes (40 balance experts and 38 endurance experts). By contrast, the data of all invited athletes ($n = 97$) were considered for the regression analyses exploring the association between balance performance and cognitive measures.

Balance experts and endurance athletes were compared to 59 sedentary participants (20–40 years of age, 43 female). Sedentary participants were recruited for a physical exercise intervention study, which will be reported elsewhere. The data recorded at baseline were included for the present study. Sedentary participants reported less than five exercise sessions a month during the last five years.

None of the participants had any history of neurological disease, and they reported no intake of antidepressant or antipsychotic medication. According to self-reporting, all participants had normal or corrected-to-normal vision and normal hearing abilities. Most of the participants (93%) held an A-level certificate or a university degree.

The local ethical board of the Faculty of Psychology and Movement Science at the University of Hamburg approved the study, and all participants gave written informed consent. Participants received course credit or monetary compensation of 16 € for participation.

2.2. Assessments

2.2.1. Balance Test

Balance performance was tested with a stability platform (Stability Platform, Modell 16,030 L, Lafayette Instrument Company, Lafayette, IN, USA). Previous studies have demonstrated a sufficient sensitivity of this method for distinguishing balance experts from professional soccer players, swimmers and non-athletes [40,41].

Participants stood barefoot on an unstable platform with a maximal deviation of 15 degrees to each side. They were instructed to place their hands on their hips, direct their gaze to a fixation cross straight ahead (eyes-open condition only) and to keep the platform in a horizontal position for as long as possible during a 30 s trial. After a practice trial, three 30 s trials with eyes open and three 30 s trials with eyes closed were run, separated by 30 s breaks. Whether participants started with eyes open or eyes closed was counterbalanced across participants. A handrail was available to prevent falls and for use during rest. When participants touched the handrail during a trial, the trial was repeated. Testing was stopped after three unsuccessful attempts. This was the case for one participant in the eyesclosed condition. A built-in digital encoder recorded the time per trial the platform was in the horizontal position ($\pm 3°$ deviation). The mean time spent in a horizontal position across trials was calculated for each participant, separately for eyes open (EO) and eyes closed (EC).

2.2.2. Perspective-Taking Abilities

The Orienting and Perspective Taking Test (OPT, [42]) was used to assesses the ability to image scenes from different viewpoints, a sub-function of spatial cognition. In this paper-pencil test, participants were shown a picture with seven objects. Their task was to imagine standing at one given object, facing a second object and indicating the direction of a third object. A circle was printed under the scene, with an arrow pointing in the direction of the object the participant was facing. Participants were instructed to draw a second arrow indicating their imagined pointing direction. They were not allowed to turn the page or their own body for viewpoint shifts. The time limit to solve 12 items was set to 5 min. Deviation errors were scored by subtracting the participants' angle estimates from the correct solutions. If a participant did not answer an item, missing values were replaced with 90°, as reported in [43]. The mean deviation across items was calculated for each participant. Smaller values represent better performance.

Data of four participants were not included in the analyses of the OPT task because two participants did not understand the instructions of the task ($n = 1$ endurance group, $n = 1$ sedentary group) and two participants (balance group) had mean deviation errors of more than three standard deviations above the group mean and were therefore excluded as outliers.

2.2.3. Executive Functions

A computer-based modified version of the Eriksen flanker task [44,45] was used to assess executive control and inhibition. Stimulus presentation and recording of responses were performed using Presentation® Software (Version 14.9, Neurobehavioral Systems, Inc., Berkeley, CA, USA). Five arrows were presented in the middle of a computer screen (white color on black background). Participants were asked to indicate the direction of the middle arrow as fast as possible by pressing a left and right button on a custom-made device. Participants responded with the index and middle fingers of the dominant hand. Each stimulus was presented for 1000 ms, and the inter-stimulus interval was set to 1000 ms. The probability of middle errors pointing to the right versus left was the same. In half of the trials, the arrow in the middle indicated the same direction as the flanker arrows (congruent trials), and in the other half of the trials, the arrow in the middle pointed in the opposite direction as the flanker arrows (incongruent trials). In total, 2 blocks of 50 trials each were run, with a short break between blocks. Congruent and incongruent trials were presented in random order within a block.

Trials with reaction times faster than 200 ms and reaction times slower than 3 standard deviations above the individual mean reaction time were discarded from further analyses. Only correct trials were considered for reaction time analyses. On average, less than 1.5% of the trials were error trials. Data of the Flanker task were missing for four participants due to technical problems ($n = 2$ balance group, $n = 1$ endurance group) and non-compliance with task instructions ($n = 1$ balance group).

2.2.4. Physical Activity Questionnaire

The German version of the "Freiburg Questionnaire of Physical Activity" (FQPA, [39]) was used to assess participants' general physical activity and their sports activities in particular. The questionnaire encompasses questions about basic physical activities (e.g., walking to work, taking the stairs, gardening), leisure-time activities (dancing, light cycling tours) and sports activities. Hours of activity per week were calculated separately for basic physical activities, leisure-time activities and sports activities. Moreover, metabolic equivalents (MET) per week were calculated to estimate the total energy expenditure associated with physical activities, using the compendium provided by Ainsworth et al. [46]. One participant did not fill in the FQPA.

2.2.5. Verbal Intelligence Test

The German "Mehrfachwahl-Wortschatztest" (MWT-B, [47]) was used to estimate participants' verbal intelligence. The MWT-B comprises 37 rows with four pseudo-words and one legal German word. Participants have to strike out the legal German word with a pencil. The item difficulty continuously increases from the first to the last row. The test score correlates with the general IQ in healthy adults [47].

2.3. Data Analysis

Balance performance and perspective-taking skills were compared between balance experts, endurance athletes and sedentary participants by means of an Analysis of Covariance (Ancova), using age and gender as covariates. The significant main effects of the factor group were followed up with simple contrasts, comparing the balance group against the endurance group and the balance group against the sedentary group.

For executive functions, a 2×3 Ancova with the within-subject factor Congruency (congruent vs. incongruent) and the between-subject factor Group (balance vs. endurance vs. sedentary) as well as the covariates age and gender was run to test for differences in reaction times in the flanker task.

Linear regression models were run to test for associations between balance performance and cognitive functions. We ran hierarchical models, entering age and gender as predictors in the first step and balance performance with eyes open in the second step, to test whether balance performance explained additional variance. Separate models with mean deviation error in the OPT and the flanker effect (reaction times in incongruent trials minus congruent trials) as dependent variable were run. Furthermore, we ran models using additional covariates, including hours of sports activities, hours of basic and leisure time activities and verbal IQ, to test whether the revealed associations were explained by the amount of physical activity and participants' IQ.

Data analyses were performed using IBM SPSS Statistics, Version 25. Figures were generated using the package gglot 2 [48] in R (Version 3.6.0, R [49]).

3. Results

3.1. Description of Participant Groups

Characteristics of participants considered for the group analyses are provided in Table 1. Sedentary participants tended to be on average younger than endurance athletes. There were significantly more female participants in the balance group and the sedentary group than in the endurance group. Thus, age and gender were included as covariates in the consecutive analyses. Sedentary participants had significantly lower scores in the verbal IQ test compared to both balance experts and endurance athletes. Verbal IQ was not available for fifteen participants ($n = 5$ did not fill in the test due to time reasons, $n = 10$ were non-native German speakers). Therefore, we reported additional analyses in the subgroup of participants for whom data on verbal IQ were available, controlling for verbal IQ. The balance experts and endurance athletes reported spending on average 8.83 h/week, CI = (7.90, 9.75) for sports activities, while sedentary participants reported 0.52 h/week, CI = (0.25, 0.79). Time spent training did not differ between balance experts and endurance athletes, $t(76) = 0.57$, $p = 0.568$, $d = 0.13$. As expected, endurance athletes practiced sports activities with higher metabolic demands than balance experts, resulting in a significant group difference in metabolic equivalents (METs) assigned to the activities reported in the FQPA, $t(76) = -3.41$, $p = 0.001$, $d = -0.77$.

Table 1. Participant characteristics (Mean, Sd) for balance experts, endurance athletes and sedentary participants.

	Balance Experts $n = 40$	Endurance Athletes $n = 38$	Sedentary Participants $n = 59$	p
Age	27.95 (8.39)	29.71 (7.07)	26.63 (5.33)	0.098 [3]
Gender (female/male)	25/15	17/21	43/16	0.020 [4]
Verbal IQ [1]	29.06 (4.15)	29.11 (3.34)	27.13 (4.23)	0.029 [3]
Self-reported sports activities [2] (hours/week)	9.06 (4.50)	8.58 (3.61)	0.52 (1.04)	<0.001 [3]
Self-reported sports activities [2] (MET)	45.28 (27.84)	67.96 (30.19)	2.48 (4.73)	<0.001 [3]
Self-reported basic and leisure time activities [2] (hours/week)	8.38 (6.96)	8.72 (5.14)	6.44 (4.59)	0.087 [3]
Self-reported basic and leisure time activities [2] (MET)	30.62 (25.11)	34.13 (19.78)	21.91 (15.25)	0.008 [3]

Note. [1] Missing data verbal IQ: $n = 7$ balance group, $n = 2$ endurance group, $n = 6$ sedentary group; [2] Missing data FQPA: $n = 1$ balance group; [3] Anova; [4] Chi-squared test.

3.2. Balance Performance

Balance performance tested with eyes open was significantly better in balance experts compared to both endurance athletes and sedentary participants, $F(2, 132) = 13.78, p < 0.001$, $\eta^2 = 0.173$, balance vs. endurance: covariate-adjusted group difference = 2.18, $p = 0.005$, 95% CI (0.69, 3.68), balance vs. sedentary: covariate-adjusted group difference = 3.56, $p < 0.001$, 95% CI (2.21, 4.90), Figure 1. Groups did not significantly differ in the eyes-closed condition, $F(2, 131) = 1.43, p = 0.244, \eta^2 = 0.021$.

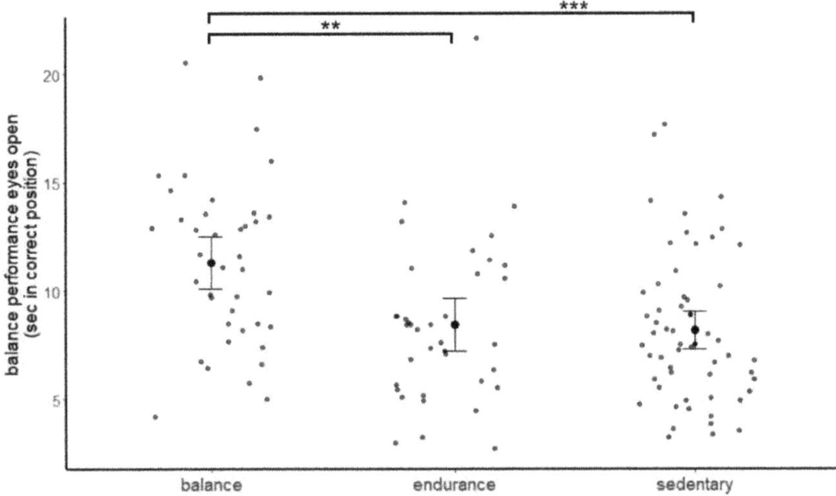

Figure 1. Mean performance on the balance platform in the eyes-open condition, separately for the balance experts, endurance athletes and sedentary participants. Error bars indicate 95% confidence intervals. Data of single participants are plotted in grey. ** $p < 0.01$, *** $p < 0.001$, post hoc contrasts.

The pattern of results was confirmed when controlling for verbal IQ, $F(2, 116) = 14.54$, $p < 0.001, \eta^2 = 0.200$, balance vs. endurance: covariate-adjusted group difference = 2.70, $p = 0.001$, 95% CI (1.16, 4.24), balance vs. sedentary: covariate-adjusted group difference = 3.88, $p < 0.001$, 95% CI (2.44, 5.32).

3.3. Perspective-Taking Abilities

Groups significantly differed in the OPT, $F(2, 128) = 10.56$, $p < 0.001$, $\eta^2 = 0.142$, with the balance group showing the smallest and the sedentary group the largest mean deviation errors, as shown in Figure 2. Planned contrast confirmed a significantly better performance for the balance group compared to the sedentary group, with a covariate-adjusted group difference = -16.23, $p < 0.001$, 95% CI $(-23.36, -9.10)$. The difference between the balance group and the endurance group was not significant, with the covariate-adjusted group difference = -6.16, $p = 0.130$, 95% CI $[-14.17, 1.84]$.

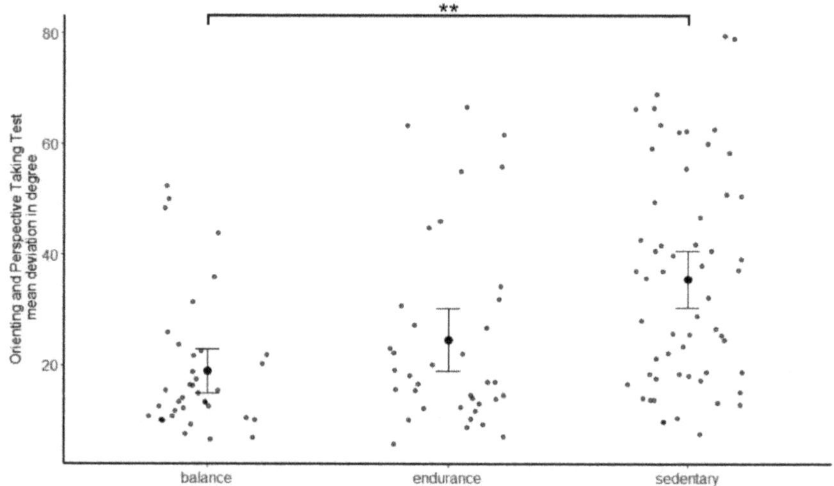

Figure 2. Mean deviation error in the Orienting and Perspective Taking Test, separately for the balance experts, endurance athletes and sedentary participants. Error bars indicate 95% confidence intervals. Data of single participants are plotted in grey. ** $p < 0.01$, post hoc contrast.

The group difference remained significant when controlling for verbal IQ, $F(2, 113) = 5.92$, $p = 0.004$, $\eta^2 = 0.095$, balance vs. sedentary group: covariate adjusted group difference = -12.40, $p = 0.001$, 95% CI $(-19.60, -5.21)$.

3.4. Executive Functions

Participants showed a reliable flanker effect, with faster reaction times in the congruent condition than in the incongruent condition: $F(1, 128) = 31.68$, $p < 0.001$, $\eta^2 = 0.198$; see Figure 3. Neither overall reaction times, Group $F(2, 128) = 0.46$, $p = 0.631$, $\eta^2 = 0.007$, nor the flanker effect differed between groups, Congruency x Group $F(2, 128) = 1.10$, $p = 0.335$, $\eta^2 = 0.017$.

Adding verbal IQ as an additional covariate did not change the pattern of results: Congruency $F(1, 113) = 19.56$, $p < 0.001$, $\eta^2 = 0.148$, Group $F(2, 113) = 0.62$, $p = 0.539$, $\eta^2 = 0.011$, Congruency x Group $F(2, 113) = 1.01$, $p = 0.367$, $\eta^2 = 0.018$.

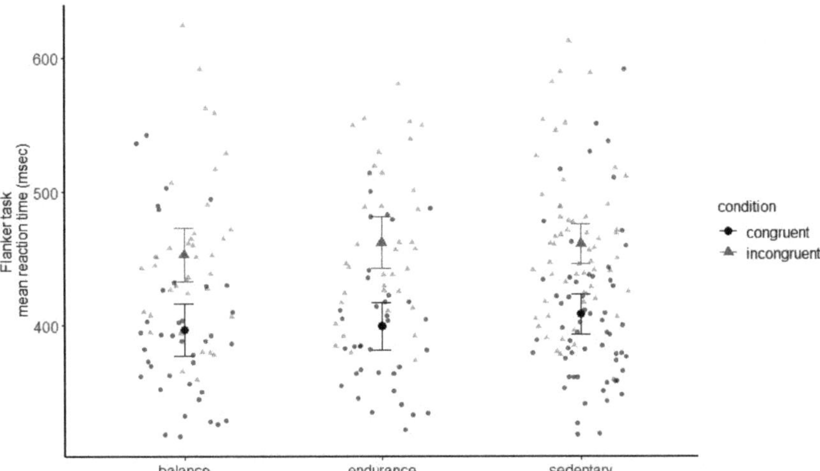

Figure 3. Mean reaction times in the flanker task, separately for congruent versus incongruent trials and for the balance experts, endurance athletes and sedentary participants. Error bars indicate 95% confidence intervals. Data of single participants are plotted in grey.

3.5. Regression Analyses

For regression analyses, data of all assessed participants were considered, including data of those who could not be unambiguously assigned to the balance group, endurance group or sedentary group. By doing so, we tested the overall association between balance expertise and cognitive function in a larger sample and ruled out the possibility that participant exclusion may have biased the reported association between balance expertise and spatial cognition. Additional regression analyses for the subsample considered for the group analyses are provided in Appendix A (Figure A1, Table A1, Table A2).

Balance performance was significantly associated with perspective-taking skills (Figure 4), such that the better the performance on the stability platform in the eyes-open condition, the lower the deviation errors in the OPT, standardized β (148) = −0.307, $p < 0.001$, adjusted for age and gender. The strength of the association was very similar when adding hours of sports activities/week as an additional covariate, standardized β (146) = −0.301, $p = 0.001$; see Table 2.

The flanker effect did not correlate significantly with balance performance, standardized β (148) = 0.025, $p = 0.773$, adjusted for age and gender, standardized β (146) = −0.027, $p = 0.760$, adjusted for age, gender and hours of sports activities; see Table 3.

We ran further regression models to test whether the reported associations could be explained by participants' basic and leisure time physical activity and verbal IQ. In one model, age, gender and the sum of basic and leisure time physical activity reported in the FQPA were entered as covariates. In a second model, age, gender and number of correct words in the MWT-B score were entered as covariates. Cognitive measures served as the dependent variable and balance performance with eyes open as the predictor. The additional analyses confirmed a specific association between balance performance and spatial cognition that could not be explained by the amount of physical activity (model 1) and verbal IQ (model 2) (all $|\beta| > 0.28$, all $p < 0.01$). That is, participants with better balance performance had lower deviation errors in the OPT. No associations between balance performance and the flanker effect were found (all $|\beta| < 0.04$, all $p > 0.7$).

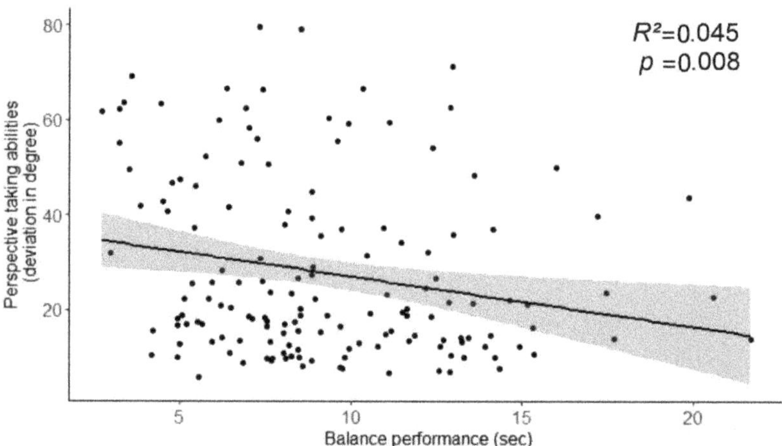

Figure 4. Correlation between balance performance in the eyes-open condition (seconds in correct position) and perspective-taking skills (OPT task, mean deviation error in degree). Confidence bands indicate 95% confidence intervals.

Table 2. Hierarchical regression model for perspective-taking abilities (OPT task).

Variable	B	95% CI for B		SE B	β	R^2	ΔR^2
		LL	*UL*				
Step 1						0.078	0.078 **
Constant	32.53	20.07	45.00	6.31			
Age	0.057	−0.38	0.50	0.22	0.02		
Gender	−5.56	−11.66	0.55	3.09	−0.15		
Hours sports/week	−0.72	−1.28	−0.16	0.28	−0.21*		
Step 2						0.151	0.073 **
Constant	53.64	36.84	70.44	8.50			
Age	−0.20	−0.64	0.24	0.22	−0.07		
Gender	−9.00	−15.17	−2.81	3.13	−0.24 **		
Hours sports/week	−0.49	−1.04	0.06	0.28	−0.14		
Stability platform eyes open	−1.53	−2.39	−0.68	0.43	−0.30 **		

Notes: CI = confidence interval, *LL* = lower limit, *UL* = upper limit, * $p < 0.05$, ** $p < 0.01$.

Table 3. Hierarchical regression model for the flanker effect (RT incongruent–RT congruent).

Variable	B	95% CI for B		SE B	β	R^2	ΔR^2
		LL	*UL*				
Step 1						0.035	0.035
Constant	44.91	28.65	61.18	8.23			
Age	0.22	−0.33	0.78	0.28	0.07		
Gender	5.39	−2.82	13.60	4.15	0.11		
Hours sports/week	0.52	−0.24	1.28	0.39	0.11		
Step 2						0.036	0.001
Constant	47.35	24.68	70.02	11.47			
Age	0.20	−0.39	0.78	0.30	0.06		
Gender	4.97	−3.70	13.64	4.39	0.10		
Hours sports/week	0.55	−0.24	1.34	0.40	0.12		
Stability platform eyes open	−0.18	−1.37	1.00	0.60	−0.03		

Notes: RT = reaction times, CI = confidence interval, *LL* = lower limit, *UL* = upper limit.

4. Discussion

In the present cross-sectional study, we provided evidence for a positive association between balance skills and visuo-spatial abilities: Participants who reported practicing balance sports for at least four hours a week during the last two years outperformed sedentary participants in the Object Perspective Taking Test (OPT). There was no significant difference in spatial abilities between endurance athletes and sedentary participants, suggesting that overall physical activity cannot explain group differences in spatial cognition. A regression analysis that included all participants additionally indicated that dynamic balance performance was associated with fewer errors in the perspective-taking task, when controlling for age, gender, verbal IQ and physical activity. The association was specific for the visuo-spatial task, with no group differences in a flanker task assessing response speed and executive functions. Moreover, individual balance performance did not correlate with executive functions.

These results are in line with a meta-analysis on mental rotation skills in motor experts [7], demonstrating overall better performance of motor experts than sedentary participants in mental rotation tasks. However, effect sizes differed considerably for athletes of different disciplines. The largest effect sizes were found for athletes practicing combat sports, followed by those engaging in gymnastics and dancing; effect sizes for endurance athletes did not differ significantly from zero. It has been argued that performing combat sports, gymnastics and dancing requires mental rotation and continuous updating of one's own and others' body position in space along all body axes [9,22], while exercise programs of endurance athletes are mostly uniform and predictable. Thus, extensive visuo-spatial processing in combination with motor practice in these disciplines might lead to improved mental rotation abilities, which transfers to better performance in visuo-spatial tasks outside of the sport context. In the present study, we excluded participants practicing combat sports, couple dance and ball sports to narrow down the possible processes that might contribute to better spatial skills in motor experts. Thus, the enhanced perspective-taking skills in the balance experts in the present study are probably less due to extensive training in anticipating the movements of an opponent, partner or object in space, but might be shaped by spatial updating and mental imagery of their own body transformations.

Furthermore, it could be speculated that the vestibular system is a mediating factor. Superior balance skills have been shown for gymnasts and dancers compared to athletes engaging in other physical activities [10,50,51]. Moreover, balance training has been shown not only to improve balance skills but to enhance performance in spatial cognition tests as well [15,16]. In the present study, we investigated the contribution of balance expertise to enhancing spatial skills in athletes by including individuals whose physical activity was associated with high demands on postural control, such as gymnastics, slacklining, acrobatics, ballet, skateboarding, bouldering, unicycling, freestyle taekwondo, dancing, trampolining, tricking and yoga. Moreover, we assessed balance performance explicitly, with a stability platform. Therefore, we demonstrate that individuals engaged in sports activities that place high demands on balance outperformed both endurance athletes and sedentary participants in terms of dynamic balance abilities. Thus, the present data suggest that balance skills should be taken into account when discussing improved visuo-spatial skills in motor experts.

Balancing requires an efficient integration of vestibular, proprioceptive, somatosensory and visual signals. There is growing evidence that the vestibular system essentially contributes to higher-order cognitive functions such as spatial orientation, memory and body self-consciousness [11,52,53]. Visuo-spatial tasks requiring mental spatial transformations are known to activate higher-order visual areas in the occipital lobe, the posterior parietal cortex and the parieto-medial temporal pathway to the hippocampus [54,55]. It should be noted that these regions partially overlap with vestibular cortical networks [11,56,57]. In brain imaging studies, long-term balance expertise and short-term balance training for several weeks have been associated with changes in grey matter volume and cortical thickness in areas receiving vestibular input, such as higher-order visual association areas [17,18], the posterior cingulate cortex [17] and the hippocampus [18,58]. Moreover, balance training has been associated with structural changes in premotor and motor cortices [17,59–61], which are known to be activated during mental rotation and mental motor imagery [55,62]. Thus, regular joined stimulation of visuo-vestibular pathways and motor cortices during balance training, which partially overlap with areas important for spatial processing, may contribute to superior performance of balance experts in the spatial task. The results can be interpreted in an embodied cognition framework, which proposes that cognitive processes are based on, or are at least moderated by, sensorimotor processes [63], probably by relying on shared neuronal representations [4]. Enhanced cognitive performance in specific cognitive domains in motor experts compared to novices, such as enhanced perspective-taking skills in balance experts, might thus be the result of long-lasting sensorimotor training [64]. Practicing physical activities with high balance needs is, however, only one factor among those linked to enhanced visuo-spatial skills. For instance, better mental rotation skills compared to controls have been reported for professional orchestral musicians [65] and users of sign language [20].

In the present study, we used a perspective-taking task to measure spatial skills, while previous studies on visuo-spatial skills in motor experts mostly assessed mental rotation abilities. The OPT measures the ability to perform egocentric spatial transformations, i.e., the ability to mentally shift one's perspective in order to judge the relative position of objects in the environment [42]. In contrast, mental rotation tasks typically measure the ability to imagine movements or the rotation of objects, which requires object-based spatial transformations but not egocentric transformations [42]. Depending on the type of the stimuli and task instructions, the distinction between perspective-taking tasks and mental rotation tasks is less clear-cut. It has been shown that body stimuli in mental rotation tasks trigger perspective-taking strategies. Steggemann et al. [9], for instance, compared experts in artistic gymnastics, aero wheel gymnastics and trampolining to athletes from disciplines that do not involve a lot of spins and turns around the body axes. In this study, the authors used three mental rotation tasks: a letter rotation task with same–mirrored judgments, a body rotation task with same-mirrored judgment and a body rotation task with left–right judgments. The motor experts showed an advantage only in the body rotation task with left–right judgments, which requires an egocentric transformation, i.e., the rotation of one's own point of view. The authors suggested that the athletes are highly trained in body transformations, even in very unfamiliar body positions, which transfers to the laboratory task of egocentric mental rotation, but not to object-based mental rotation. However, other studies have reported better performance in object-based mental rotation in motor experts [21,23]. Thus, it is still an open question whether there are specific effects of balance expertise on egocentric spatial transformations.

Superior performance of the balance experts in the dynamic balance task compared to endurance athletes and sedentary participants was only found for the eyes-open condition, not for the eyes-closed condition. The balance experts regularly perform their sports with eyes open, making use of visual cues to detect motion displacements and to stabilize posture. Dordevic et al. [58] reported better balance performance in professional ballet dancers compared to age-matched controls in both eyes-open and eyes-closed conditions, but effect sizes were larger for the eyes-open condition. Sedentary participants showed an increase in balance performance after a complex balance training only in the eyes-open testing conditions [66]. Taken together, these results suggest that the effects of motor training on balance skills are more pronounced for the sensory condition that athletes usually experience during their regular training.

We hypothesized that balance training is specially associated with visuo-spatial abilities. Physical exercise is defined as an activity that is "planned, structured, repetitive, and purposeful in the sense that the improvement or maintenance of one or more components of physical fitness is the objective" [67] (pp. 52–53). Physical exercise is a subcategory of physical activity, defined as "any bodily movement produced by skeletal muscles that requires energy expenditure" [67] (pp. 52–53). Both physical exercise and overall physical activity have been linked to enhanced cognitive performance in epidemiological studies [31,32]. In the present study, participants' amounts of physical exercise training as well as their basic and leisure time physical activity was assessed with a questionnaire. The data showed that the balance group and the endurance group did not differ in the amount of physical exercise nor physical activity in everyday life. Furthermore, the difference between athletes and sedentary participants was much more pronounced for physical exercise (8 h per week) compared to the group difference in basic and leisure time physical activity (2 h per week). Adding basic and leisure time physical activity as additional covariates in the models did not reduce the association between balance skills and visuo-spatial abilities. Therefore, we concluded that there is a specific association between practicing balance sports and visuo-spatial processing that cannot be accounted for by the amount of time spent physically exercising and overall physical activity.

As hypothesized, superior performance of the balance group compared to sedentary participants was specific for the visuo-spatial task and was not observed for executive functions. This is in line with findings of a recent balance training study, which found effects of the balance training compared to a relaxation control training on spatial cognition and memory, but not for executive functions as assessed with the Stroop task [16]. In the present study, the endurance group also did not differ from the sedentary group in executive functions. This was expected on the basis of studies reporting improved executive functions in physically active people compared to sedentary individuals [68,69] and on the basis of training studies proposing beneficial effects of aerobic exercise, particularly on executive functions [34,35,70]. Age has been shown to moderate the influence of exercise on executive functions, with the beneficial effects of aerobic exercise increasing with age in adulthood [71]. It has been suggested that cognitive functions are most sensitive to sensorimotor experiences in phases during which they undergo developmental changes, such as executive functions in childhood and old age [72–75]. Participants in the present study were between 18 and 50 years of age. In this age range, executive functions are rather stable and at their functional peak [76,77] and thus maybe less likely to be affected by the habitual level of physical exercise.

Studying balance experts allows exploring the effects of regular sensorimotor experience accumulated over years, which is hardly possible to investigate in randomized controlled trials. However, data of the present study are correlational only and do not allow causal interpretations. Reverse causality might account for the reported associations with visuo-spatial abilities influencing whether or not individuals start practicing balance sports and the proficiency level they reach. We showed that balance performance explained variance in visuo-spatial skills that could not be accounted for by participants' age, gender, verbal IQ and the amount of physical activity. Moreover, participants in the present sample had an overall high level of education, making it unlikely that the reported effects could be explained by group differences in general intelligence. Nevertheless, further variables might have contributed to the association between balance skills and perspective-taking abilities. For instance, playing action video games has been shown to improve spatial cognition [78], which was not assessed in the present study. We measured balance performance with a standardized protocol on a stability platform, but the amount of endurance exercise and participants' overall physical activity was assessed via self-reporting only, which may be prone to higher measurement errors and biases [79]. Adding standardized assessments of cardiovascular fitness and using wearable activity trackers to draw samples of participants' physical activity in everyday life might be fruitful approaches to disentangle the contribution of different forms of physical activity to cognitive functions in athletes and sedentary participants in future studies.

The athletes recruited for the present study trained at least four hours per week in their discipline over the past two years. Although some participants reported taking part in sports competitions, this was not an inclusion criterion, and we did not systematically assess how many were engaging in competitions and at which level. Recruiting elite athletes and using standardized procedures for defining expertise level [80] might have increased the power of detecting specific associations between balance training and cognitive functions.

The positive effects of physical exercise on cognition have not only been reported after months or years of training, but after a single bout of cardiovascular exercise [33,81]. One might hypothesize that acute and chronic effects of physical exercise add up when performed regularly and over long periods of time. Studies on acute effects of balance training, however, are lacking. There are some recent results suggesting that the acute cognitive effects of balance exercise do not differ from those after a cardiovascular exercise session [82]. These results should be confirmed in further studies including a control group without training.

Perspective-taking abilities in the present study were assessed using a paper-pencil test, while executive functions were measured with a computer-based flanker task. One might argue that the difference in the assessment modalities might have caused the reported dissociation between cognitive domains. Computerized tests of cognitive function provide very precise measures of reaction times and are therefore more sensitive to individual differences in cognitive performance than paper and pencil tests [83,84]. Thus, if anything, we would have expected more of a correlation between balance expertise and executive functions, which we did not find.

5. Conclusions

In the present cross-sectional study, we demonstrate that high balance skills are associated with higher spatial cognitive abilities and that this association cannot be explained by time spent on physical exercise or overall verbal IQ. Moreover, a similar association was not found for executive functions. Practicing balance sports such as gymnastics, acrobatics, dancing and slacklining requires complex body transformations in space and an efficient integration of vestibular, proprioceptive and visual signals. We speculate that the specific skills acquired during balance training transfer to higher spatial cognitive functions in balance experts, probably based on shared neural circuits. From an applied perspective, incorporating balance tasks into physical exercise programs might be a promising approach to increase spatial cognition in athletes and in individuals with deficits in spatial cognition, such as aging populations or patients suffering from neuro-degenerative diseases.

Author Contributions: Conceptualization, K.H. and B.R.; methodology, K.H., A.-K.R. and L.A.K.; software, A.-K.R.; formal analysis, K.H.; investigation, A.-K.R. and L.A.K.; writing—original draft preparation, K.H.; writing—review and editing, A.-K.R., L.A.K. and B.R.; visualization, K.H.; supervision, K.H. and B.R.; funding acquisition, B.R. and K.H. All authors have read and agreed to the published version of the manuscript.

Funding: This research was funded by a grant from the German Research Foundation [DFG Ro 2625/10-1] to Brigitte Röder and a grant from the Center for a Sustainable University, Universität Hamburg, to Kirsten Hötting.

Institutional Review Board Statement: The study was conducted according to the guidelines of the Declaration of Helsinki and approved by the Ethics Committee of the Faculty of Psychology and Movement Science at the University of Hamburg (protocol code 2017_97, date of approval: 5 April 2017).

Informed Consent Statement: Informed consent was obtained from all participants involved in the study.

Data Availability Statement: The data presented in this study are available on request from the corresponding author. The data are not publicly available due to privacy concerns.

Acknowledgments: We thank Steven Behn, Antonio Garcia, Luba Lindt, Silke Lippok, Janne Nold and Annika Struck for their help with data acquisition and data entry.

Conflicts of Interest: The authors declare no conflict of interest. The funders had no role in the design of the study; in the collection, analyses, or interpretation of data; in the writing of the manuscript, or in the decision to publish the results.

Appendix A. Regression Analyses: Subsample Considered for Group Analyses

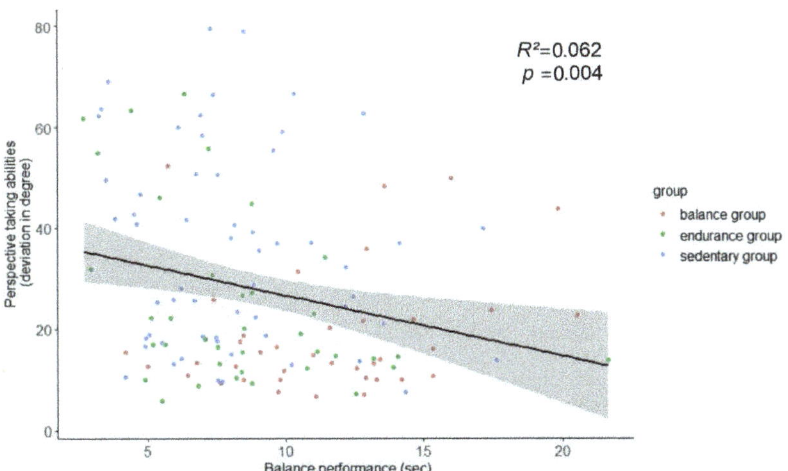

Figure A1. Correlation between the balance performance in the eyes-open condition (seconds in correct position) and perspective-taking skills (OPT task, mean deviation error in degree). N = 133 (n = 38 balance group, n = 37 endurance group, n = 58 sedentary group). Confidence bands indicate 95% confidence intervals.

Table A1. Hierarchical regression model for perspective-taking abilities (OPT task).

Variable	B	95% CI for B		*SE* B	β	R^2	ΔR^2
		LL	*UL*				
Step 1						0.081	0.081 *
Constant	31.14	17.57	44.70	6.85			
Age	0.092	−0.39	0.58	0.25	0.03		
Gender	−3.61	−10.19	2.97	3.33	−0.10		
hours sports/week	−0.88	−1.28	−0.15	0.32	−0.25 **		
Step 2						0.170	0.089 **
Constant	55.18	36.91	73.46	9.24			
Age	−0.24	−0.74	0.26	0.25	−0.08		
Gender	−7.28	−13.86	−0.70	3.33	−0.19 *		
Hours sports/week	−0.59	−1.21	0.02	0.31	−0.17		
Stability platform eyes open	−1.66	−2.55	−0.77	0.45	−0.34 ***		

Note: CI = confidence interval, *LL* = lower limit, *UL* = upper limit, * $p < 0.05$, ** $p < 0.01$, *** $p < 0.001$.

Table A2. Hierarchical regression model for the flanker effect (RT incongruent–RT congruent).

Variable	B	95% CI for B		*SE* B	β	R^2	ΔR^2
		LL	*UL*				
Step 1						0.046	0.046
Constant	47.17	29.89	64.45	8.73			
Age	0.11	−0.49	0.72	0.31	0.03		
Gender	5.81	−2.91	14.54	4.41	0.12		
Hours sports/week	0.68	−0.16	1.53	0.42	0.14		
Step 2						0.047	0.001
Constant	49.67	25.42	73.91	12.25			
Age	0.08	−0.56	0.72	0.32	0.02		
Gender	5.40	−3.81	14.60	4.65	0.11		
Hours sports/week	0.72	−0.16	1.60	0.44	0.15		
Stability platform eyes open	−0.18	−1.41	1.05	0.62	−0.03		

Note: RT = reaction times, CI = confidence interval, *LL* = lower limit, *UL* = upper limit.

References

1. Münte, T.F.; Altenmüller, E.; Jäncke, L. The musician's brain as a model of neuroplasticity. *Nat. Rev. Neurosci.* **2002**, *3*, 473–478. [CrossRef]
2. Röder, B.; Rösler, F. The principle of brain plasticity. In *Principles of Learning and Memory*; Kluwe, R.H., Lüer, G., Eds.; Birkhäuser Verlag: Basel, Germany, 2002.
3. Ericsson, K.A.; Krampe, R.T.; Tesch-Römer, C. The role of deliberate practice in the acquisition of expert performance. *Psychol. Rev.* **1993**, *100*, 363–406. [CrossRef]
4. Yarrow, K.; Brown, P.; Krakauer, J.W. Inside the brain of an elite athlete: The neural processes that support high achievement in sports. *Nat. Rev. Neurosci.* **2009**, *10*, 585–596. [CrossRef]
5. Land, M.F.; McLeod, P. From eye movements to actions: How batsmen hit the ball. *Nat. Neurosci.* **2000**, *3*, 1340–1345. [CrossRef]
6. Aglioti, S.M.; Cesari, P.; Romani, M.; Urgesi, C. Action anticipation and motor resonance in elite basketball players. *Nat. Neurosci.* **2008**, *11*, 1109–1116. [CrossRef] [PubMed]
7. Voyer, D.; Jansen, P. Motor expertise and performance in spatial tasks: A meta-analysis. *Hum. Mov. Sci.* **2017**, *54*, 110–124. [CrossRef]
8. Wheeler, M.; Clark, A. Culture, embodiment and genes: Unravelling the triple helix. *Philos. Trans. R. Soc. B Biol. Sci.* **2008**, *363*, 3563–3575. [CrossRef]
9. Steggemann, Y.; Engbert, K.; Weigelt, M. Selective effects of motor expertise in mental body rotation tasks: Comparing object-based and perspective transformations. *Brain Cogn.* **2011**, *76*, 97–105. [CrossRef] [PubMed]
10. Hrysomallis, C. Balance ability and athletic performance. *Sports Med.* **2011**, *41*, 221–232. [CrossRef]
11. Ehitier, M.; Ebesnard, S.; Smith, P.F. Vestibular pathways involved in cognition. *Front. Integr. Neurosci.* **2014**, *8*, 59. [CrossRef]
12. Brandt, T.; Schautzer, F.; Hamilton, D.A.; Brüning, R.; Markowitsch, H.J.; Kalla, R.; Darlington, C.; Smith, P.; Strupp, M. Vestibular loss causes hippocampal atrophy and impaired spatial memory in humans. *Brain* **2005**, *128*, 2732–2741. [CrossRef]
13. Dilda, V.; MacDougall, H.G.; Curthoys, I.S.; Moore, S.T. Effects of galvanic vestibular stimulation on cognitive function. *Exp. Brain Res.* **2012**, *216*, 275–285. [CrossRef]
14. Mast, F.W.; Merfeld, D.M.; Kosslyn, S.M. Visual mental imagery during caloric vestibular stimulation. *Neuropsychologia* **2006**, *44*, 101–109. [CrossRef]
15. Dordevic, M.; Hökelmann, A.; Müller, P.; Rehfeld, K.; Müller, N.G. Improvements in orientation and balancing abilities in response to one month of intensive slackline-training. A randomized controlled feasibility study. *Front. Hum. Neurosci.* **2017**, *11*, 55. [CrossRef] [PubMed]
16. Rogge, A.-K.; Röder, B.; Zech, A.; Nagel, V.; Hollander, K.; Braumann, K.-M.; Hötting, K. Balance training improves memory and spatial cognition in healthy adults. *Sci. Rep.* **2017**, *7*, 1–10. [CrossRef]
17. Rogge, A.-K.; Röder, B.; Zech, A.; Hötting, K. Exercise-induced neuroplasticity: Balance training increases cortical thickness in visual and vestibular cortical regions. *NeuroImage* **2018**, *179*, 471–479. [CrossRef]
18. Hüfner, K.; Binetti, C.; Hamilton, D.A.; Stephan, T.; Flanagin, V.L.; Linn, J.; Labudda, K.; Markowitsch, H.; Glasauer, S.; Jahn, K.; et al. Structural and functional plasticity of the hippocampal formation in professional dancers and slackliners. *Hippocampus* **2010**, *21*, 855–865. [CrossRef] [PubMed]
19. Emmorey, K.; Kosslyn, S.M.; Bellugi, U. Visual imagery and visual-spatial language: Enhanced imagery abilities in deaf and hearing ASL signers. *Cognition* **1993**, *46*, 139–181. [CrossRef]
20. Stroh, A.-L.; Rösler, F.; Röder, B. The interaction of the visuo-spatial and the vestibular system depends on sensory experience in development. *Neuropsychologia* **2021**, *152*, 107736. [CrossRef]

21. Jansen, P.; Lehmann, J. Mental rotation performance in soccer players and gymnasts in an object-based mental rotation task. *Adv. Cogn. Psychol.* **2013**, *9*, 92–98. [CrossRef] [PubMed]
22. Moreau, D. The role of motor processes in three-dimensional mental rotation: Shaping cognitive processing via sensorimotor experience. *Learn. Individ. Differ.* **2012**, *22*, 354–359. [CrossRef]
23. Ozel, S.; LaRue, J.; Molinaro, C. Relation between sport activity and mental rotation: Comparison of three groups of subjects. *Percept. Mot. Ski.* **2002**, *95*, 1141–1154. [CrossRef] [PubMed]
24. Schmidt, M.; Egger, F.; Kieliger, M.; Rubeli, B.; Schuler, J.; Information, R. Gymnasts and orienteers display better mental rotation performance than nonathletes. *J. Individ. Differ.* **2016**, *37*, 1–7. [CrossRef]
25. Churchill, J.D.; Galvez, R.; Colcombe, S.; Swain, R.; Kramer, A.; Greenough, W.T. Exercise, experience and the aging brain. *Neurobiol. Aging* **2002**, *23*, 941–955. [CrossRef]
26. Etnier, J.L.; Nowell, P.M.; Landers, D.M.; Sibley, B.A. A meta-regression to examine the relationship between aerobic fitness and cognitive performance. *Brain Res. Rev.* **2006**, *52*, 119–130. [CrossRef]
27. Sibley, B.A.; Etnier, J.L. The relationship between physical activity and cognition in children: A meta-analysis. *Pediatr. Exerc. Sci.* **2003**, *15*, 243–256. [CrossRef]
28. Kao, S.-C.; Westfall, D.R.; Parks, A.C.; Pontifex, M.B.; Hillman, C.H. Muscular and aerobic fitness, working memory, and academic achievement in children. *Med. Sci. Sports Exerc.* **2017**, *49*, 500–508. [CrossRef]
29. Ishihara, T.; Miyazaki, A.; Tanaka, H.; Matsuda, T. Identification of the brain networks that contribute to the interaction between physical function and working memory: An fMRI investigation with over 1,000 healthy adults. *NeuroImage* **2020**, *221*, 117152. [CrossRef]
30. Willey, J.Z.; Gardener, H.; Caunca, M.R.; Moon, Y.P.; Dong, C.; Cheung, Y.K.; Sacco, R.L.; Elkind, M.S.; Wright, C.B. Leisure-time physical activity associates with cognitive decline. *Neurology* **2016**, *86*, 1897–1903. [CrossRef]
31. Hillman, C.H.; Erickson, K.I.; Kramer, A. Be smart, exercise your heart: Exercise effects on brain and cognition. *Nat. Rev. Neurosci.* **2008**, *9*, 58–65. [CrossRef]
32. Kramer, A.F.; Colcombe, S. Fitness effects on the cognitive function of older adults: A meta-analytic study—revisited. *Perspect. Psychol. Sci.* **2018**, *13*, 213–217. [CrossRef] [PubMed]
33. Roig, M.; Nordbrandt, S.; Geertsen, S.S.; Nielsen, J.B. The effects of cardiovascular exercise on human memory: A review with meta-analysis. *Neurosci. Biobehav. Rev.* **2013**, *37*, 1645–1666. [CrossRef] [PubMed]
34. Kramer, A.F.; Hahn, S.; Cohen, N.J.; Banich, M.; McAuley, E.; Harrison, C.R.; Chason, J.; Vakil, E.; Bardell, L.; Boileau, R.A.; et al. Ageing, fitness and neurocognitive function. *Nat. Cell Biol.* **1999**, *400*, 418–419. [CrossRef] [PubMed]
35. Colcombe, S.; Kramer, A.F. Fitness effects on the cognitive function of older adults: A meta-analytic study. *Psychol. Sci.* **2003**, *14*, 125–130. [CrossRef]
36. Voelcker-Rehage, C.; Godde, B.; Staudinger, U.M. Cardiovascular and coordination training differentially improve cognitive performance and neural processing in older adults. *Front. Hum. Neurosci.* **2011**, *5*, 26. [CrossRef] [PubMed]
37. Ludyga, S.; Gerber, M.; Pühse, U.; Looser, V.N.; Kamijo, K. Systematic review and meta-analysis investigating moderators of long-term effects of exercise on cognition in healthy individuals. *Nat. Hum. Behav.* **2020**, *4*, 603–612. [CrossRef]
38. Faul, F.; Erdfelder, E.; Lang, A.-G.; Buchner, A. G*Power 3: A flexible statistical power analysis program for the social, behavioral, and biomedical sciences. *Behav. Res. Methods* **2007**, *39*, 175–191. [CrossRef]
39. Frey, I.; Berg, A.; Grathwohl, D.; Keul, J. Freiburger fragebogen zur körperlichen aktivität-entwicklung, prüfung und anwendung. *Int. J. Public Health* **1999**, *44*, 55–64. [CrossRef]
40. Davlin, C.D. Dynamic balance in high level athletes. *Percept. Mot. Ski.* **2004**, *98*, 1171–1176. [CrossRef]
41. Kioumourtzoglou, E.; Derri, V.; Mertzanidou, O.; Tzetzis, G. Experience with perceptual and motor skills in rhythmic gymnastics. *Percept. Mot. Ski.* **1997**, *84*, 1363–1372. [CrossRef]
42. Hegarty, M. A dissociation between mental rotation and perspective-taking spatial abilities. *Intelligence* **2004**, *32*, 175–191. [CrossRef]
43. Tarampi, M.R.; Heydari, N.; Hegarty, M. A tale of two types of perspective taking. *Psychol. Sci.* **2016**, *27*, 1507–1516. [CrossRef] [PubMed]
44. Colcombe, S.J.; Kramer, A.F.; Erickson, K.I.; Scalf, P.; McAuley, E.; Cohen, N.J.; Webb, A.; Jerome, G.J.; Marquez, D.X.; Elavsky, S. Cardiovascular fitness, cortical plasticity, and aging. *Proc. Natl. Acad. Sci. USA* **2004**, *101*, 3316–3321. [CrossRef]
45. Eriksen, B.A.; Eriksen, C.W. Effects of noise letters upon identification of a target letter in a nonsearch task. *Percept. Psychophys.* **1974**, *16*, 143–149. [CrossRef]
46. Ainsworth, B.E.; Haskell, W.L.; Herrmann, S.D.; Meckes, N.; Bassett, D.R.; Tudor-Locke, C.; Greer, J.L.; Vezina, J.; Whitt-Glover, M.C.; Leon, A.S. 2011 Compendium of physical activities. *Med. Sci. Sports Exerc.* **2011**, *43*, 1575–1581. [CrossRef]
47. Lehrl, S.; Triebig, G.; Fischer, B. Multiple choice vocabulary test MWT as a valid and short test to estimate premorbid intelligence. *Acta Neurol. Scand.* **1995**, *91*, 335–345. [CrossRef]
48. Wickham, H. *ggplot2: Elegant Graphics for Data Analysis*; Springer: New York, NY, USA, 2016.
49. R Core Team. A Language and Environment for Statistical Computing. Available online: https://www.R-project.org/ (accessed on 22 October 2021).
50. Gautier, G.; Thouvarecq, R.; LaRue, J. Influence of experience on postural control: Effect of expertise in gymnastics. *J. Mot. Behav.* **2008**, *40*, 400–408. [CrossRef]

51. Gerbino, P.G.; Griffin, E.D.; Zurakowski, D. Comparison of standing balance between female collegiate dancers and soccer players. *Gait Posture* **2007**, *26*, 501–507. [CrossRef]
52. Bigelow, R.T.; Agrawal, Y. Vestibular involvement in cognition: Visuospatial ability, attention, executive function, and memory. *J. Vestib. Res.* **2015**, *25*, 73–89. [CrossRef]
53. Smith, P.F.; Darlington, C.L.; Zheng, Y. Move it or lose it—Is stimulation of the vestibular system necessary for normal spatial memory? *Hippocampus* **2009**, *20*, 36–43. [CrossRef] [PubMed]
54. Kravitz, D.J.; Saleem, K.S.; Baker, C.; Mishkin, M. A new neural framework for visuospatial processing. *Nat. Rev. Neurosci.* **2011**, *12*, 217–230. [CrossRef] [PubMed]
55. Zacks, J.M. Neuroimaging Studies of Mental Rotation: A Meta-analysis and review. *J. Cogn. Neurosci.* **2008**, *20*, 1–19. [CrossRef] [PubMed]
56. Cullen, K.E. The vestibular system: Multimodal integration and encoding of self-motion for motor control. *Trends Neurosci.* **2012**, *35*, 185–196. [CrossRef] [PubMed]
57. Lopez, C.; Blanke, O. The thalamocortical vestibular system in animals and humans. *Brain Res. Rev.* **2011**, *67*, 119–146. [CrossRef]
58. Dordevic, M.; Schrader, R.; Taubert, M.; Müller, P.; Hökelmann, A.; Müller, N.G. Vestibulo-hippocampal function is enhanced and brain structure altered in professional ballet dancers. *Front. Integr. Neurosci.* **2018**, *12*, 50. [CrossRef]
59. Dordevic, M.; Taubert, M.; Müller, P.; Kaufmann, J.; Hökelmann, A.; Müller, N.G. Brain gray matter volume is modulated by visual input and overall learning success but not by time spent on learning a complex balancing task. *J. Clin. Med.* **2018**, *8*, 9. [CrossRef]
60. Hänggi, J.; Koeneke, S.; Bezzola, L.; Jäncke, L. Structural neuroplasticity in the sensorimotor network of professional female ballet dancers. *Hum. Brain Mapp.* **2009**, *31*, 1196–1206. [CrossRef]
61. Taubert, M.; Draganski, B.; Anwander, A.; Müller, K.; Horstmann, A.; Villringer, A.; Ragert, P. Dynamic properties of human brain structure: Learning-related changes in cortical areas and associated fiber connections. *J. Neurosci.* **2010**, *30*, 11670–11677. [CrossRef]
62. Ptak, R.; Schnider, A.; Fellrath, J. The dorsal frontoparietal network: A core system for emulated action. *Trends Cogn. Sci.* **2017**, *21*, 589–599. [CrossRef]
63. Raab, M.; Araújo, D. Embodied cognition with and without mental representations: The case of embodied choices in sports. *Front. Psychol.* **2019**, *10*, 1825. [CrossRef]
64. Beilock, S.L. Beyond the playing field: Sport psychology meets embodied cognition. *Int. Rev. Sport Exerc. Psychol.* **2008**, *1*, 19–30. [CrossRef]
65. Sluming, V.; Brooks, J.; Howard, M.; Downes, J.J.; Roberts, N. Broca's area supports enhanced visuospatial cognition in orchestral musicians. *J. Neurosci.* **2007**, *27*, 3799–3806. [CrossRef]
66. Rogge, A.-K.; Hötting, K.; Nagel, V.; Zech, A.; Hölig, C.; Röder, B. Improved balance performance accompanied by structural plasticity in blind adults after training. *Neuropsychologia* **2019**, *129*, 318–330. [CrossRef] [PubMed]
67. Word Health Organization. *Global Recommendations on Physical Activity for Health*; WHO Press: Geneva, Switzerland, 2010.
68. Voelcker-Rehage, C.; Godde, B.; Staudinger, U.M. Physical and motor fitness are both related to cognition in old age. *Eur. J. Neurosci.* **2010**, *31*, 167–176. [CrossRef] [PubMed]
69. Hayes, S.M.; Forman, D.E.; Verfaellie, M. Cardiorespiratory fitness is associated with cognitive performance in older but not younger adults. *J. Gerontol. B Psychol. Sci. Soc. Sci.* **2014**, *71*, 474–482. [CrossRef]
70. Best, J.R. Effects of physical activity on children's executive function: Contributions of experimental research on aerobic exercise. *Dev. Rev.* **2010**, *30*, 331–351. [CrossRef] [PubMed]
71. Stern, Y.; MacKay-Brandt, A.; Lee, S.; McKinley, P.; McIntyre, K.; Razlighi, Q.; Agarunov, E.; Bartels, M.; Sloan, R.P. Effect of aerobic exercise on cognition in younger adults. *Neurology* **2019**, *92*, 905–916. [CrossRef]
72. Schäfer, S.; Huxhold, O.; Lindenberger, U. Healthy mind in healthy body? A review of sensorimotor–cognitive interdependencies in old age. *Eur. Rev. Aging Phys. Act.* **2006**, *3*, 45–54. [CrossRef]
73. Barenberg, J.; Berse, T.; Dutke, S. Executive functions in learning processes: Do they benefit from physical activity? *Educ. Res. Rev.* **2011**, *6*, 208–222. [CrossRef]
74. Hötting, K.; Röder, B. Beneficial effects of physical exercise on neuroplasticity and cognition. *Neurosci. Biobehav. Rev.* **2013**, *37*, 2243–2257. [CrossRef]
75. Park, D.C.; Reuter-Lorenz, P. The adaptive brain: Aging and neurocognitive scaffolding. *Annu. Rev. Psychol.* **2009**, *60*, 173–196. [CrossRef] [PubMed]
76. Cepeda, N.J.; Kramer, A.F.; de Sather, J.C.M.G. Changes in executive control across the life span: Examination of task-switching performance. *Dev. Psychol.* **2001**, *37*, 715–730. [CrossRef] [PubMed]
77. West, R.L. An application of prefrontal cortex function theory to cognitive aging. *Psychol. Bull.* **1996**, *120*, 272–292. [CrossRef] [PubMed]
78. Dale, G.; Joessel, A.; Bavelier, D.; Green, C.S. A new look at the cognitive neuroscience of video game play. *Ann. N.Y. Acad. Sci.* **2020**, *1464*, 192–203. [CrossRef]
79. Ainsworth, B.E.; Caspersen, C.J.; Matthews, C.E.; Mâsse, L.C.; Baranowski, T.; Zhu, W. Recommendations to improve the accuracy of estimates of physical activity derived from self report. *J. Phys. Act. Health* **2012**, *9*, S76–S84. [CrossRef]

80. Swann, C.; Moran, A.; Piggott, D. Defining elite athletes: Issues in the study of expert performance in sport psychology. *Psychol. Sport Exerc.* **2015**, *16*, 3–14. [CrossRef]
81. Chang, Y.; Labban, J.; Gapin, J.; Etnier, J. The effects of acute exercise on cognitive performance: A meta-analysis. *Brain Res.* **2012**, *1453*, 87–101. [CrossRef]
82. Formenti, D.; Cavaggioni, L.; Duca, M.; Trecroci, A.; Rapelli, M.; Alberti, G.; Komar, J.; Iodice, P. Acute effect of exercise on cognitive performance in middle-aged adults: Aerobic versus balance. *J. Phys. Act. Health* **2020**, *17*, 773–780. [CrossRef]
83. Germine, L.; Reinecke, K.; Chaytor, N. Digital neuropsychology: Challenges and opportunities at the intersection of science and software. *Clin. Neuropsychol.* **2019**, *33*, 271–286. [CrossRef]
84. Kessels, R.P.C. Improving precision in neuropsychological assessment: Bridging the gap between classic paper-and-pencil tests and paradigms from cognitive neuroscience. *Clin. Neuropsychol.* **2018**, *33*, 357–368. [CrossRef] [PubMed]

Review

Concurrent Performance of Executive Function during Acute Bouts of Exercise in Adults: A Systematic Review

Kefeng Zheng [1], Liye Zou [2], Gaoxia Wei [3,4] and Tao Huang [1,*]

1   Department of Physical Education, Shanghai Jiao Tong University, Shanghai 200240, China; z757068@sjtu.edu.cn
2   Exercise Psychophysiology Laboratory, Institute of KEEP Collaborative Innovation, School of Psychology, Shenzhen University, Shenzhen 518060, China; liyezou123@gmail.com
3   Key Laboratory of Mental Health, Institute of Psychology, Chinese Academy of Sciences, Beijing 100864, China; weigx@psych.ac.cn
4   Department of Psychology, University of Chinese Academy of Sciences, Beijing 100049, China
*   Correspondence: taohuang@sjtu.edu.cn; Tel.: +86-21-54747672

Citation: Zheng, K.; Zou, L.; Wei, G.; Huang, T. Concurrent Performance of Executive Function during Acute Bouts of Exercise in Adults: A Systematic Review. *Brain Sci.* **2021**, *11*, 1364. https://doi.org/10.3390/brainsci11101364

Academic Editors: Filipe Manuel Clemente and Ana Filipa Silva

Received: 2 August 2021
Accepted: 13 October 2021
Published: 17 October 2021

Publisher's Note: MDPI stays neutral with regard to jurisdictional claims in published maps and institutional affiliations.

Abstract: The purpose of the study was to systematically review the evidence on the effects of an acute bout of exercise on concurrent performance of core executive function (EF) during exercise in adults. Four electronic databases (i.e., PubMed, Web of Science, PsycINFO, and SportDiscus) were searched from inception dates to 30 December 2020. The literature searches were conducted using the combinations of two groups of relevant items related to exercise and executive function. Articles were limited to human studies in adults. The search process, study selection, data extraction, and study quality assessments were carried out independently by two researchers. A total of 4899 studies were identified. Twenty-two studies met our inclusion criteria. Of the 42 reported outcomes in the 22 studies, 13 (31%) of the 42 outcomes showed that core EF performance was enhanced during exercise and 14 (33%) found that core EF performance did not differ from control conditions. Fifteen (36%) found that core EF performance was impaired. Notably, improved EF performances tend to be observed during moderate-intensity exercise, whereas impaired EF performances were more likely to be observed at vigorous-high intensity. The review suggests mixed findings regarding the effects of an acute bout of exercise on concurrent performance of core EF. Exercise intensity seems to influence the effects. The underlying neural mechanisms remain to be elucidated.

Keywords: brain; exercise; exercise intensity; executive function

1. Introduction

Accumulating evidence indicates that exercise is not only beneficial to physical health but also to brain and cognitive function [1–3], and that the exercise-induced positive effects seem to be significantly larger for executive function (EF) compared to other cognitive subdomains [4,5]. EF refers to a subset of higher-order mental skills, primarily encompassing inhibitory control, working memory, and cognitive flexibility. These three core subdomains have attracted substantial attention from the research community, especially its association with school-based academic performance and future career success [6,7]. Observational studies have shown that higher levels of physical activity or greater aerobic fitness are associated with better EF-related performance among various age groups [8–10]. Chronic exercise intervention studies further support that exercise could promote cognitive (and brain) development in children and attenuate the progress of age-related cognitive decline [11,12]. In addition, a growing body of studies have also examined the effects of acute exercise on EF [13,14]. Specifically, the concurrent and subsequent performance of EF were both affected by an acute bout of exercise [15]. Several hypotheses have proposed in the literature to explain the mechanisms underlying the effects of exercise on cognitive performance, such as the cardiorespiratory hypothesis, neurogenesis increase hypothesis, synaptic plasticity increase hypothesis, catecholamines increase hypothesis, and cognitive enrichment hypothesis [16,17].

Individuals often confront a situation wherein cognitive skills and physical activities are simultaneously needed [18], which seems to be simulated by the concurrent measures of cognitive performance during exercise. For example, in sports and military activities, successful performances partly depend on the ability to simultaneously handle physical and cognitive loads [19,20]. To this end, it is of great importance to elucidate how the concurrent performance of EF is modulated during exercise. However, some studies show that EF performances appear to be impaired [21–23] or improved [15,24] during an acute bout of exercise, whereas others did not observe any influences [25,26]. Collectively, there is a lack of consensus with regard to the changes in core EF during exercise.

A previous review assessed the effects of acute bouts of exercise on cognitive performance during and after an exercise session [27]. They concluded that submaximal aerobic exercise with a duration up to 60 min facilitated specific aspects of information processing, and that extended exercise negatively influenced information processing and memory function. In a later systematic review with meta-analysis, Lambourne and Tomporowski [28] quantitatively analyzed the effects of acute exercise on cognitive function. They found that cognitive performance declined during the initial 20 min and then facilitated after 20 min of exercise. Similar views have been presented in the review by Chang et al. [29], indicating that the time of cognitive test administration during exercise significantly influenced the outcome, such that effects in the first 20 min of exercise were negligible or negative, and effects after 20 min of exercise were positive. Furthermore, Schmit and Brisswalter [30] proposed a fatigue-based neurocognitive perspective of EF during exercise. They demonstrated that EF performance during relatively long exercise would be dynamic rather than steady and that exercise intensity may not be the most crucial factor to explain EF performance. However, in the aforementioned reviews, the included studies examined performance of a diversity of cognitive skills during exercise. Meanwhile, only a limited number of the included studies investigated the effects of acute exercise on core EF while exercising and highlighted the lack of consensus. Obviously, the inconsistent findings regarding EF performance during exercise warrant further investigations.

Taken together, we identified two gaps in the literature. First, there is no systematic review that has exclusively evaluated the core EF performance during acute bouts of exercise. Second, factors that may moderate the concurrent EF performance during exercise are not clear. With these thoughts in mind, the current study aims to systematically review the evidence on the effects of an acute bout of exercise on concurrent core EF performance during exercise in adults.

2. Methods

The systematic review was performed according to the guidelines from the Preferred Reporting Items for Systematic Reviews and Meta-Analyses [31].

2.1. Data Sources and Search Strategy

The electronic databases PubMed, Web of Science, SportDiscus, and PsycINFO were searched for relevant articles. Articles were retrieved from inception dates to 30 December 2020. The literature searches were conducted using combinations of two groups of relevant items: ("exercise" OR "aerobic exercise" OR "acute exercise") AND ("executive function" OR "cognitive control" OR "inhibitory control" OR "working memory" OR "cognitive flexibility").

2.2. Study Selection

Two authors independently performed the literature searches. Upon performing the computerized searches, the article titles and abstracts were reviewed in order to identify potentially relevant articles. All potential and relevant articles were retrieved and reviewed at the full text level. In addition, the bibliographies of the included studies were further screened for missing relevant studies. Any disagreements about the study selection were discussed among the authors until a consensus was reached.

2.3. Inclusion/Exclusion Criteria

Studies were included if they: (1) published in peer-reviewed journals with full text available in English; (2) investigated core executive function performance during an acute bout of exercise in adults (18–65 years); and (3) employed an experimental design with a comparison to a no-exercise control group/condition. Studies without predefined exercise intensity (e.g., self-paced cycling or walking) were excluded. Studies conducted in a special condition (e.g., severe hypoxia or breakfast omission) were also excluded.

2.4. Methodological Quality of Included Studies

Two authors independently evaluated the methodological quality of the included studies. The methodological quality of each included study was evaluated using the Physiotherapy Evidence Database (PEDro) scale [32,33]. This scale consists of 11 items to assess the methodological quality: eligibility criteria, randomization, concealed allocation, similar baseline, blinding of all subjects, blinding of all therapists, blinding of all assessors, more than 85% retention, intention to treat analysis, between/within group comparison, and point measures and measures of variability. If an item was described absently or unclearly, the article would be given 0 points; if an item was described clearly, the article would be given 1 point. When disagreements of rating between the two reviewers occurred, they discussed and re-evaluated the discrepant results together until reaching a consensus. Total score was calculated, with a higher score indicating better methodological quality. Given that some of the studies employed a within-subject design, these studies were automatically awarded a point for the baseline comparability item. The item for blinded subjects is given 0 points for all studies, since the blinding of participants is not possible in exercise behavioral intervention studies.

2.5. Data Extraction of Included Studies

Detailed information of the included studies was extracted, including the first author, methodological quality, participants description, study design, exercise protocol, time of EF test administration and duration, EF task and subdomain(s), and main results. If study findings suggest divergent effects (e.g., shorter RT but impaired RA) of exercise on EF performance, this systematic review retrieved RT as a measure of EF performance.

3. Results

3.1. Study Selection and Study Characteristics

Figure 1 depicts the flow chart of the article selection process. The computerized searches identified 4892 articles from the four electronic databases and seven were identified through screening the references in the relevant articles.

After removing duplicates and irrelevant articles, 3624 articles were eligible for further screening. After screening via title and abstract, 35 were identified as potentially relevant, and the full text articles were reviewed. Among them, 13 articles were ineligible as they employed a self-paced exercise protocol (e.g., self-paced cycling), were conducted in a special condition (e.g., severe hypoxia or breakfast omission), or did not employ a comparison to a no-exercise control group/condition. Thus, 22 articles met our inclusion criteria and were included for the qualitative research. Notably, due to the lack of relevant data and the diversity of the experimental protocols (e.g., exercise protocol, time points of EF measurement, and nature of the cognitive tasks), it is not feasible to quantitatively synthesize the findings using a meta-analytic method.

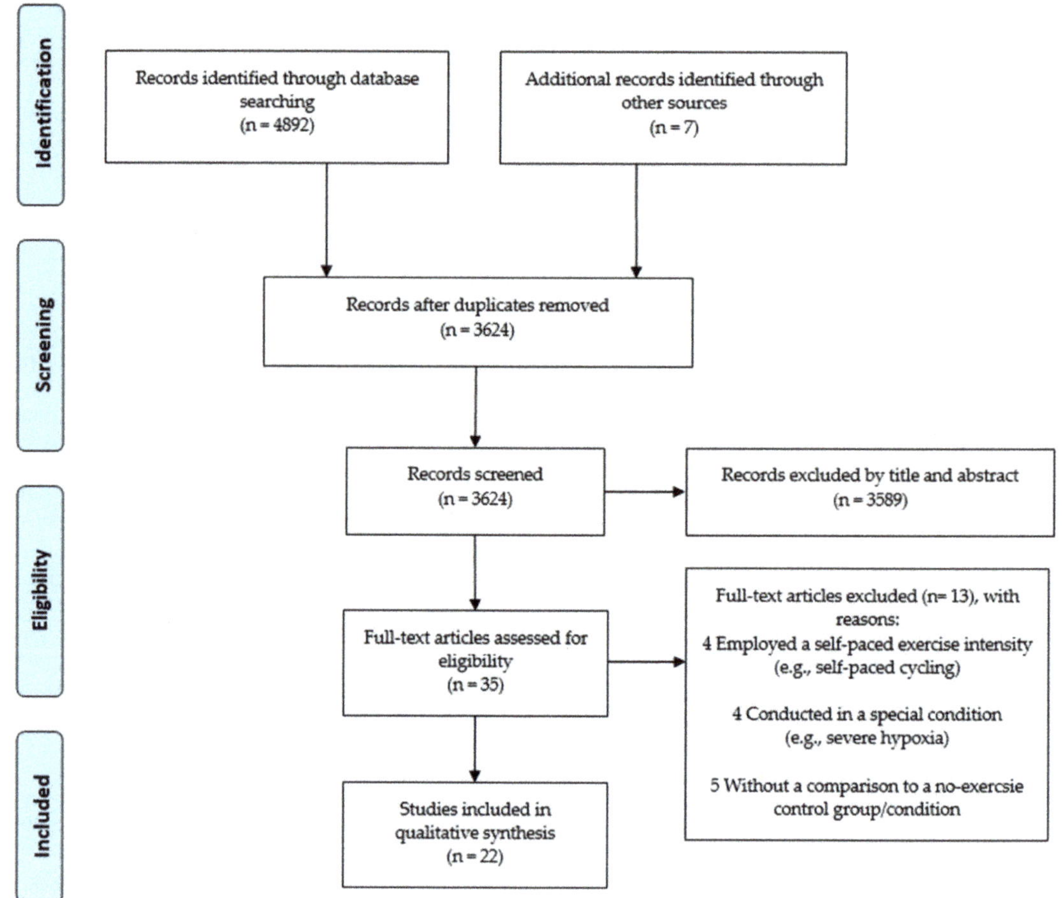

Figure 1. Flow chart of study selection.

Collectively, in the 22 studies, a total of 590 participants were included, with sample sizes of each individual study ranging from 7 to 120. A total of 13 EF tasks were used. The included studies employed light to high intensities based on classifications of aerobic exercise intensity by Norton et al. [34]. The exercise duration lasted from 7 to 65 min. The study characteristics are summarized in Table 1.

Table 1. Characteristics of the included studies.

Study (Authors, Methodological Quality)	Participants Description	Study Design	Exercise Protocol	Time of EF Test Administration and Duration	EF Task	EF Domain	Results
Audiffren et al. [22] 5/11	Female: 21.11 ± 1.05 Male: 21.14 ± 0.69 (N = 18) 9M/18	Within-subject design	Cycling at 90% VT (moderate intensity); exercise duration (40 min)	Intermittent assessment (a total of 5 times)	RNG	Working memory; inhibitory control	Inhibitory control was impaired while working memory did not differ from the control conditions. The EF modulation can be interpreted as a change in strategy.
Davranche et al. [35] 7/11	30 ± 8 (N = 14) 11M/14	Within-subject design	Cycling at 50% MAP (moderate intensity); Exercise duration (two periods of 20 min cycling)	Four blocks of task trials were performed during a first 15-min period and another four blocks were performed during a second 15-min period.	Flanker task	Inhibitory control	The task performance was not different from the control condition.
Davranche and McMorris [20] 7/11	32 ± 9 (N = 12) 8M/12	Within-subject design	Cycling at VT intensity (moderate intensity); exercise duration (20 min)	The Simon task began at the end of the 3-min warm-up period and performed within the remaining 17-min period.	Simon task	Inhibitory control	According to the Simon effect, inhibitory control (RT) was impaired.
Del Giorno et al. [36] 7/11	20.2 ± 1.1 (N = 30) 17M/30	Within-subject design	Cycling at 75% VT and VT intensity (light and moderate intensity); exercise duration (30 min)	Cognitive tests began at 20 min following the onset of exercise, lasting for approximately 4 min.	CPT; WCST	Inhibitory control; cognitive flexibility	The performance of two tasks (RA) was impaired during exercise at both light and moderate intensities.
Dietrich and Sparling [21] 7/11							
Exp.1	23.7 ± 9.4 (N = 24) 24M/24	Between-subject design	Cycling or running at 70–80% HR$_{max}$ (vigorous intensity); exercise duration (50 min)	Cognitive tests began after 25 min of exercise, lasting for approximately 10 min.	WCST	Cognitive flexibility	For the WCST, the exercise group made significantly more errors compared to the control group.
Exp.2	25.1 ± 6.3 (N = 8) 8M/8	Within-subject design	Running at vigorous intensity; exercise duration (65 min)	After 25 min of exercise, lasting for approximately 28 min.	PASAT	Working memory	For the PASAT, the exercise condition resulted in significantly more errors than the control condition.
Dodwell et al. [37] 5/11	24.5 ± 2.6 (N = 18) 10M/18	Within-subject design	Cycling or running at 65% HRR intensity (vigorous intensity)	The Retro-cue task began following 5–10 min of warm-up period, included 4 blocks of 96 trials.	Retro-cue task	Working memory	RT was facilitated in the exercise condition compared to the control condition.

Table 1. *Cont.*

Study (Authors, Methodological Quality)	Participants Description	Study Design	Exercise Protocol	Time of EF Test Administration and Duration	EF Task	EF Domain	Results
Joyce et al. [15] 7/11	23 ± 2 (N = 10) 7M/10	Within-subject design	Cycling at 40% MAP (moderate intensity); exercise duration (30 min)	The Stop-signal task was performed whilst cycling after a 4-min warm-up period and lasted approximately 22 min.	Stop-signal task	Inhibitory control	Inhibitory control was improved during exercise (shorter RT without a change in RA).
Joyce et al. [26] 7/11	23 ± 2 (N = 12) 3M/12	Within-subject design	Cycling at 65% of HR_{max} (moderate intensity); exercise duration (30 min)	The Simon task was performed after 5-min warm-up period and lasted approximately 23 min.	Simon Task	Inhibitory control	According to the Simon effect, inhibitory control was unchanged during exercise.
Komiyama et al. [38] 6/11	21.5 ± 3.5 (N = 13) 13M/13	Within-subject design	Cycling at 50% VO_{2max} (moderate intensity); exercise duration (20 min)	The EF tasks were started after a 5-min warm-up period.	Spatial DR task; Go/No-Go task	Working memory; inhibitory control	The task performance (RT) was improved during exercise without sacrificing RA.
Komiyama et al. [39] 6/11	23.0 ± 2.3 (N = 16) 16M/16	Within-subject design	Cycling at heart rate of 140 beats/min (moderate intensity); exercise duration (30 min)	Intermittent assessment (a total of 2 times)	Spatial DR task; Go/No-Go task	Working memory; inhibitory control	RA was not changed in the Spatial DR task; RT was shorter without sacrificing RA in the Go/No-Go task.
Komiyama et al. [23] 6/11	22.1 ± 1.7 (N = 17) 17M/17	Within-subject design	Cycling at 50%$VO_{2\ peak}$ (moderate intensity) for 8 min; thereafter, participants cycled at 80% $VO_{2\ peak}$ (vigorous intensity) for an additional 8 min.	Participant performed the EF tasks 3 min after commencing each workload.	Spatial DR task; Go/No-Go task	Working memory; inhibitory control	RA of the tasks was impaired during vigorous-intensity exercise, whereas it was not changed during moderate-intensity exercise; RT was not changed during both intensity exercises.
Lambourne et al. [25] 7/11	21.1 ± 1.7 (N = 19) 8M/19	Within-subject design	Cycling at 90% VT intensity (moderate intensity); exercise duration (40 min)	Intermittent assessment (a total of 5 times)	PASAT	Working memory	RA of the task in the exercise condition did not differ from the control condition.
Lucas et al. [40] 6/11	24 ± 5 (N = 13) 7M/13	Within-subject design	Cycling at 30% followed by 70% of HRR (light and vigorous intensity); exercise duration (two 8-min bouts of cycling)	The Stroop task involved 2 blocks of 20 trials.	Stroop task	Inhibitory control	RT was facilitated during exercise. Vigorous-intensity exercise led to greater improvement compared to light-intensity exercise.

Table 1. *Cont.*

Study (Authors, Methodological Quality)	Participants Description	Study Design	Exercise Protocol	Time of EF Test Administration and Duration	EF Task	EF Domain	Results
Martins et al. [24] 7/11							
Exp. 1	20.50 ± 0.89 (N = 24) 24M/24	Between-subject design	Cycling at moderate intensity; (short duration)	Four blocks lasting approximately 8 min.	PASAT	Working memory	RA of the task was improved during moderate-intensity exercise.
Exp. 2	19.57 ± 0.83 (N = 120) 55M/120	Mixed Multi-factorial experimental design	Cycling at light and moderate intensity; (short duration)	Two blocks lasting approximately 16 min.	Sternberg task	Working memory	Light and moderate intensity exercise lowered the response latency slopes, resulting in improved working memory.
McMorris et al. [41] 7/11	24.32 ± 7.10 (N = 24) 24M/24	Within-subject design	Cycling at 50% and 80% MAP (moderate and vigorous intensity); exercise duration (15 min or until voluntary exhaustion)	Intermittent assessment (a total of 3 times)	Flanker task	Inhibitory control	Vigorous-intensity exercise impaired RT, but moderate-intensity exercise did not change the task performance.
Ogoh et al. [42] 6/11	20.4 ± 0.6 (N = 7) 7M/7	Within-subject design	Cycling at heart rate of 140 beats/min (moderate intensity); exercise duration (50 min)	Intermittent assessment (a total of 4 times)	Stroop task	Inhibitory control	RT was facilitated during exercise without any loss of performance accuracy.
Olson et al. [43] 7/11	20.4 ± 2.0 (N = 27) 16M/27	Within-subject design	Cycling at 40% and 60% VO_{2peak} (light and moderate intensity); exercise duration (31 min)	Intermittent assessment (a total of 3 times)	Flanker task	Inhibitory control	RA was impaired during both light and moderate intensity exercise, but RT was facilitated during moderate-intensity exercise.
Pontifex and Hillman [44] 7/11	20.2 ± 1.6 (N = 41) 15M/41	Within-subject design	Cycling at 60% of HR_{max} (moderate intensity); exercise duration (approximately 11 min)	The Simon task was performed after 5 min of exercise, lasing for approximately 6.5 min.	Flanker task	Inhibitory control	Exercise did not affect RT but showed a decrease in RA for incongruent trials, resulting in impaired inhibitory control.
Schmit et al. [45] 7/11	22.1 ± 0.6 (N = 15) 10M/15	Within-subject design	Cycling at 85% MAP until exhaustion; exercise duration (approximately 7 min)	Participants performed the Flanker task until exhaustion.	Flanker task	Inhibitory control	RT was facilitated during exercise in the initial stage and remained unaltered in the final stage.
Smith et al. [46] 7/11	28 ± 5 (N = 15) 6M/15	Within-subject design	Running at moderate and high intensity; exercise duration (10 min)	The EF task was performed during the last 2 min of exercise.	Go/No-Go task	Inhibitory control	RT was impaired during high-intensity exercise, whereas it was not changed during moderate-intensity exercise.

Table 1. *Cont.*

Study (Authors, Methodological Quality)	Participants Description	Study Design	Exercise Protocol	Time of EF Test Administration and Duration	EF Task	EF Domain	Results
Stone et al. [19] 5/11	19.6 ± 2 (N = 13) 8M/13	Within-subject design	Conducted at an exercise intensity in an incremental manner; the average duration was between 20–24 min.	The OWAT test was administered throughout the entirety of the graded exercise test.	OWAT	Cognitive flexibility	RA was not changed at an intensity from 20% to 80% HRR, where it was impaired from 80% to 100% HRR.
Wang et al. [47] 7/11	20.51 ± 1.99 (N = 80) 49M/80	Between-subject design	Cycling at 30%, 50%, and 80% HRR (light, moderate, and vigorous intensity); exercise duration (30 min)	The WCST was performed 6 min after exercise onset.	WCST	Cognitive flexibility	Cognitive flexibility (WCST indices) was impaired in the group of vigorous intensity, whereas it was not changed in groups of light and moderate intensity compared to the control group.

M, male; Exp, experiment; WCST, Wisconsin card sorting task; PASAT, paced auditory serial addition task; RNG, random number generation; CPT, contingent continuous performance task; Spatial DR, spatial delayed response; OWAT: operator workload assessment tool; HRR, heart rate range; MAP, maximal aerobic power; PPO, peak power output; VT, ventilatory threshold; RT, reaction time; RA, response accuracy.

3.2. Study Quality

The methodological quality scores are presented in Table 1. The average score of the methodological quality of the 22 studies are 6.5, with scores ranging from 5 to 7 (see Supplementary Table S1 for details).

3.3. Study Findings

In the 22 studies, 42 outcomes are reported. Overall, 13 (31%) of these 42 outcomes showed that EF performance was enhanced, whereas 14 (33%) indicated no change in this outcome. Fifteen (36%) found that EF performance was impaired.

Seven of the 42 outcomes, within six studies, employed a light-intensity exercise protocol. Among them, two outcomes demonstrated that EF was improved [24,40], three outcomes found that exercise did not affect EF performance [19,43,47], and another two outcomes found that EF performance was declined [36]. Twenty-four of the 42 outcomes, within 18 studies, employed moderate-intensity exercise protocol. Eight outcomes observed that EF performance was improved [15,24,38,39,42,43]. Eleven outcomes found that EF performance during exercise remained unaltered [19,22,23,25,26,35,39,41,46,47]. Five outcomes found that EF performance was deteriorated [20,22,36,44]. Eleven of the 42 outcomes, within nine studies, employed a vigorous- to high-intensity protocol, of which three found that EF performance was improved [37,40,45]. Eight showed that EF performance was declined [19,21,23,41,46,47]. According to the findings, impaired EF performance most likely occurs during vigorous- to high-intensity exercise. In contrast, the unaltered or even facilitated outcomes of concurrent EF performance are predominantly observed during moderate-intensity exercise. A summary of the outcomes across the varying exercise intensities is presented in Table 2.

The exercise duration of the 22 studies ranged from 7 to 65 min. Six of the twenty-two studies examined the temporal dynamic of EF during relatively long, steady-state exercise (30 min or longer). Two of the six studies found that the improvement of EF performance was greater in the terminal stage compared with earlier stages [42,43]. One study found that the shift towards a less effortful strategy was more pronounced in the first stage of the exercise bout than later [22], which means that inhibitory control was compromised during the initial stage of exercise. Another three studies showed that EF remained unaltered during the entire steady-state exercise session [25,39,45].

Table 2. Summary of the core EF performance across the varying exercise intensities.

Intensity	Facilitation	No Effect	Impairment
Light	🟢🟡	🟢🔴🔴	🟢🔴
Moderate	🟢🟢🟢🟢🟢🟢🟡🟡🟡	🟢🟢🟢🟢🟢🟡🟡🟡🟡🔴🔴	🟢🟢🟢🟢🔴
Vigorous-high	🟢🟢🟡		🟢🟢🟢🟡🟡🔴🔴🔴

The three exercise intensities included: (1) light-intensity exercise ($40 < 55\%HR_{max}$, $20 < 40\%HRR$, $20 < 40\%VO_{2max}$) vs. control, (2) moderate-intensity exercise ($55 < 70\%HR_{max}$, $40 < 60\%HRR$, $40 < 60\%VO_{2max}$) vs. control, (3) vigorous-high intensity exercise ($\geq 70\%HR_{max}$, $\geq 60\%HRR$, $\geq 60\%VO_{2max}$) vs. control. Green 🟢: inhibitory control, Yellow 🟡: working memory, Red 🔴: cognitive flexibility.

Four studies examined EF at several time points (either during or after the first 20 min relative to the onset of exercise). A study by Olson et al. [43] found a significant decline in EF performance either during or after the first 20 min relative to exercise onset. In contrast, one study demonstrated that EF performance was improved after the first 20 min and remained stable during first 20 min relative to the onset of exercise [42]. This is different to the findings of Komiyama et al. [39], who found that EF performance was facilitated both during and after the first 20 min of exercise. In addition, Lambourne et al. [25] revealed that the time points had no significant effect on concurrent EF performance.

Five of the 22 studies investigated cerebral oxygenation during exercise using near-infrared spectroscopy (NIRS). One of the five studies found that EF performance was not associated with changes in cerebral oxygenation sampled at the prefrontal cortex (PFC) [38]. Two showed that EF performance was unchanged or even improved despite a decrease in cerebral oxygenation [40,45]. One study observed that decline in EF performance was related to attenuated cortex oxygenation [19]. Another study revealed that improvement in EF performance was associated with higher cerebral oxygenation, but the change of cerebral oxygenation was not significant [39]. Four studies (two of the four studies also used NIRS) used transcranial Doppler to assess cerebral blood flow (CBF) velocity. They found that improvements or impairments in EF during exercise are not directly related to alterations in CBF [23,38,40,42].

4. Discussion

4.1. Main Findings

The current study critically reviewed the evidence regarding how the concurrent performance of core EF is affected during an acute exercise session in adults. The available studies revealed mixed findings. Collectively, the findings of the current review indicate that exercise intensity is a potential factor influencing the core EF performance while simultaneously performing exercise. Improved EF performances tend to be observed during moderate-intensity exercise, whereas impaired EF performances during exercise were more likely to be observed at vigorous-to high-intensity.

4.2. Role of Exercise Intensity

Although the included studies depicted a picture of mixed findings, exercise intensity seems to moderate the effects of exercise on concurrent EF performance. Most outcomes of these studies found that EF performance during light to moderate intensity exercise remained unchanged or even improved. With regard to the studies that employed a vigorous- to high-intensity protocol, most of outcomes found that EF performance was impaired. Meanwhile, most of the facilitating outcomes were observed during moderate-intensity exercise. The findings are inconsistent with previous studies by Lambourne and Tomporowski [28] and Chang et al. [29], who suggested that exercise intensity might not be a crucial factor influencing the exercise–cognition relationship. The plausibility of our findings can be partly explained by the inverted-U shape relationship between arousal and mental information processing [27,48,49]. Moderate-intensity exercise may lead to a moderate level of arousal [50], thereby facilitating the EF performance during exercise.

According to the reticular-activating hypofrontality model (RAH) proposed by Dietrich and Audiffren [51], it is hypothesized that, within the context of cognitive and physical demands, exercise engages arousal mechanism in the reticular-activating system and disengages the higher-order functions of the prefrontal cortex partly due to the limited resources. As the findings in this review found, the transient hypofrontality most likely occurs during vigorous- to high-intensity exercise. In contrast to the inhibition predicted by the RAH model, these findings suggest that, during moderate-intensity exercise, EF performance can be either unchanged or even facilitated.

4.3. Role of Participants' Physical Fitness and Exercise Mode

In addition to exercise intensity, previous studies suggest that factors such as physical fitness level of participants and exercise mode may moderate the exercise-cognitive function relationship. In the review by Chang et al. [29], physical fitness level was found to be a moderator of the relationship. Positive effects were evident for high-fitness participants, while negative effects were observed for low-fitness counterparts. Unfortunately, the included studies of this review did not examine the potential moderating effects of physical fitness. It is, therefore, not possible to clarify whether the effects of an acute bout of exercise on the performance of core EF during exercise differ among participants with different fitness levels. Most of the included studies employed a cycling protocol. It is also not possible to clarify the moderating effects of exercise mode. However, one included study by Dodwell et al. [37] found that the concurrent performance of working memory was improved during running but not during cycling. Therefore, future studies may further investigate the moderating effects of physical fitness level and exercise mode.

4.4. Time Point of the EF Task Administration

Four of the included studies examined the temporal change of EF during exercise and the direction of EF change is mixed [25,39,42,43]. The current findings suggest that the time point of the EF task administration did not significantly moderate the effects of concurrent EF performance. Another two studies measured EF at 20 min following the onset of exercise and found that EF performance was impaired [21,36]. However, these findings are in contrast with a previous review by Lambourne and Tomporowski [28], which suggests that improved cognitive performance occurred when measured after the first 20 min relative to the onset of exercise. Unlike the studies by Lambourne and Tomporowski [28] and Chang et al. [29], which reviewed acute exercise on both higher-order cognitive processes and basic-cognitive processes, the present review focused solely on core EF. Therefore, the inconsistent findings may be driven in part by the nature of cognitive tasks.

This systematic review also sheds light on the importance of considering EF performance as a dynamic process during relatively long and steady-state exercise. For the study conducted by Schmit et al. [37] aimed at investigating inhibitory control during steady-state exercise to exhaustion, EF was measured at an initial and a terminal period of the exercise session. They found that EF was enhanced during the first part of the exercise session, while no sign of deficit was observed shortly before exhaustion. However, individuals' susceptibility to making errors was increased during the terminal period. Based on those findings, Schmit and Brisswalter [28] proposed a fatigue-based neurocognitive perspective of EF during prolonged, steady-state exercise. From this perspective, EF performance during exercise would be dynamic rather than steady (i.e., positively then negatively impacted by exercise). Apart from that, Olson et al. [43] also examined temporal dynamic of EF during steady-state exercise. EF was assessed at three different time points during the stage of a 31-min cycle. Behavioral findings revealed impaired response accuracy, while faster reaction time was observed at 25 min relative to earlier time points. Although the findings are mixed, it is possible that the performance of EF are dynamic rather than steady during steady-state exercise.

4.5. Neural Physiological Responses

In order to elucidate the neural physiological responses, several included studies examined the relationship between CBF or cerebral hemodynamic responses and EF performance during exercise. CBF did not relate to the changes in EF during exercise in all of the included studies [23,38,40,42]. One study using NIRS found that the changes in cerebral hemodynamic responses were not directly related to the changes of EF [38]. However, other studies found a negative or positive link between cerebral hemodynamic responses and EF performance [19,39,40,45]. Overall, the conflicting results from NIRS studies might be due to the different experimental protocols. Moreover, the insufficient spatial resolution of NIRS may partly explain the inconsistent findings.

4.6. Strengths and Limitations

The findings of this systematic review have some potential practical implications, since individuals often confront a situation wherein cognitive skills and physical activities are simultaneously needed. It is of importance to understand how the concurrent EF performance is modulated during exercise. Unlike other similar reviews, this study focused solely on concurrent performance of core EF during an acute bout of exercise in adults. To some extent, it eliminates the confounding effects from a diverse cognitive process and helps draw a clearer picture of the concurrent performance of core EF during exercise. However, the review must be interpreted within the context of its potential limitations. First, the included studies employed a diversity of experimental protocols (e.g., exercise protocol, EF test administration and duration, cognitive tasks). The diversity and lack of relevant data make it impossible to perform a meta-analysis. Second, although the outcomes measures consist of three core EFs in this review, the insufficient number of studies limits the possibility of clarifying the potential domain-specific effects on core EF. Lastly, the language of included studies was limited to English, which may lead to the omitting of potential studies published in other languages.

4.7. Perspectives and Future Direction

Considering the findings presented in this systematic review, future studies are needed to investigate the concurrent EF performance during exercise in individuals with different fitness levels. It is also interesting to clarify the effects of different exercise modes (e.g., cycling or running) on concurrent EF performance. Furthermore, the possible temporal dynamic of EF at different time points is worth investigating during relatively long exercise. Although the findings tend to suggest that exercise intensity may be a potential factor to moderate the concurrent EF performance during exercise, the underlying mechanisms and neural responses are not clear. Future neuroimaging studies with more rigorous designs are needed to look into the neural responses, thereby elucidating the neural mechanisms of the observed findings.

5. Conclusions

The current review suggests mixed findings of the concurrent performance of core EF during exercise in adults. Exercise intensity seems to be a potential factor influencing the effects. The underlying neural mechanisms remain to be elucidated.

Supplementary Materials: The following are available online at https://www.mdpi.com/article/10.3390/brainsci11101364/s1, Table S1: Methodological quality scores.

Author Contributions: K.Z.: Conceptualization, Methodology, Validation, Formal analysis, Investigation, Data curation, Writing—original draft, Writing—review and editing, Visualization. L.Z.: Conceptualization, Methodology, Investigation, Writing—review and editing. G.W.: Methodology, Investigation, Writing—review and editing. T.H.: Conceptualization, Methodology, Investigation, Writing—original draft, Writing—review and editing, Resources, Visualization, Supervision, Project administration. All authors approved the final version. All authors have read and agreed to the published version of the manuscript.

Funding: This research received no external funding.

Institutional Review Board Statement: Not applicable.

Informed Consent Statement: Not applicable.

Data Availability Statement: The data used to support the findings of this study are included within the article.

Acknowledgments: Thanks to Shiyuan Li for reading the manuscript and providing feedback.

Conflicts of Interest: The authors declare no conflict of interest.

References

1. Stillman, C.M.; Esteban-Cornejo, I.; Brown, B.; Bender, C.M.; Erickson, K.I. Effects of Exercise on Brain and Cognition Across Age Groups and Health States. *Trends Neurosci.* **2020**, *43*, 533–543. [CrossRef]
2. Ludyga, S.; Gerber, M.; Pühse, U.; Looser, V.N.; Kamijo, K. Systematic review and meta-analysis investigating moderators of long-term effects of exercise on cognition in healthy individuals. *Nat. Hum. Behav.* **2020**, *4*, 603–612. [CrossRef]
3. Hillman, C.H.; Erickson, K.I.; Kramer, A. Be smart, exercise your heart: Exercise effects on brain and cognition. *Nat. Rev. Neurosci.* **2008**, *9*, 58–65. [CrossRef]
4. Kramer, A.; Erickson, K.I. Capitalizing on cortical plasticity: Influence of physical activity on cognition and brain function. *Trends Cogn. Sci.* **2007**, *11*, 342–348. [CrossRef]
5. Chaddock, L.; Pontifex, M.; Hillman, C.; Kramer, A.F. A Review of the Relation of Aerobic Fitness and Physical Activity to Brain Structure and Function in Children. *J. Int. Neuropsychol. Soc.* **2011**, *17*, 975–985. [CrossRef] [PubMed]
6. Diamond, A. Executive functions. *Annu. Rev. Psychol.* **2013**, *64*, 135–168. [CrossRef] [PubMed]
7. Magalhães, S.; Carneiro, L.; Limpo, T.; Filipe, M. Executive functions predict literacy and mathematics achievements: The unique contribution of cognitive flexibility in grades 2, 4, and 6. *Child Neuropsychol.* **2020**, *26*, 934–952. [CrossRef] [PubMed]
8. Lin, J.; Wang, K.; Chen, Z.; Fan, X.; Shen, L.; Wang, Y.; Yang, Y.; Huang, T. Associations Between Objectively Measured Physical Activity and Executive Functioning in Young Adults. *Percept. Mot. Ski.* **2017**, *125*, 278–288. [CrossRef] [PubMed]
9. Cox, E.P.; O'Dwyer, N.; Cook, R.; Vetter, M.; Cheng, H.L.; Rooney, K.; O'Connor, H. Relationship between physical activity and cognitive function in apparently healthy young to middle-aged adults: A systematic review. *J. Sci. Med. Sport* **2016**, *19*, 616–628. [CrossRef]
10. Huang, T.; Tarp, J.; Domazet, S.L.; Thorsen, A.K.; Froberg, K.; Andersen, L.B.; Bugge, A. Associations of Adiposity and Aerobic Fitness with Executive Function and Math Performance in Danish Adolescents. *J. Pediatr.* **2015**, *167*, 810–815. [CrossRef]
11. Lautenschlager, N.T.; Cox, K.L.; Flicker, L.; Foster, J.K.; Van Bockxmeer, F.M.; Xiao, J.; Greenop, K.R.; Almeida, O.P. Effect of physical activity on cognitive function in older adults at risk for Alzheimer disease—A randomized trial. *JAMA* **2008**, *300*, 1027–1037. [CrossRef]
12. Xue, Y.; Yang, Y.; Huang, T. Effects of chronic exercise interventions on executive function among children and adolescents: A systematic review with meta-analysis. *Br. J. Sports Med.* **2019**, *53*, 1397–1404. [CrossRef] [PubMed]
13. Moreau, D.; Chou, E. The Acute Effect of High-Intensity Exercise on Executive Function: A Meta-Analysis. *Perspect. Psychol. Sci.* **2019**, *14*, 734–764. [CrossRef] [PubMed]
14. Ludyga, S.; Gerber, M.; Brand, S.; Holsboer-Trachsler, E.; Pühse, U. Acute effects of moderate aerobic exercise on specific aspects of executive function in different age and fitness groups: A meta-analysis. *Psychophysiology* **2016**, *53*, 1611–1626. [CrossRef] [PubMed]
15. Joyce, J.; Graydon, J.; McMorris, T.; Davranche, K. The time course effect of moderate intensity exercise on response execution and response inhibition. *Brain Cogn.* **2009**, *71*, 14–19. [CrossRef] [PubMed]
16. Agbangla, N.F.; Maillot, P.; Vitiello, D. Mini-Review of Studies Testing the Cardiorespiratory Hypothesis With Near-Infrared Spectroscopy (NIRS): Overview and Perspectives. *Front. Neurosci.* **2021**, *15*, 699948. [CrossRef] [PubMed]
17. Dustman, R.E.; Ruhling, R.O.; Russell, E.M.; Shearer, D.E.; Bonekat, H.; Shigeoka, J.W.; Wood, J.S.; Bradford, D.C. Aerobic exercise training and improved neuropsychological function of older individuals. *Neurobiol. Aging* **1984**, *5*, 35–42. [CrossRef]
18. Davranche, K.; Audiffren, M. Facilitating effects of exercise on information processing. *J. Sports Sci.* **2004**, *22*, 419–428. [CrossRef]
19. Stone, B.L.; Beneda-Bender, M.; Mccollum, D.L.; Sun, J.; Shelley, J.H.; Ashley, J.D.; Fuenzalida, E.; Kellawan, J.M. Understanding cognitive performance during exercise in Reserve Officers' Training Corps: Establishing the executive function-exercise intensity relationship. *J. Appl. Physiol.* **2020**, *129*, 846–854. [CrossRef]
20. Davranche, K.; McMorris, T. Specific effects of acute moderate exercise on cognitive control. *Brain Cogn.* **2009**, *69*, 565–570. [CrossRef]
21. Dietrich, A.; Sparling, P.B. Endurance exercise selectively impairs prefrontal-dependent cognition. *Brain Cogn.* **2004**, *55*, 516–524. [CrossRef]
22. Audiffren, M.; Tomporowski, P.D.; Zagrodnik, J. Acute aerobic exercise and information processing: Modulation of executive control in a Random Number Generation task. *Acta Psychol.* **2009**, *132*, 85–95. [CrossRef]

23. Komiyama, T.; Tanoue, Y.; Sudo, M.; Costello, J.T.; Uehara, Y.; Higaki, Y.; Ando, S. Cognitive Impairment during High-Intensity Exercise: Influence of Cerebral Blood Flow. *Med. Sci. Sports Exerc.* **2019**, *52*, 561–568. [CrossRef]
24. Martins, A.Q.; Kavussanu, M.; Willoughby, A.; Ring, C. Moderate intensity exercise facilitates working memory. *Psychol. Sport Exerc.* **2013**, *14*, 323–328. [CrossRef]
25. Lambourne, K.; Audiffren, M.; Tomporowski, P.D. Effects of Acute Exercise on Sensory and Executive Processing Tasks. *Med. Sci. Sports Exerc.* **2010**, *42*, 1396–1402. [CrossRef] [PubMed]
26. Joyce, J.; Smyth, P.J.; Donnelly, A.E.; Davranche, K. The Simon Task and Aging: Does Acute Moderate Exercise Influence Cognitive Control? *Med. Sci. Sport Exerc.* **2014**, *46*, 630–639. [CrossRef] [PubMed]
27. Tomporowski, P.D. Effects of acute bouts of exercise on cognition. *Acta Psychol.* **2002**, *112*, 297–324. [CrossRef]
28. Lambourne, K.; Tomporowski, P. The effect of exercise-induced arousal on cognitive task performance: A meta-regression analysis. *Brain Res.* **2010**, *1341*, 12–24. [CrossRef] [PubMed]
29. Chang, Y.; Labban, J.; Gapin, J.; Etnier, J. The effects of acute exercise on cognitive performance: A meta-analysis. *Brain Res.* **2012**, *1453*, 87–101. [CrossRef] [PubMed]
30. Schmit, C.; Brisswalter, J. Executive functioning during prolonged exercise: A fatigue-based neurocognitive perspective. *Int. Rev. Sport Exerc. Psychol.* **2018**, *13*, 21–39. [CrossRef]
31. Moher, D.; Liberati, A.; Tetzlaff, J.; Altman, D.G.; Grp, P. Preferred Reporting Items for Systematic Reviews and Me-ta-Analyses: The PRISMA Statement. *PLoS Med.* **2009**, *89*, 873–880.
32. Elkins, M.R.; Moseley, A.M.; Sherrington, C.; Herbert, R.D.; Maher, C.G. Growth in the Physiotherapy Evidence Database (PEDro) and use of the PEDro scale. *Br. J. Sports Med.* **2012**, *47*, 188–189. [CrossRef] [PubMed]
33. PEDro Scale. Available online: https://pedro.org.au/english/resources/pedro-scale/ (accessed on 1 March 2021).
34. Norton, K.; Norton, L.; Sadgrove, D. Position statement on physical activity and exercise intensity terminology. *J. Sci. Med. Sport* **2010**, *13*, 496–502. [CrossRef]
35. Davranche, K.; Hall, B.; McMorris, T. Effect of Acute Exercise on Cognitive Control Required during an Eriksen Flanker Task. *J. Sport Exerc. Psychol.* **2009**, *31*, 628–639. [CrossRef] [PubMed]
36. Del Giorno, J.M.; Hall, E.E.; O'Leary, K.C.; Bixby, W.R.; Miller, P.C. Cognitive function during acute exercise: A test of the transient hypofrontality theory. *J. Sport Exerc. Psychol.* **2010**, *32*, 312–323. [CrossRef]
37. Dodwell, G.; Müller, H.J.; Töllner, T. Electroencephalographic evidence for improved visual working memory performance during standing and exercise. *Br. J. Psychol.* **2018**, *110*, 400–427. [CrossRef] [PubMed]
38. Komiyama, T.; Katayama, K.; Sudo, M.; Ishida, K.; Higaki, Y.; Ando, S. Cognitive function during exercise under severe hypoxia. *Sci. Rep.* **2017**, *7*, 10000. [CrossRef]
39. Komiyama, T.; Sudo, M.; Higaki, Y.; Kiyonaga, A.; Tanaka, H.; Ando, S. Does moderate hypoxia alter working memory and executive function during prolonged exercise? *Physiol. Behav.* **2015**, *139*, 290–296. [CrossRef]
40. Lucas, S.J.; Ainslie, P.N.; Murrell, C.J.; Thomas, K.N.; Franz, E.A.; Cotter, J.D. Effect of age on exercise-induced alterations in cognitive executive function: Relationship to cerebral perfusion. *Exp. Gerontol.* **2012**, *47*, 541–551. [CrossRef]
41. McMorris, T.; Davranche, K.; Jones, G.; Hall, B.; Corbett, J.; Minter, C. Acute incremental exercise, performance of a central executive task, and sympathoadrenal system and hypothalamic-pituitary-adrenal axis activity. *Int. J. Psychophysiol.* **2009**, *73*, 334–340. [CrossRef]
42. Ogoh, S.; Tsukamoto, H.; Hirasawa, A.; Hasegawa, H.; Hirose, N.; Hashimoto, T. The effect of changes in cerebral blood flow on cognitive function during exercise. *Physiol. Rep.* **2014**, *2*, e12163. [CrossRef]
43. Olson, R.; Chang, Y.-K.; Brush, C.J.; Kwok, A.N.; Gordon, V.X.; Alderman, B.L. Neurophysiological and behavioral correlates of cognitive control during low and moderate intensity exercise. *NeuroImage* **2016**, *131*, 171–180. [CrossRef] [PubMed]
44. Pontifex, M.; Hillman, C.H. Neuroelectric and behavioral indices of interference control during acute cycling. *Clin. Neurophysiol.* **2007**, *118*, 570–580. [CrossRef] [PubMed]
45. Schmit, C.; Davranche, K.; Easthope, C.S.; Colson, S.S.; Brisswalter, J.; Radel, R. Pushing to the limits: The dynamics of cognitive control during exhausting exercise. *Neuropsychologia* **2015**, *68*, 71–81. [CrossRef] [PubMed]
46. Smith, M.; Tallis, J.; Miller, A.; Clarke, N.; Guimarães-Ferreira, L.; Duncan, M. The effect of exercise intensity on cognitive performance during short duration treadmill running. *J. Hum. Kinet.* **2016**, *51*, 27–35. [CrossRef] [PubMed]
47. Wang, C.-C.; Chu, C.-H.; Chu, I.-H.; Chan, K.-H.; Chang, Y.-K. Executive function during acute exercise: The role of exercise intensity. *J. Sport Exerc. Psychol.* **2013**, *35*, 358–367. [CrossRef] [PubMed]
48. Brisswalter, J.; Collardeau, M.; René, A. Effects of Acute Physical Exercise Characteristics on Cognitive Performance. *Sports Med.* **2002**, *32*, 555–566. [CrossRef]
49. Davey, C. Physical exertion and mental performance. *Appl. Ergon.* **1974**, *5*, 171. [CrossRef]
50. Kamijo, K.; Nishihira, Y.; Hatta, A.; Kaneda, T.; Kida, T.; Higashiura, T.; Kuroiwa, K. Changes in arousal level by differential exercise intensity. *Clin. Neurophysiol.* **2004**, *115*, 2693–2698. [CrossRef] [PubMed]
51. Dietrich, A.; Audiffren, M. The reticular-activating hypofrontality (RAH) model of acute exercise. *Neurosci. Biobehav. Rev.* **2011**, *35*, 1305–1325. [CrossRef]

Systematic Review

Depressive Symptoms and Burnout in Football Players: A Systematic Review

Hugo Sarmento [1,2,*], Roberta Frontini [2], Adilson Marques [3], Miguel Peralta [3], Nestor Ordoñez-Saavedra [4], João Pedro Duarte [1,2], António Figueiredo [1,2], Maria João Campos [1,2] and Filipe Manuel Clemente [5]

1 University of Coimbra, Research Unit for Sport and Physical Activity, Faculty of Sport Sciences and Physical Education, 3040-256 Coimbra, Portugal; joaopedromarquesduarte@gmail.com (J.P.D.); afigueiredo@fcdef.uc.pt (A.F.); mjcampos@fcdef.uc.pt (M.J.C.)
2 Center for Innovative Care and Health Technology, Polytechnic of Leiria, 2411-901 Leiria, Portugal; roberta_frontini@hotmail.com
3 CIPER, Faculdade de Motricidade Humana, Universidade de Lisboa, 1499-002 Lisbon, Portugal; adncmpt@gmail.com (A.M.); miguel.peralta14@gmail.com (M.P.)
4 Faculty of Health Sciences, Sports Science Program, University of Applied and Environmental Sciences, Bogota 111166, Colombia; sportnestor@gmail.com
5 Escola Superior Desporto e Lazer, Instituto Politécnico de Viana do Castelo, 4960-320 Viana do Castelo, Portugal; filipe.clemente5@gmail.com
* Correspondence: hugo.sarmento@uc.pt

Citation: Sarmento, H.; Frontini, R.; Marques, A.; Peralta, M.; Ordoñez-Saavedra, N.; Duarte, J.P.; Figueiredo, A.; Campos, M.J.; Clemente, F.M. Depressive Symptoms and Burnout in Football Players: A Systematic Review. *Brain Sci.* **2021**, *11*, 1351. https://doi.org/10.3390/brainsci11101351

Academic Editor: Mario Luciano

Received: 4 August 2021
Accepted: 12 October 2021
Published: 14 October 2021

Publisher's Note: MDPI stays neutral with regard to jurisdictional claims in published maps and institutional affiliations.

Abstract: The purpose of this article was to systematically review and organise the available literature devoted to the topic of depressive symptoms and burnout in football players. A systematic search was conducted in Web of Science, Scopus, SPORTdiscus, PubMed, and Psychinfo for articles published up to June 2020. The searches yielded 1589 articles, and after the screening process, a total of 18 studies met the eligibility criteria and were included for review. Playing position and conflicts with coach/management seems to have a direct influence on the prevalence of depressive symptoms in current players as do the injuries and life events of former players. During the pre-competition phase, most of the athletes displayed reduced rates, indicating burnout. An exploration of the mental health of football players will help to create models of care and guide professionals so that they may help players achieve better performance while also having better wellbeing. Understanding how to prevent and cope with the emotional wellbeing of football players will be possible to guide players and coaches.

Keywords: soccer; sports; depressive symptoms; mental health

1. Introduction

Football is one of the world's most popular sports. Millions of amateur and professional players are involved. For a player to become an elite performer, they need to have exceptional skills and abilities involving the investment of large amounts of time, effort, and dedication [1].

Football is linked to a multitude of different emotions, some of them experienced very markedly both by fans and players. Hence, there has recently been an interest in understanding the impact of the game on the physical and mental health of fans and players [2].

This is particularly important considering that football puts intense mental pressure on the players, which may increase their susceptibility to certain mental health problems [3]. Losing a game may result in distress, disappointment, sadness [2] and, ultimately, depressive symptoms or even burnout.

Depression is a disturbance of mood characterized by constant feelings of sadness, hopelessness, despair, and a loss of interest in formerly appreciated activities [4]. It may be accompanied by a lack of any desire to move and, thus, with reduced physical activity [5]. A symptom can be defined as the "patient's perception of an abnormal physical, emotional, or cognitive state" [6] and the depressive symptoms can vary from mild to severe (e.g., feeling sad, changes in appetite, trouble sleeping, loss of energy, feeling worthless or guilty, etc.). According to the Diagnostic and Statistical Manual (DSM) of Mental Disorders (DSM-5), there are several depressive disorders which are characterized by the occurrence of a sad, empty, or irritable mood, and both somatic and cognitive fluctuations that, for the purpose of diagnosis, must significantly affect the individual's capacity to function [7]. Worldwide, nearly 98.7 million people suffer from depression [4]. Past literature has found that elite sportspeople, such as football players, may present higher levels of depression compared to the general population [8,9]. However, there is little empirical data regarding the mechanisms of depression in football athletes. [10].

The cause of depression and the appearance of depressive symptoms are not fully understood. Apart from biological and genetic predisposition [9], several possible psychological reasons could trigger a previous susceptibility. For footballers, these reasons could include the intense mental demands and the enormous pressure of this particular sport [11], the higher standards of performance, the responsibility of being part of a team, or the fact that, usually, players spend much time away from family and friends [9]. Staying away from home may increase feelings of loneliness and lack of social support which, ultimately, can be related to depressive symptoms [12]. Injuries may also be frequent in elite athletes, and can also play an important role in depression [9]. One must also consider the impact of negative media content [13]. Thus, understanding depression in athletes is crucial not only because of personal suffering, which sometimes leads to desperate acts such as suicide [9], but also because depression is linked to more non-adherence and more dropouts in sport and physical exercise [5].

Burnout is a state of mental, emotional, and physical exhaustion due to a constant commitment to ambitious goals [14]. It has been studied extensively in sports, specifically considering that intensive training has been related to higher levels of burnout [15]. However, burnout has been linked not only to injuries but also to low perceptions of ability in athletes [16]. Thus, higher levels of training, constant injuries and not feeling up to the challenge [15], as well as an inability to get the physical and mental recovery needed may lead to burnout [17].

Whereas burnout and depression share several common features, such as loss of interest, loss of energy or fatigue [7], there has been some discussion whether they are different constructs [18]. Research has found a positive correlation between burnout and depression [19]. This is one of the reasons why most prevention programs with elite athletes focus both on depression and burnout [10]. Although they are usually studied together, recent research recognizes they are distinct concepts [18]. However, the relationship between the two is usually acknowledged and depression in elite athletes seems to be related to several sport-specific mechanisms such as significant stress [20]. In recent years, several studies attempted to better understand the possibly of a biological signature for burnout considering the global research on burnout-depression overlap [21]. Several other studies searched for the psychological mechanisms underlying burnout and depression, suggesting the importance of personality and the importance of the environment where the person is situated [22].

It is also worth mentioning that both athlete burnout and depression are often conceptualized in a stress-based model [10]. In fact, burnout has often been considered as part of depression [23]. It is also important to note that there are no diagnostic criteria to identify and diagnose burnout [24]. Considering the clear association between these two constructs [18], the fact that many football athletes suffer from both burnout and depression and the fact that mechanisms underlying the two variables are not yet understood [10], these two variables should be studied together whenever possible, which may help to clarify the association between the two. Although there have been important systematic reviews regarding the physical health of football players, little research has been devoted to their mental health [25]. Therefore, this review aimed to systematically review and organize the available literature dedicated to the topic of depression and burnout among football players.

2. Materials and Methods

This systematic review was conducted according to PRISMA (Preferred Reporting Items for Systematic Reviews and Meta-analyses) guidelines [26]. The study is registered in the International Prospective Register of Systematic Reviews (INPLASY ID 202080074) and DOI number is 10.37766/inplasy2020.8.0074.

2.1. Search Strategy: Databases and Eligibility Criteria

A systematic review strategy was conducted according to PRISMA guidelines [26]. Electronic databases (Web of Science, Scopus, SPORTdiscus, PubMed, and Psychinfo) were searched (5 October 2021) for relevant publications before and up to 30 June 2020. A research strategy was performed through Mesh terms obtained at the MeSH Database, leading to the following search strings used (depress* OR "depressive disorder*" OR "depressive symptom*" OR burnout OR "mental health" OR "emotional" OR "emotional depression*") AND (Soccer OR football). The publications included met the following criteria: (1) to be performed with adult male/female amateur/professional players; (2) contained relevant data concerning depression and/or burnout; and (3) produced relevant data concerning prevalence, treatment, diagnosis, of depression/burnout. Studies were excluded if: (1) they were written in a language other than English; (2) they were editorials, review articles, conference abstracts, books, or book chapters; and (3) they were not subject to peer review. Two reviewers (RF and FC) then independently screened the titles and abstracts of all retrieved studies and determined the eligibility of the potentially relevant full-text articles. If the decision of eligibility was not unanimous, a third reviewer was consulted (HS) to evaluate the identified articles and to reaching a final consensus on inclusion.

2.2. Extraction of Data

A data extraction sheet, adapted from the Cochrane Consumers and Communication Review Group's data extraction template [27], was used to assess inclusion requirements, and was subsequently tested on ten randomly selected studies (i.e., pilot testing). Similar to what was reported above, this process was conducted by two independent reviewers (RF, FC). Any disagreement regarding study eligibility was resolved by a third reviewer (HS). The data extracted from the eligible studies was grouped into three categories: (1) general study descriptors (e.g., authors, year of publication and study design); (2) description of the study population (e.g., sample size, age, gender, country, and competitive level), and (3) data concerning the qualitative synthesis (e.g., outcomes, instruments used to evaluate the symptoms, and main results).

2.3. Methodological Quality

An appraisal tool to assess the quality of cross-sectional studies (AXIS) was used to classify the methodological quality of the articles [28]. Additionally, the critical appraisal skills programme checklists were used according to the study design, namely the checklist for: (1) Randomised Controlled Trials; (2) Cohort Studies; and (3) Qualitative studies (http://www.casp-uk.net/ accessed on 15 September 2021).

All the articles related to burnout were cross-sectional. Of the 11 articles related to depression, seven were cross-sectional. The scale includes 20 items, in which one is related to the introduction, 10 are related to methods, five are related to results, two are related to discussion, and two consider other factors. Two of the authors (FMC and HS) independently screened and rated the included full articles. The agreement of both authors was tested using the k agreement rate. The Cohen's kappa coefficient (k) was executed and revealed a k agreement of k = 0.94.

3. Results

3.1. Study Identification and Selection

The searching of databases identified an initial 2730 titles. These studies were then exported to reference manager software (EndNoteTM X9, Clarivate Analytics, Philadelphia, PA, USA). Duplicates (1388 references) were subsequently removed either automatically or manually. The remaining 1342 articles were screened for their relevance based on titles and abstracts, resulting in the removal of a further 1157 studies. The full texts of the remaining 185 articles were examined diligently. After reading full texts, a further 167 studies were excluded owing to several reasons including a lack of relevance to the research topic (n = 121), the fact that they were conference abstracts (n = 27), they presented data from other sports (n = 14), and they were written in languages other than English (n = 5). Following this trimming, 18 articles were accepted for the systematic review (Figure 1).

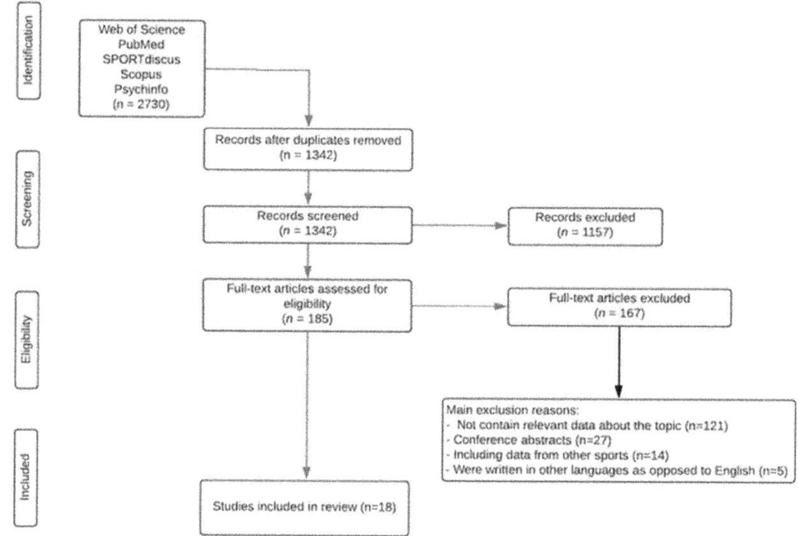

Figure 1. PRISMA flow diagram highlighting the selection process for the studies included in the current systematic review.

3.2. Quality Assessment

The results of the methodological assessment can be found in Table 1. The overall methodological quality of the studies included in this review, demonstrates that a set of components exists that should be improved in future studies. Approximately 78% of the studies don't justify the sample size. In 50% of the reviewed papers, measures were not taken to categorise non-responders. Additionally, most of the papers fail in providing information about non-responders.

3.3. General Description of the Studies

The characterization of the included studies can be found in Table 2. Among the included studies related to depression and depressive symptoms (n = 11), four of them included former players, three included youth players, five included elite/professional players, and two of them included amateur players. Seven of the studies exclusively analysed men, two included both men and women, and two analysed women only. Considering the studies on depression and depressive symptoms, the majority assumed a cross-sectional design (n = 7). Considering the studies on burnout (n = 7), five of them were focused on elite or professionals and two on youth players. All of the studies (*n* = 7) on burnout were conducted on men and had a cross-sectional study design.

Table 1. Quality assessment of individual studies.

Study	1	2	3	4	5	6	7	8	9	10	11	12	13	14	15	16	17	18	19	20
Depression/Depressive symptoms																				
Norouzi et al. [29]	Y	Y	Y	N	N	N	Y	Y	Y	Y	Y	Y	Y	–	–	–	–	–	–	–
Junge and Prinz [30]	Y	Y	N	Y	Y	Y	N	Y	Y	Y	Y	Y	N	N	Y	Y	Y	N	N	Y
Smith et al. [31]	Y	Y	Y	Y	N	N	Y	Y	Y	Y	Y	Y	Y	Y	–	–	–	–	–	–
Olmedilla et al. [32]	Y	Y	N	Y	Y	Y	N	Y	Y	Y	Y	Y	N	N	Y	Y	Y	Y	N	N
Jensen et al. [33]	Y	Y	N	Y	Y	Y	N	Y	Y	Y	Y	Y	N	N	Y	Y	Y	Y	N	N
Wood et al. [34]	Y	Y	Y	Y	Y	Y	Y	Y	Y	Y	–	–	–	–	–	–	–	–	–	–
Van Ramele et al. [35]	Y	Y	Y	Y	N	N	Y	Y	Y	Y	Y	Y	Y	Y	–	–	–	–	–	–
Sanders and Stevinson [36]	Y	Y	N	Y	Y	Y	N	Y	Y	Y	Y	Y	N	N	Y	Y	Y	Y	N	Y
Prinz et al. [37]	Y	Y	N	Y	Y	Y	N	Y	Y	Y	Y	Y	N	N	Y	Y	Y	Y	N	Y
Junge and Feddermann-Demont [38]	Y	Y	Y	Y	Y	Y	Y	Y	Y	Y	Y	Y	N	Y	Y	Y	Y	N	N	Y
Gouttebarge et al. [39]	Y	Y	Y	Y	Y	Y	Y	Y	Y	Y	Y	Y	N	N	Y	Y	Y	Y	N	Y
Burnout																				
Yildiz [40]	Y	Y	N	Y	Y	Y	Y	Y	Y	Y	Y	Y	N	N	Y	Y	Y	Y	N	N
Verardi et al. [41]	Y	Y	N	Y	Y	Y	N	N	Y	Y	Y	Y	N	N	Y	Y	Y	N	N	Y
Hill [42]	Y	Y	N	Y	Y	Y	Y	N	Y	Y	Y	Y	Y	Y	Y	Y	Y	Y	N	Y
Curran et al. [43]	Y	Y	N	Y	Y	Y	Y	Y	Y	Y	Y	Y	N	N	Y	Y	Y	Y	N	Y
Tabei et al. [44]	Y	Y	Y	Y	Y	Y	Y	Y	Y	Y	Y	Y	N	N	Y	Y	Y	Y	N	Y
Yildiz [45]	Y	Y	N	Y	Y	Y	Y	Y	Y	Y	Y	Y	Y	N	Y	Y	Y	N	N	N
Curran et al. [46]	Y	Y	N	Y	Y	Y	Y	Y	Y	Y	Y	Y	Y	N	Y	Y	Y	Y	N	Y

Abbreviation: Y, yes; N, no.

Table 2. Characterization of the included studies.

Study	TS	N	CL	Sex	Age	Years of Experience	Country
Depression/Depressive symptoms							
Norouzi et al., 2020 [29]	RCT	40	Retired	Men	34.1 ± 1.7	N.R.	Iran
Junge and Prinz, 2019 [30]	CS	290	Elite and amateur	Women	<20 to >26	N.R.	Germany
Smith et al., 2018 [31]	Cohort	108	Youth	Men	16.2 ± 1.8	3.65	United Kingdom
Olmedilla et al., 2018 [32]	CS	187	Amateur levels	Men and women	22.1 ± 4.7	N.R.	Spain
Jensen et al., 2018 [33]	CS	323	Junior and professional	Men	22.1 ± 5.2	N.R.	Denmark and Sweden
Wood et al., 2017 [34]	DQ	7	Professional	Men	NR	N.R.	England
Van Ramele et al., 2017 [35]	Cohort	194	Retired	Men	35	12 *	International
Sanders and Stevinson, 2017 [36]	CS	307	Retired	Men	46.8 ± 15.7	6.7 *	United Kingdom
Prinz et al., 2016 [37]	CS	157	Elite	Women	33.0 ± 6.25	N.R.	Germany
Junge and Feddermann-Demont, 2016 [38]	CS	471	Elite and youth	Men and women	League men: 24.8 ± 2.3; U-21 men: 18.4 ± 1.2; League women: 21.0 ± 3.8	N.R.	Switzerland
Gouttebarge et al., 2015 [39]	CS	253	Current and retired	Men	Current: 27 ± 5; Retired: 36 ± 5	9 and 12 *	International
Burnout							
Yildiz, 2015 [40]	CS	102	Professional	Men	25.55	6.72 *	Turkey
Verardi et al., 2015 [41]	CS	134	Professional and amateur	Men	22.8 ± 4.0 and 17.1 ± 0.8	NR	Brazil
Hill, 2013 [42]	CS	171	Elite	Men	16.17 ± 1.57	4.35	England
Curran, et al., 2013 [43]	CS	173	Elite	Men	15.46 ± 1.47	9.45	England
Tabei, et al., 2012 [44]	CC	98	Youth	Men	20.25 ± 1.20	13.22	England and Japan
Yildiz, 2011 [45]	CS	150	Elite	Men	25.7 ± 4.40	7.07	Turkey
Curran, et al., 2011 [46]	CS	149	Youth	Men	16.2 ± 2.00	9.1	England

Abbreviation: TS, Type of study; N, number of participants; CL, Competitive level; RCT, randomized clinical trial; CS, cross-sectional; DQ, descriptive qualitative; NR, non-reported; *: professional years. U-: under.

3.3.1. Depression and Depressive Symptoms

The qualitative synthesis of the studies related to depression and depressive symptoms can be found in Table 3. A description of the prevalence of depressive symptoms was the purpose most commonly observed among the studies ($n = 6$). An establishment of the relationship between depression and depressive symptoms and factors that contribute to the occurrence of symptoms was also recurrent among the studies ($n = 3$). One study tested the effect of a controlled intervention to reduce depressive symptoms. Considering the instruments to assess the main outcome, the Center of Epidemiologic Studies Depression Scale was the most used ($n = 5$), followed by the 12-item General Health Questionnaire ($n = 2$).

Table 3. A qualitative synthesis of the studies related to depression and depressive symptoms.

Study	Aim	Outcomes	Instrument	Main Results
Norouzi et al., 2020 [29]	Test the efficacy of a mindfulness-based stress reduction program on depression symptoms	Depressive symptoms	Montgomery-Åsberg Depression Rating Scale	Depressive symptoms significantly decreased after an intervention (22.9 to 9.4 points) and the values in follow-up also remain low (12.4 points). The intervention group had significant benefits compared to the active control group.
Junge and Prinz, 2019 [30]	Describe the prevalence and risk factors of depression	Depressive symptoms	Center of Epidemiologic Studies Depression Scale	The prevalence of depressive symptoms was 31% among the participants. Additionally, 14% of the players revealed severe symptoms of depression. Despite 16% declared the need for clinical support, only 1/3 have reported that are under treatment or counselling.
Smith et al., 2018 [31]	Test relationships between depressive symptoms, burnout, and perfectionism	Depressive symptoms	Center of Epidemiologic Studies Depression Scale	Socially prescribed perfectionism did not predict depressive symptoms. However, depressive symptoms did predict an increase in socially prescribed perfectionism over time.
Olmedilla et al., 2018 [32]	Analyse the post-injury impact on depression	Depressive symptoms	Depression, Anxiety and Stress Scale—21 Items	No significant differences were found between sexes. Depression symptoms were not significantly different between injured and non-injured players.
Jensen et al., 2018 [33]	Analyse the relationship between perfectionism and depressive symptoms	Depressive symptoms	Center of Epidemiologic Studies Depression Scale	A prevalence of 16.7% was found among the participants. Depression was not correlated with age. However, significantly greater values of depression were found in youth than in professionals.
Wood et al., 2017 [34]	Describe the lived experiences of mental health difficulties	*	*	Survival terms emerged from the interviews. Injury and transition were related to mental health difficulties.
Van Ramele et al., 2017 [35]	Analyse the incidence of anxiety/depressive symptoms	Depressive symptoms	12-item General Health Questionnaire	Common mental disorders ranged between 11 and 29% during 12-month. Players with life events showed a higher risk of experiencing mental disorders.
Sanders and Stevinson, 2017 [36]	Analyse the relationships between career-ending injury, chronic pain, athletic identity and depressive symptoms	Depressive symptoms	Short Depression-Happiness Scale	Retired players with depressive symptoms were more likely to cite injury as retirement reasons. The injury was the greater determinant to explain the depressive symptoms in retired players.

Table 3. *Cont.*

Study	Aim	Outcomes	Instrument	Main Results
Prinz et al., 2016 [37]	Analyse depressive symptoms during and after career	Depressive symptoms	Modified Centre of Epidemiologic Studies Depression Scale	Almost 1/3 of the participants had symptoms of major depression at least once during their career. Average depression scores were different between playing positions and levels of play. Conflicts with coach/management were frequently stated as a reason for lows in mood.
Junge and Feddermann-Demont, 2016 [38]	Analyse the prevalence of depression	Depressive symptoms	Centre of Epidemiologic Studies Depression Scale	Players had a similar prevalence of depressive symptoms to the general population, despite under-21 reported higher prevalence. Symptoms of severe depression were identified in an average of a player per team. Age, sex, playing position, level of play, and a current injury resulted in significant differences in depressive symptoms.
Gouttebarge et al., 2015 [39]	Analyse the prevalence of anxiety/depression	Anxiety/depressive symptoms	12-item General Health Questionnaire	The prevalence of mental health problems achieved 26 and 39% for current and former players, respectively. The low social support and recent live events were cited as main reasons to justify the mental health problems.

* qualitative methodology (interview).

The prevalence of depressive symptoms found in male and female players as well as in current and former players assumes an important relevance for public health [26–28]. Playing position [28,29], and conflicts with coach/management [27] seems to have a direct influence on the prevalence of depressive symptoms in current players, along with injury episodes [30] or live events [28,31] in former players. Mindfulness-based stress reduction programs seem to have a positive effect on retired players [32].

Since terminology is important in the studies developed in this context, Table 4 summarizes the terms adopted by the authors in different studies as well the measures scales used (Table 4).

Table 4. Definitions used in the depression variable and evaluation instrument tools.

Study	Depression Variable	Measurement
Norouzi et al. [29]	- The authors define the variable as "depressive symptoms". However, during the paper, they also use the term depression. - The sum scores can be interpreted as follows (Ahmadpanah et al., 2016): 0–6 points: no depression; 7–19 points: mild depression; 20–34 points: moderate depression; >34 points: severe depression.	- Montgomery-Åsberg Depression Rating Scale (MADRS)
Junge and Prinz [30]	- The authors define the variable as "depressive symptoms". However, during the paper, they also use the term depression. - The questionnaire also included questions on frequency of intake of medication for depression.	- The Center for Epidemiologic Studies Depression Scale (CES-D)
Smith et al. [31]	- The authors define the variable as "depressive symptoms". However, during the paper, they also use the term depression.	- The Center for Epidemiologic Studies Depression Scale (CES-D)
Olmedilla et al. [32]	- Depression	- DASS-21

Table 4. *Cont.*

Study	Depression Variable	Measurement
Jensen et al. [33]	- The authors define the variable as "depressive symptoms". However, during the paper, they also use the term depression. - A cutoff score of 16 to define a clinical, significant level of depression was used.	- The Center for Epidemiologic Studies Depression Scale (CES-D)
Wood et al. [34]	- Depression	- Qualitative research design using interpretative phenomenological - analysis (IPA)
Van Ramele et al. [35]	- Depression	- General Health Questionnaire (GHQ-12)
Sanders and Stevinson [36]	- The authors define the variable as "depressive symptoms". - Total possible scores range from 0 to 18, with lower scores indicating greater depression.	- Short Depression-Happiness Scale (SDHS; Joseph, Linley, Harwood, Lewis, & McCollam, 2004).
Prinz et al. [37]	- Severity of depression symptoms - Depression	- The Center for Epidemiologic Studies Depression Scale (CES-D) - PHQ-2
Junge and Feddermann-Demont [38]	- The authors define the variable as "depressive symptoms".	- The Center for Epidemiologic Studies Depression Scale (CES-D)
Gouttebarge et al. [39]	- Depression	- General Health Questionnaire

3.3.2. Burnout

The analysis of burnout in football players has been on the agenda of sport scientists in the last decade (Table 5). All of the studies reviewed are developed in the context of male football, integrating both senior and youth levels as well as the amateur versus professional context. The Athlete Burnout Questionnaire has been used extensively in this context. Only one study used a different scale to analyse burnout symptoms [45]. In general, the reviewed studies always sought to establish relationships between burnout and other moderating variables (e.g., perfectionism, bullying, psychological need satisfaction, organizational stressors, self-determined motivation, and leader-member exchange).

Table 5. A qualitative synthesis of the studies related to burnout.

Study	Aim	Outcomes	Instrument	Main Results
Yildiz, 2015 [40]	Analyse the effect of burnout and bullying in professional players.	Burnout symptoms	Athlete Burnout Questionnaire (ABQ)	Bullying had the strongest and statistically significant direct influence on three dimensions of burnout. (1) reduced sense of accomplishment, (2) emotional/physical exhaustion, and (3) devaluation.
Verardi et al., 2015 [41]	Identify and interpret the occurrence of symptoms associated with burnout syndrome during the pre-competition.	Burnout symptoms	Athlete Burnout Questionnaire (ABQ)	The incidence, and consequently the vulnerability, to burnout, were identified in a portion of the athletes during the pre-competition phase.
Hill, 2013 [42]	Examine the interactive effects of dimensions of perfectionism in predicting symptoms of athlete burnout.	Burnout symptoms	Athlete Burnout Questionnaire (ABQ)	Pure personal perfectionism provided some, albeit limited, protection from burnout in comparison with non-perfectionism. Also, pure evaluative concerns perfectionism, as opposed to mixed perfectionism, emerged as the most debilitating in terms of burnout symptoms.
Curran et al., 2013 [43]	Examine the mediating role of psychological need satisfaction in relationships between types of passion for sport and athlete burnout.	Burnout symptoms and basic psychological need satisfaction	Athlete Burnout Questionnaire (ABQ) and different scales to measure basic psychological need satisfaction	An inverse relationship between harmonious passion and burnout can be explained by higher levels of psychological need satisfaction. However, this was not the case for obsessive passion, which was not associated with psychological need satisfaction or most symptoms of athlete burnout.
Tabei et al., 2012 [44]	Investigate the relationship between organizational stressors in sport and athlete burnout.	Burnout symptoms and organizational stressors	Athlete Burnout Questionnaire (ABQ) and interview	Results revealed multiple demands linked to the dimensions of athlete burnout and identified specific organizational-related issues that players associated with the incidence of burnout.

Table 5. *Cont.*

Study	Aim	Outcomes	Instrument	Main results
Yildiz, 2011 [45]	determine whether leader-member exchange quality affects burnout in professional footballers.	Burnout symptoms	An abbreviated version of the 10-item burnout scale	The results demonstrated that the quality of leader-member exchange significantly and inversely influenced burnout of professional footballers
Curran et al., 2011 [46]	To examine the relationship between forms of passion and whether these relationships are mediated by self-determined motivation.	Burnout symptoms, self-determined motivation, and passion	Athlete Burnout Questionnaire (ABQ), Sport Motivation Scale, Passion Scale	The results suggest that harmonious passion may offer some protection from burnout for athletes due to higher levels of self-determined motivation.

4. Discussion

4.1. Depression and Depressive Symptoms

The prevalence of depressive symptoms was analysed in current and former players (both men and women). Depressive symptoms varied from 16.7% [27] to 39% [33]. Severe symptoms were found in between 14% [30] and 33% [37] of players with depressive symptoms. Understanding the prevalence of depressive symptoms is important for public health. The prevalence found in this review suggests that depressive symptoms are more prevalent in football players when compared to the general population [9]. However, studying the prevalence of depression may be very difficult since the definition of depression ranges from episodes of unhappiness to persistent mood changes [47]. In the case of current players, reviewed studies suggested the influence of playing position as a possible cause for variation in depressive symptoms [37,38]. Thus, and considering that some research highlighted that some playing positions (such as goalkeepers) used to have higher levels of depressive symptoms, future studies should try to better explore the role of playing position. Tables 2 and 3 present more detailed information regarding the samples assessed as well as the instruments used to assess depressive symptoms. It is important to state that the variability in prevalence estimates may be due to different assessment methods, times or even samples (e.g., difference prevalence in gender etc.) [10]. Systematic reviews such as the one we present may bring some light and help to better understand what has been made in this field.

Some of the studies have tried to justify the causes of such prevalence. In the case of former players, the main reason to justify depressive symptoms was having to retire because of injury [30] or live events [35,39]. In the case of current players, some reports suggested the influence of conflicts with the coach/management [37], or live events [39]. Interestingly, in one study, no differences were found in depressive symptoms between injured and non-injured players [32]. Nonetheless, injury has been highlighted in past literature as being linked with depression [9]. Thus, we acknowledge the need of future studies to continue exploring the link between injuries and depressive symptoms. Moreover, when a player gets injured, it must be a priority to connect with that player and try to understand if he/she needs some extra support to deal with any subsequent mental issues related to the injury. Regarding former players, it is also important that, at the end of their career, they receive support in terms of a retirement plan.

A randomized clinical trial [29] tested the effects of a mindfulness-based stress reduction programs on retired players. The intervention revealed a significant benefit compared to the control group. Additionally, the effects continued through to the follow-up. Cognitive-behavioural therapies are still the most used in sports psychology interventions and football interventions [48]. Mindfulness strategies, focusing on the present in a nonjudgmental way [49], may be helpful. However, future studies should try to understand if other interventions and strategies might be effective in order to find the best methods to prevent and alleviate depressive symptoms in football players. Although many studies have already discussed some of the most effective interventions in terms of fighting depression and depressive symptoms, football players are a specific population, and their uniqueness should be considered.

Finally, a result worth noting relates to sex differences. In one study, no significant differences were found between sexes [23]. However, Junge and Feddermann-Demont [29] reported significant differences between sexes, with females presenting higher prevalence of depression. This result is also present in other populations [50] and it is not exclusive to football players.

4.2. Burnout

The reviewed studies demonstrated the existence of a great dispersion of objectives. In this sense, the study of Verardi et al. [41] demonstrated that during the pre-competition phase, most of the athletes displayed reduced rates, indicating burnout. Additionally, it is important to note that professional players achieved maximum average scores related to burnout in the three dimensions, which should be taken into consideration in future studies.

The other studies included in this review always sought to establish relationships between burnout and other moderating variables, namely perfectionism [42], bullying [40], psychological need satisfaction [43], organizational stressors [44], self-determined motivation [46], and leader-member exchange [45]. Football players have very busy lives, with different competitions sometimes in different parts of the world. Players should take some time to switch off and both physically and mentally detach from the game to prevent burnout symptoms [9]. Considering the association that burnout may have with so many different variables, it is important to better understand the mechanisms underlying them. Coaches, with their privileged position, should be alert to earlier signs of burnout, such as physical difficulties, sadness, or even fatigue. In some cases it may be possible to schedule individualized training sessions. Working closely with a psychologist may also be important. Coaches might identify players needing attention to a sports psychologist. The sports psychologist could then work with the player in order to help him/her overcome any difficulties.

Specific techniques may apply, depending on the particularities of burnout. Nonetheless, typical techniques that may help reduce anxiety may also be used with these players, such as: imagery techniques, relaxation, and problem-solving techniques. Considering the association found between anxiety and burnout [51], it is possible that techniques to lower anxiety in football players may be effective in helping players deal with burnout as well. One important specific finding of one of the papers was the mediating role of psychological need satisfaction in the relationship between harmonious passion and dimensions of athlete burnout. This suggests that harmonious passion may protect athletes from the development of athlete burnout through psychological need satisfaction [43]. This finding has many important practical implications that are discussed in the paper, such as the importance of promoting sporting atmospheres emphasizing harmonious tendencies. Creating a supporting environment may also be a protective factor against bullying, which has also been found to be associated with player burnout [40].

It is important to note that from the studies selected for the review that met the inclusion criteria, only one study [31] referred to both depression and burnout. This is also an important result. Considering the importance that researchers have attributed to both variables [23] and the fact that prevention programs in elite athletes usually highlight the fact that it is crucial to consider both variables [10], the results of the present study underline that there is still work to be done in the area. In fact, the literature has already pointed to the importance of studying both variables, especially considering that that mechanisms underlying the two variables are not yet understood [10]. The results of the present study reinforce that studies underlying the connection between the two variables are still scarce.

4.3. Recommendations and Future Directions

The studies found in this review revealed some interesting results in terms of practical implications and possible future directions in research. However, there is still a long way to go.

First of all, the results highlight the importance of talking about depression and depressive symptoms. Depression is a serious disease, and the results of this review suggest that the prevalence of depressive symptoms in football players is very high. It is imperative to talk about depression, the symptoms, and the consequences, as well as the early signs of burnout. Improving open communication between players and football staff may be important. It is essential not to hide it, especially considering that depression and depressive symptoms can often be linked to irreversible issues such as suicide [52]. However, there is still a lot of stigma surrounding depression. Many people are afraid to seek help, concluding that they will be judged as weak. Mental health is usually relegated to another dimension when compared to physical health. Thus, an athlete seeking help for a physical condition is usually not stigmatized the same way they feel they might be for seeking help for mental health problems. Educational programs at the beginning of a career, and possibly at the end of a career, might help players to understand the early signs of the disease, the places where they may seek help, and the best times to do so.

Moreover, clinical sport psychological approaches are still scarce [11]. Considering the prevalence of depression and burnout in football players, it is of the utmost importance to incorporate psychologists in multidisciplinary teams working with these professionals. Only a sports psychologist has the necessary tools to help overcome depressive symptoms and burnout symptoms. Thus, multidisciplinary teams, with a psychologist, would provide a balance between physical and emotional components, using important tools and techniques that would help to prevent burnout and depression in players [17]. It is important to note that psychologists may work with the team, but also in particular cases with particular players. A counselling-based session may, for instance, be needed in particular cases and for specific players. In fact, due to the prevalence revealed in the studies, psychotherapeutic support during one's football career may be needed. Although coaches are not experts at diagnosing symptoms, considering their privileged position and proximity to players, they should be alert for possible symptoms and early signs. A routine screening of mental health problems (and specifically a quick screening of depression, depressive symptoms and burnout), performed by a psychologist, may also be important. Future studies should also try to understand if players are usually referred to psychologists/psychiatrists to decrease possible suffering and, if so, the strategies that were used (and to what level of effectiveness).

The study's result demonstrated the existence of a great dispersion of objectives on studies. Thus, more studies are needed. It is imperative to develop a comprehensive understanding of the mental health of football players and to create models of care and management which will have an impact on the performance of players and their wellbeing [25]. By better understanding how to manage the emotional wellbeing of football players, it will be possible to guide all sports staff, players and coaches alike [53]. Coping skills should also be researched extensively in terms of working toward a better understanding of optimal educational intervention.

Another important recommendation relates to social support. Considering the nature of football, and the constant travel and possible disillusions related to the game, social support may be an important preventive strategy. Social support may also be an important protective factor [13], helping players who are dealing with critical problems such as injury and loneliness [9,12], and minimizing important negative behaviours such as bullying.

Finally, future investigations should take into account the level of professionalism of footballers. The evolutionary tendencies of the game (e.g., compressed schedules and multiple games in short amounts of time) can have a significant influence on depressive symptoms and burnout.

Future studies should consider not only the use of self-reports but also more accurate methodologies. Self-report measures are of utmost importance in screening and measuring progress. However, when we are interested in studying the diagnosis of diseases such as depression it is important to include a professional structured clinical interview by professional psychiatrists/psychologists and to have a confirmed diagnosis.

A possible limitation of this systematic review is that it only includes studies in English from the Web of Science, Scopus, SPORTdiscus, PubMed, and Psychinfo databases. Additionally, the fact that the selected studies are mostly cross-sectional may result in a limitation of the existing research to date, along with the level of evidence (as explained in Section 3.2). Additionally, the lack of homogeneous studies about the topic under study should be considered a limitation that future research should take into account. It is also important to note that the use of self-reported scales do not allow the generalization of some of the results.

5. Conclusions

Although there are many studies in this area, little attention has been paid to the psychological component. When looking for information about depression, and depressive symptoms or burnout and sport, there is a clear primacy with respect to the beneficial effects of sports and exercise on symptoms of depression and anxiety, while there is little on the levels of depression and burnout of players. Recently, however, there has been a greater openness to this topic.

The reviewed studies showed that depressive symptoms in football players are more prevalent when compared to the general population. During the pre-competition phase, most of the athletes displayed reduced rates of indicating burnout. Considering the constant travel and possible disillusions related to the game, social support might be an important preventive strategy and protective factor. This type of intervention would help players dealing with critical problems such as injury and loneliness, while serving to minimize a host of negative behaviours. By better understanding how to prevent and cope with the emotional wellbeing of football players, it will be possible to guide all sports staff.

Author Contributions: Conceptualization, H.S, A.F, A.M. and F.M.C.; methodology, H.S, F.M.C., M.J.C. and M.P.; writing—original draft preparation, R.F., J.P.D., N.O.-S. and M.P.; writing—review and editing, H.S., F.M.C., A.F. and A.M.; supervision, H.S. and A.M. All authors have read and agreed to the published version of the manuscript.

Funding: H.S. gratefully acknowledges the support of a Spanish government subproject Integration ways between qualitative and quantitative data, multiple case development, and synthesis review as the main axis for an innovative future in physical activity and sports research [PGC2018-098742-B-C31] (Ministerio de Economía y Competitividad, Programa Estatal de Generación de Conocimiento y Fortalecimiento Científico y Tecnológico del Sistema I + D + i), that is part of the coordinated project New approach of research in physical activity and sport from mixed methods perspective (NARPAS_MM) [SPGC201800X098742CV0]. H.S., A.F., M.J.C and J.P.D gratefully acknowledges the support of FCT (uid/dtp/04213/2020). Authors also thanks Bruce Jones for reviewing the document.

Conflicts of Interest: The authors declare no conflict of interest.

References

1. Sarmento, H.; Anguera, M.T.; Pereira, A.; Araújo, D. Talent Identification and Development in Male Football: A Systematic Review. *Sports Med.* **2018**, *48*, 907–931. [CrossRef] [PubMed]
2. Heun, R.; Pringle, A. Football does not improve mental health: A systematic review on football and mental health disorders. *Glob. Psychiatry* **2018**, *1*, 25–38. [CrossRef]
3. Hughes, L.; Leavey, G. Setting the bar: Athletes and vulnerability to mental illness. *Br. J. Psychiatry* **2012**, *200*, 95–96. [CrossRef]
4. Richardson, K.; Barkham, M. Recovery from depression: A systematic review of perceptions and associated factors. *J. Ment. Health* **2020**, *29*, 103–115. [CrossRef] [PubMed]

5.   Hanin, J.; Ekkekakis, P. Emotions in sport and exercise settings. In *Routledge Companion to Sport and Exercise Psychology*; Papaioannou, A., Hackfort, D., Eds.; Routledge: New York, NY, USA, 2014.

6.   Bekelman, D.B.; Havranek, E.P.; Becker, D.M.; Kutner, J.S.; Peterson, P.N.; Wittstein, I.S.; Gottlieb, S.H.; Yamashita, T.E.; Fairclough, D.L.; Dy, S.M. Symptoms, Depression, and Quality of Life in Patients With Heart Failure. *J. Card. Fail.* **2007**, *13*, 643–648. [CrossRef] [PubMed]

7.   Association, A.P. *Diagnostic and Statistical Manual of Mental Disorders: DSM-5*, 5th ed.; American Psychiatry Publishing: Arlington, VA, USA, 2013.

8.   Brewer, B.W.; Petrie, T. Psychopathology in sport and exercise. In *Exploring Sport and Exercise Psychology*, 3rd ed.; Van Raalte, J., Brewer, B.W., Eds.; American Psychological Association: Washington, DC, USA, 2014.

9.   Pruna, R.; Badhur, K. Depression in Football. *J. Nov. Physiother.* **2016**, *6*, 6. [CrossRef]

10.   Nixdorf, I.; Bechmann, J.; Nixdorf, R. Preventing depression and burnout in youth football. In *Football Psychology: From Theory to Practice*; Konter, T.M., Beckmann, E., Loughead, J., Eds.; Routledge: London, UK, 2019; pp. 337–351.

11.   Frank, R.; Nixford, I.; Beckam, J. Depression among Elite Athletes: Prevalence and Psychological Factors. *Dtsch. Z. Sportmed.* **2015**, *64*, 320–326.

12.   Donohue, B.; Chow, G.M.; Pitts, M.; Loughran, T.; Schubert, K.N.; Gavrilova, Y.; Allen, D.N. Piloting A Family-Supported Approach to Concurrently Optimize Mental Health and Sport Performance in Athletes. *Clin. Case Stud.* **2015**, *14*, 159–177. [CrossRef]

13.   Kristiansen, E.; Roberts, G.C.; Sisjord, M.K. Coping with negative media content: The experiences of professional football goalkeepers. *Int. J. Sport Exerc. Psychol.* **2011**, *9*, 295–307. [CrossRef]

14.   Freudenberger, H. *Burnout*; Doubleday: New York, NY, USA, 1980.

15.   Goodger, K.; Kenta, G. Professional practice in sport psychology: A review. In *Professional Practice Issues in Athlete Burnout*; Routledge: London, UK, 2012.

16.   Lemyre, P.N.; Hall, H.K.; Roberts, G.C. A social cognitive approach to burnout in elite athletes. *Scand. J. Med. Sci. Sports* **2008**, *18*, 221–234. [CrossRef]

17.   Nixdorf, I.; Nixdorf, R.; Hautzinger, M.; Beckmann, J. Prevalence of Depressive Symptoms and Correlating Variables among German Elite Athletes. *J. Clin. Sport Psychol.* **2013**, *7*, 313–326. [CrossRef]

18.   Koutsimani, P.; Montgomery, A.; Georganta, K. The relationship between burnout, depression, and anxiety: A systematic review and meta-analysis. *Front Psychol.* **2019**, *10*, 284. [CrossRef]

19.   Bianchi, R.; Laurent, E. Emotional information processing in depression and burnout: An eye-tracking study. *Eur. Arch. Psychiatry Clin. Neurosci.* **2015**, *265*, 27–34. [CrossRef]

20.   Nixdorf, I.; Frank, R.; Beckmann, J. An Explorative Study on Major Stressors and Its Connection to Depression and Chronic Stress among German Elite Athletes. *Adv. Phys. Educ.* **2015**, *5*. [CrossRef]

21.   Bianchi, R.; Schonfeld, I.S.; Laurent, E. Biological research on burnout-depression overlap: Long-standing limitations and on-going reflections. *Neurosci. Biobehav. Rev.* **2017**, *83*, 238–239. [CrossRef]

22.   Kaschka, W.P.; Korczak, D.; Broich, K. Burnout: A fashionable diagnosis. *Dtsch. Arztebl. Int.* **2011**, *108*, 781–787. [CrossRef]

23.   Bianchi, R.; Schonfeld, I.S.; Laurent, E. Burnout-depression overlap: A review. *Clin. Psychol. Rev.* **2015**, *36*, 28–41. [CrossRef] [PubMed]

24.   Bakusic, J.; Schaufeli, W.; Claes, S.; Godderis, L. Stress, burnout and depression: A systematic review on DNA methylation mechanisms. *J. Psychosom. Res.* **2017**, *92*, 34–44. [CrossRef]

25.   Rice, S.M.; Purcell, R.; De Silva, S.; Mawren, D.; McGorry, P.D.; Parker, A.G. The Mental Health of Elite Athletes: A Narrative Systematic Review. *Sports Med.* **2016**, *46*, 1333–1353. [CrossRef] [PubMed]

26.   Moher, D.; Liberati, A.; Tetzlaff, J.; Altman, D.G. Preferred reporting items for systematic reviews and meta-analyses: The PRISMA statement. *BMJ (Clin. Res. Ed.)* **2009**, *339*, b2535. [CrossRef] [PubMed]

27.   Group, Cochrane Consumers and Communication. Data Extraction Template for Included Studies. 2016. Available online: https://cccrg.cochrane.org/author-resources (accessed on 11 October 2021).

28.   Downes, M.J.; Brennan, M.L.; Williams, H.C.; Dean, R.S. Development of a critical appraisal tool to assess the quality of cross-sectional studies (AXIS). *BMJ Open* **2016**, *6*, e011458. [CrossRef]

29.   Norouzi, E.; Gerber, M.; Masrour, F.F.; Vaezmosavi, M.; Puhse, U.; Brand, S. Implementation of a mindfulness-based stress reduction (MBSR) program to reduce stress, anxiety, and depression and to improve psychological well-being among retired Iranian football players. *Psychol. Sport Exerc.* **2020**, *47*, 101636. [CrossRef]

30.   Junge, A.; Prinz, B. Depression and anxiety symptoms in 17 teams of female football players including 10 German first league teams. *Br. J. Sports Med.* **2019**, *53*, 471–477. [CrossRef]

31.   Smith, E.; Hill, A.; Hall, H. Perfectionism, Burnout, and Depression in Youth Soccer Players: A Longitudinal Study. *J. Clin. Sport Psychol.* **2018**, *12*, 179–200. [CrossRef]

32.   Olmedilla, A.; Ortega, E.; Robles-Palazon, F.J.; Salom, M.; Garcia-Mas, A. Healthy Practice of Female Soccer and Futsal: Identifying Sources of Stress, Anxiety and Depression. *Sustainability* **2018**, *10*, 2268. [CrossRef]

33.   Jensen, S.N.; Ivarsson, A.; Fallby, J.; Dankers, S.; Elbe, A.M. Depression in Danish and Swedish elite football players and its relation to perfectionism and anxiety. *Psychol. Sport Exerc.* **2018**, *36*, 147–155. [CrossRef]

34. Wood, S.; Harrison, L.K.; Kucharska, J. Male professional footballers' experiences of mental health difficulties and help-seeking. *Physician Sportsmed.* **2017**, *45*, 120–128. [CrossRef] [PubMed]
35. Van Ramele, S.; Aoki, H.; Kerkhoffs, G.; Gouttebarge, V. Mental health in retired professional football players: 12-month incidence, adverse life events and support. *Psychol. Sport Exerc.* **2017**, *28*, 85–90. [CrossRef]
36. Sanders, G.; Stevinson, C. Associations between retirement reasons, chronic pain, athletic identity, and depressive symptoms among former professional footballers. *Eur. J. Sport Sci.* **2017**, *17*, 1311–1318. [CrossRef]
37. Prinz, B.; Dvořák, J.; Junge, A. Symptoms and risk factors of depression during and after the football career of elite female players. *BMJ Open Sport Exerc. Med.* **2016**, *2*, e000124. [CrossRef]
38. Junge, A.; Feddermann-Demont, N. Prevalence of depression and anxiety in top-level male and female football players. *BMJ Open Sport Exerc. Med.* **2016**, *2*, e000087. [CrossRef] [PubMed]
39. Gouttebarge, V.; Frings-Dresen, M.H.; Sluiter, J.K. Mental and psychosocial health among current and former professional footballers. *Occup. Med. Lond.* **2015**, *65*, 190–196. [CrossRef] [PubMed]
40. Yildiz, S.M. The relationship between bullying and burnout an empirical investigation of Turkish professional football players. *Sport Bus. Manag. Int. J.* **2015**, *5*, 6–20. [CrossRef]
41. Verardi, C.E.L.; Nagamine, K.K.; Domingos, N.A.M.; De Marco, A.; Miyazaki, M. Burnout and pre-competition: A study of its occurrence in brazilian soccer players. *Rev. Psicol. Deporte* **2015**, *24*, 259–264.
42. Hill, A.P. Perfectionism and Burnout in Junior Soccer Players: A Test of the 2 x 2 Model of Dispositional Perfectionism. *J. Sport Exerc. Psychol.* **2013**, *35*, 18–29. [CrossRef] [PubMed]
43. Curran, T.; Appleton, P.R.; Hill, A.P.; Hall, H.K. The mediating role of psychological need satisfaction in relationships between types of passion for sport and athlete burnout. *J. Sports Sci.* **2013**, *31*, 597–606. [CrossRef] [PubMed]
44. Tabei, Y.; Fletcher, D.; Goodger, K. The Relationship between Organizational Stressors and Athlete Burnout in Soccer Players. *J. Clin. Sport Psychol.* **2012**, *6*, 146–165. [CrossRef]
45. Yildiz, S.M. Relationship between leader-member exchange and burnout in professional footballers. *J. Sports Sci.* **2011**, *29*, 1493–1502. [CrossRef]
46. Curran, T.; Appleton, P.R.; Hill, A.P.; Hall, H.K. Passion and burnout in elite junior soccer players: The mediating role of self-determined motivation. *Psychol. Sport Exerc.* **2011**, *12*, 655–661. [CrossRef]
47. Biddle, S.J.H.; Mutrie, N. *Psychology of Physical Activity: Determinants, Well-Being and Interventions*; Routledge: New York, NY, USA, 2001.
48. Nesti, M. *Psychology in Football: Working with Elite and Professional Players*; Routledge: Oxon, UK, 2010.
49. Kabat-Zinn, J. Mindfulness-Based Interventions in Context: Past, Present, and Future. *Clin. Psychol. Sci. Pract.* **2003**, *10*, 144–156. [CrossRef]
50. Llorente, J.M.; Olivan-Blazquez, B.; Zuniga-Anton, M.; Masluk, B.; Andres, E.; Garcia-Campayo, J.; Magallon-Botaya, R. Variability of the prevalence of depression in function of sociodemographic and environmental factors: Ecological model. *Front Psychol.* **2018**, *9*, 2182. [CrossRef] [PubMed]
51. Wiggins, M.S.; Lai, C.; Deiters, J.A. Anxiety and burnout in female collegiate ice hockey and soccer athletes. *Percept. Mot. Ski.* **2005**, *101*, 519–524. [CrossRef] [PubMed]
52. Reng, R. *A Life Too Short: The Tragedy of Robert Enke*; Yellow Jersey: Munich, Germany, 2011.
53. Nicholls, A.R.; Polman, R.C.J. Coping in sport: A systematic review. *J. Sports Sci.* **2007**, *25*, 11–31. [CrossRef] [PubMed]

Article

Impact of Long-Rope Jumping on Monoamine and Attention in Young Adults

Masatoshi Yamashita [1,2,*] and Takanobu Yamamoto [2]

[1] Graduate School of Advanced Integrated Studies in Human Survivability, Kyoto University, Kyoto 606-8306, Japan
[2] Department of Psychology, Tezukayama University, Nara 631-8585, Japan; takaoxford@gmail.com
* Correspondence: myamashita.fatiguepsychology@gmail.com

Abstract: Previous research has shown that rope jumping improves physical health; however, little is known about its impact on brain-derived monoamine neurotransmitters associated with cognitive regulation. To address these gaps in the literature, the present study compared outcomes between 15 healthy participants (mean age, 23.1 years) after a long-rope jumping exercise and a control condition. Long-rope jumping also requires co-operation between people, attention, spatial cognition, and rhythm sensation. Psychological questionnaires were administered to both conditions, and Stroop task performance and monoamine metabolite levels in the saliva and urine were evaluated. Participants performing the exercise exhibited lower anxiety levels than those in the control condition. Saliva analyses showed higher 3-methoxy-4-hydroxyphenylglycol (a norepinephrine metabolite) levels, and urine analyses revealed higher 3-methoxy-4-hydroxyphenylglycol and 5-hydroxyindoleacetic acid (a serotonin metabolite) levels in the exercise condition than in the control. Importantly, urinary 5-hydroxyindoleacetic acid level correlated with salivary and urinary 3-methoxy-4-hydroxyphenylglycol levels in the exercise condition. Furthermore, cognitive results revealed higher Stroop performance in the exercise condition than in the control condition; this performance correlated with salivary 3-methoxy-4-hydroxyphenylglycol levels. These results indicate an association between increased 3-methoxy-4-hydroxyphenylglycol and attention in long-rope jumping. We suggest that long-rope jumping predicts central norepinephrinergic activation and related attention maintenance.

Keywords: long-rope jumping; attention; 3-methoxy-4-hydroxyphenylglycol; 5-hydroxyindoleacetic acid

Citation: Yamashita, M.; Yamamoto, T. Impact of Long-Rope Jumping on Monoamine and Attention in Young Adults. *Brain Sci.* **2021**, *11*, 1347. https://doi.org /10.3390/brainsci11101347

Academic Editor: Filipe Manuel Clemente

Received: 7 September 2021
Accepted: 8 October 2021
Published: 13 October 2021

Publisher's Note: MDPI stays neutral with regard to jurisdictional claims in published maps and institutional affiliations.

1. Introduction

Exercise has beneficial effects on psychophysiological well-being and brain function [1,2]. However, a monotonous or complicated exercise training program may not be the most enjoyable or easily performed exercise. In a society suffering from increasing health issues, identifying simple and enjoyable exercises that can effectively mitigate mental illness-related cognitive and brain dysfunction is important.

Being a simple and aerobic rhythmic exercise, the rope jumping program was introduced as a teaching tool for improving physical health in Japanese elementary schools in the early 1900s. Rope jumping co-ordinates movements of the upper and lower body to maintain balance and rhythm. Pitreli and O'Shea reported that rope jumping combines the angular momentum of the rope and vertical displacement of the body and involves upper and lower synchrony where positioning and timing are critical [3]. Given that time perception is involved in the processing of timing information [4], the rope jumping skill may be associated with superior cognitive functions, including spatial-temporal perception. A previous study reported that young adults who engaged in rope jumping had higher levels of selective attention than young adults who did not exercise [5]. Rope jumping improves C-reactive protein levels and bone density in adolescents [6,7]. However, little is known about how rope jumping affects monoaminergic activity related to cognitive performance.

The monoamine neurotransmitters norepinephrine, serotonin, and dopamine have been implicated in various cognitive functions [8,9]. Norepinephrine is involved in the regulation of the dorsal and ventral attention networks and thereby in reorienting and switching attention [10,11]. Additionally, some positron emission tomography studies have reported associations between dopamine or serotonin availability and attention in healthy individuals [12,13]. These findings indicate that monoamines are critically linked to selective attention. However, investigations of monoamines have been limited to neuroimaging, and little is known about the association between direct monoamine levels and cognitive function after rope jumping. Does rope jumping, which enhances attention in young adults, change monoamine levels in the central nervous system (CNS)?

Biochemical studies have shown an association between monoamine levels in the peripheral nervous system (PNS) and the CNS [14,15]. Although 3-methoxy-4-hydroxyphenylglycol (MHPG; a norepinephrine metabolite) levels in the plasma and cerebrospinal fluid (CSF) were related to the central norepinephrinergic system according to some studies [16,17], several other studies failed to demonstrate this association [18,19]. This discrepancy may be associated with the stress of invasive blood and/or CSF extraction. In contrast, salivary MHPG (sMHPG), delivered via blood circulation, is a non-invasive marker for detecting changes in the central norepinephrinergic system, as sMHPG has been reported to correlate with plasma and CSF MHPG levels [20,21]. Additionally, a growing body of evidence indicates that sMHPG levels can predict cognitive performance and mental health associated with central norepinephrinergic activity [22,23]. Moreover, it has been reported that urinary MHPG (uMHPG), which does not require invasive sampling, originates in the CNS [24,25]. Urinary homovanillic acid (uHVA; a dopamine metabolite) and 5-hydroxyindoleacetic acid (u5-HIAA; a serotonin metabolite) are transported from the brain via the organic anion transporter 3 system (OAT 3) in the blood-brain barrier [26,27]. Moreover, the reduction of brain serotonin levels was correlated with lower u5-HIAA levels [14]. Such monoamine metabolites in the saliva and urine can serve as non-invasive and useful markers for detecting central monoamine activity in humans.

The effect of rope jumping exercise on cognitive regulation-relevant monoamine neurotransmitters has received scant attention, and little is known about monoamine metabolite levels in the saliva and urine after performing rope jumping exercise. However, given the effect of rope jumping exercise on attention performance and physiological function, it is important to clarify whether rope jumping exercise also affects the association between monoamines and attention. To determine this association, we used long-rope jumping as an experimental condition, as it is a rhythmic exercise requiring considerable co-operation, attention, spatial cognition, and rhythm sensation when performed with multiple persons.

The present study addresses two research questions. First, does the rhythmic exercise of long-rope jumping enhance monoamine metabolite levels? Second, if improvement is observed with the rhythmic exercise, does it lead to a corresponding improvement in attention performance as measured by the Stroop task? To address these questions, we compared monoamine metabolites levels in saliva and urine and Stroop performance between long-rope jumping and non-exercise conditions. Based on the monoamine metabolite differences between conditions, we further examined the associations between monoamine metabolite levels and Stroop performance.

2. Materials and Methods

2.1. Participants

The Psychological Research Ethics Committee of Tezukayama University approved the protocol (approval number 27−15), and all study participants provided informed consent. Twenty-four university students (10 women and 14 men, aged 20–33 years) participated in this study; they were recruited from Tezukayama University. Individuals with a history of neurological, cardiovascular, or psychiatric illness, smoking, or use of drugs (e.g., tranquilizers or hypnotics) were excluded. Since previous studies have reported associations between menstruation and monoamine metabolism and physical activity-related physiological systems [28–30], five women were also excluded from the analysis because of ongoing menstruation and/or physical deconditioning. Four men were excluded because of physical deconditioning. The final analysis included 15 healthy students (5 women and 10 men; mean age = 23.1 years, *SD* = 3.4; mean education = 16.6 years, *SD* = 2.2).

2.2. Experimental Conditions

We selected long-rope jumping as the rhythmic exercise condition (EC). The rhythmic exercise performed by four persons per group consisted of seven sessions of jumping for 2 min, with 1 min of rest after each session. The rope made one rotation per second, and a metronome was used to measure this rate. Moreover, the mean intensity in long-rope jumping was 56.7% (mean heart rate during rope-jumping = 109.3 bpm, *SE* = 4.5; mean maximum heart rate = 192.7 bpm, *SE* = 0.7), indicating mild rhythmic exercise for maintaining health. The subjects had free access to water during the rhythmic exercise. In contrast, the control condition (CC) was set as a period of relaxation in a private room for 30 min. During the relaxation period, the participants were allowed to drink only water, and were not allowed to use the phone, read, exercise, or sleep. The present study was a within-subjects counterbalanced design (Figure 1). First, all participants were randomly assigned to either of the two conditions (control or exercise). After experiencing either of the two conditions, the participants were treated with the unexperienced condition 1 month later.

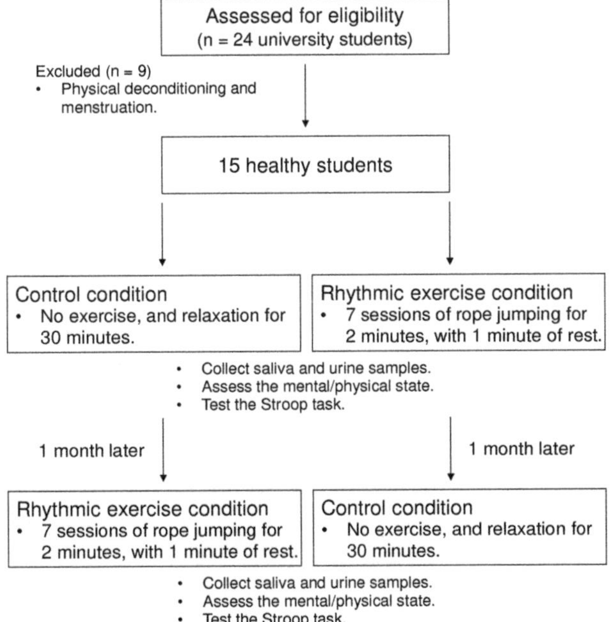

Figure 1. Study flow diagram.

2.3. Determination of MHPG in Saliva

Saliva was collected by placing two swabs on the sublingual gland in both conditions. Saliva was immediately transferred to a polypropylene tube with 2.5% perchloric acid. The tube was centrifuged at 10,000× *g* for 10 min at 4 °C. The obtained supernatant was stored at −78 °C until assayed by high performance liquid chromatography (HPLC; Nanospace SI-2 3001, Shiseido, Tokyo, Japan) equipped with an electrochemical detector (Nanospace SI-2 3005, Shiseido, Japan). The supernatant was directly injected into the

HPLC system. sMHPG concentration was measured using HPLC with an electrochemical detector (voltage: 700 mV) and a chromate recorder (C-R8A, Shimadzu Co, Kyoto, Japan). The mobile phase consisted of 15% methanol in a solution (pH 4.13) containing 30 mM citric acid, 10 mM disodium hydrogen phosphate, 0.5 mM sodium octyl sulphate, 50 mM sodium chloride, and 0.05 mM ethylenediaminetetraacetic acid. This was pumped through a 5-μM C_{18} column (150 mm $\times$ 4.6 mm; TSK gel, ODS-80TM, Tosoh, Tokyo, Japan) at a flow rate of 0.7 mL/min. sMHPG retention time was approximately 6 min. However, s5-HIAA and sHVA were not detected in either condition.

2.4. Determination of Monoamine Metabolites in Urine

The collected urine in both conditions was diluted with 6.7 mM hydrochloric acid and 2.5% perchloric acid. The mixture was centrifuged at 10,000$\times$ g for 10 min at 4 °C to separate albumin. The obtained supernatant was stored at -78 °C until the HPLC assay. The mobile phase, column, flow rate, and detection system were the same as those used for sMHPG determination. The retention times of uMHPG, u5-HIAA, and uHVA were 6, 12, and 18 min, respectively.

2.5. Psychological Measurements

The profile of mood states (POMS) is a 30-item questionnaire used to measures six basic mood states on a 5-point Likert scale (0 = not at all to 4 = very): tension (e.g., "I feel fidgety"), depression (e.g., "I feel dark"), anger (e.g., "I feel intense anger"), vigor (e.g., "I have vigour"), fatigability (e.g., "I was tired"), and confusion (e.g., "I have not gathered my thoughts"). In addition, the social behavioral questionnaire included 17 question items that measure two basic social dimensions on a 5-point Likert scale (0 = not at all to 4 = very): individual (e.g., "I value my own individuality") and social (e.g., "I stay coordinated with other people") orientations.

2.6. Cognitive Measurements

Selective attention and cognitive flexibility were measured to compare the rhythmic EC and CC. The cognitive test consisted of the classical version of the Stroop task as previously reported [31]. Subjects were asked to name the ink color in congruent words (e.g., the word red written in red ink) and incongruent words (e.g., the word blue written in green ink, or the word yellow written in red ink). Under both congruent and incongruent conditions, the words were printed on an A3 size paper with a font size of 16 for each word. The participants were asked to answer with the word or color on the paper as quickly and accurately as possible for 60 seconds/task, and the number of correct responses was recorded. To offset the effect of task order, the order of tasks was counterbalanced for all participants.

2.7. Procedure

The participants were instructed to refrain from consuming alcohol, coffee, fish, red beef, and blue cheese for at least 24 h before the experiment. In addition, they were asked to fast with water during the 90-min period before saliva and urine collection. On the day of the experiment, participants urinated, brushed their teeth with water, and then rested in a private room for 30 min. In the CC, saliva and urine were collected 30 min after the relaxation period. Next, the mental/physical state of the participants was assessed using questionnaires, and their cognitive function was tested using the Stroop task. In the EC, saliva and urine were collected 30 min after completing the rhythmic exercise. Next, questionnaires were administered to determine the mental and physical state of the participants, and the Stroop task was administered to determine their cognitive function.

2.8. Statistical Analyses

Psychological, biochemical, and cognitive data were compared between the rhythmic EC and CC using the paired sample *t*-test in IBM-SPSS, version 25 (IBM Corp., Armonk, NY, USA). However, multiple tests may induce a type I error for overestimating significant effects under no-correction, or a type II error for underestimating significant effects under conservative correction such as the Bonferroni methods. The resampling method (e.g., permutation) can be used to estimate adjusted p-values while avoiding parametric assumptions about the joint distribution of the test statistics [32,33]. We conducted a permutation test using MATLAB R2020a (The Mathworks Inc., Natick, MA, USA) for the validation of the original exercise effects detected under the uncorrected α-level threshold. For each experimental data, all samples were randomized together and resampled to obtain a dummy t-value. This procedure was repeated 10,000 times for each of the 8 psychological, 4 biochemical, and 2 cognitive data. We pooled a total of t-values (80,000 t-values: 10,000 resampling × 8 psychological data, 40,000 t-values: 10,000 resampling × 4 biochemical data, 20,000 t-values: 10,000 resampling × 2 cognitive data) and created a unique permutation t-distribution to obtain a single adjusted α-level threshold (the top five percentile ranks in the distribution) of each t-value in the psychological, biochemical, and cognitive data.

Finally, correlations were calculated using the Pearson correlation coefficient in IBM-SPSS. For these multiple coefficients, validation tests for correlations were performed in a permutation test using MATLAB R2020a. To examine the correlation in a given pair of variables (e.g., urinary 5-HIAA and salivary MHPG), a dummy coefficient was obtained by correlating the two variables randomly across participants. This procedure was repeated 10,000 times for each of the 16 correlations. We pooled a total of 160,000 dummy coefficients (10,000 resampling × 16 correlations) and created a unique permutation coefficient distribution to obtain a single adjusted α-level threshold (the top five percentile ranks in the distribution). For all analyses, $p < 0.05$ was considered statistically significant.

3. Results

3.1. Psychological Scores

The psychological data are shown in Table 1. There was a significant difference between the two conditions in the POMS anxiety score ($t_{(14)} = 5.28$, $p < 0.001$, $d = 0.96$). The t-values in the POMS anxiety score were higher than the adjusted significance level threshold ($t_{(14)} = 3.24$) obtained in the permutation test. In contrast, there were no significant differences in POMS depression, anger, vigor, fatigability and confusion scores, and individual and social orientation between the two conditions. These results indicate that, compared with the CC, the rhythmic exercise of long-rope jumping reduced anxiety levels.

Table 1. Psychological results in both conditions.

Scale	Control	Exercise	p-Value
POMS anxiety	7.7 (1.2)	4.0 (0.7)	<0.001
POMS depression	3.9 (0.8)	2.9 (0.7)	0.267
POMS anger	1.8 (0.5)	2.1 (1.1)	0.832
POMS vigour	7.9 (1.1)	7.7 (0.9)	0.909
POMS fatigability	10.1 (0.8)	11.5 (1.1)	0.257
POMS confusion	6.2 (0.7)	5.0 (0.8)	0.132
Individual orientation	23.9 (0.7)	25.1 (0.8)	0.098
Social orientation	33.7 (1.2)	34.9 (1.5)	0.342

Parameters are indicated as the mean (*SE*). *p*-values are from *t*-tests on condition differences. POMS, profile of mood states; *SE*, standard error.

3.2. Rope Jumping Exercise-Related Monoamine Metabolite Changes and Cognitive Effects

Biochemical HPLC analyses showed significantly higher sMHPG levels in the EC than in the CC (Figure 2A: $t_{(14)} = 2.48$, $p = 0.027$, $d = 0.61$). As Figure 3 shows, most individuals had an increased sMHPG ratio after rhythmic exercise. The condition differences in urinary monoamine are shown in Figure 2B–D. Compared with the CC, rhythmic exercise caused a significant increase in uMHPG (Figure 2B: $t_{(14)} = 3.18$, $p = 0.007$, $d = 1.15$) and u5-HIAA levels (Figure 2C: $t_{(14)} = 2.55$, $p = 0.023$, $d = 0.79$). As Figures 4 and 5 show, most individuals showed an increased uMHPG and u5-HIAA ratio after rhythmic exercise. These *t*-values were higher than the adjusted significance level threshold ($t_{(14)}$ =2.47) obtained in the permutation test. In contrast, there were no significant between-condition differences in uHVA levels (Figure 2D: $t_{(14)} = 0.86$, $p = 0.406$, $d = 0.27$). As Figure 6 shows, most individuals did not show differences in uHVA levels between CC and EC. These results indicate that the rhythmic exercise of long-rope jumping increased the s/uMHPG and u5-HIAA levels.

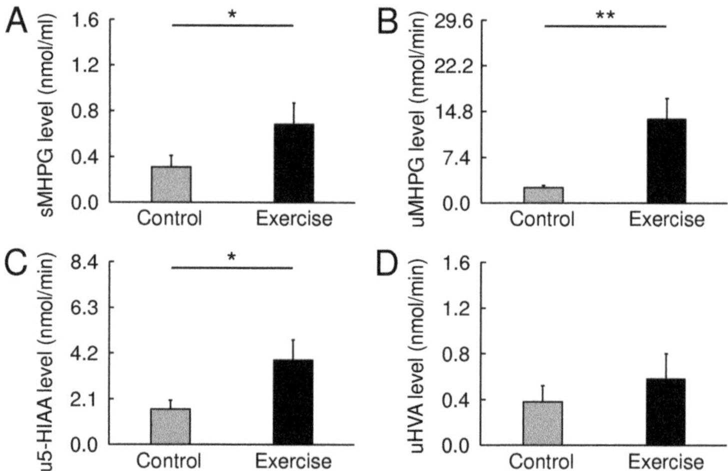

Figure 2. Change in monoamine metabolite levels stimulated by rhythmic rope jumping exercise. (**A**) Compared with the control condition, rhythmic exercise increased salivary 3-methoxy-4-hydroxyphenylglycol levels 30 min after exercise. (**B,C**) rhythmic exercise participants showed increased urinary 3-methoxy-4-hydroxyphenylglycol and 5-hydroxyindoleacetic acid levels when compared with participants in the control condition. (**D**) There was no significant difference between conditions in urinary homovanillic acid level. Parameters are indicated as mean (*SE*). * $p < 0.05$, ** $p < 0.01$. sMHPG, salivary 3-methoxy-4-hydroxyphenylglycol; uMHPG, urinary 3-methoxy-4-hydroxyphenylglycol; u5-HIAA, urinary 5-hydroxyindoleacetic acid; uHVA, urinary homovanillic acid; *SE*, standard error.

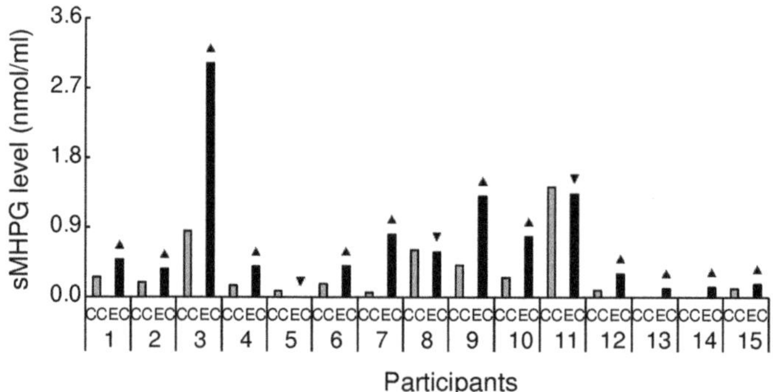

Figure 3. Increased and decreased ratio of salivary 3-methoxy-4-hydroxyphenylglycol level between conditions for each participant. Despite the large interindividual variance, 12 out of 15 participants had an increased ratio of 3-methoxy-4-hydroxyphenylglycol level during exercise compared with the control condition. sMHPG, salivary 3-methoxy-4-hydroxyphenylglycol; CC, control condition; EC, exercise condition; ▲, increase ratio in EC compared with CC; ▼, decrease ratio in EC compared with CC.

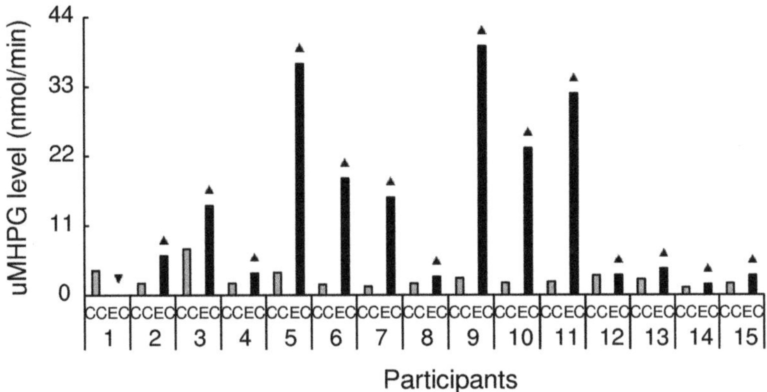

Figure 4. Increased and decreased ratio of urinary 3-methoxy-4-hydroxyphenylglycol level between conditions for each participant. Despite the large interindividual variance, 14 out of 15 participants had an increased ratio of 3-methoxy-4-hydroxyphenylglycol level during exercise compared with the control condition. uMHPG, urinary 3-methoxy-4-hydroxyphenylglycol; CC, control condition; EC, exercise condition; ▲, increase ratio in EC compared with CC; ▼, decrease ratio in EC compared with CC.

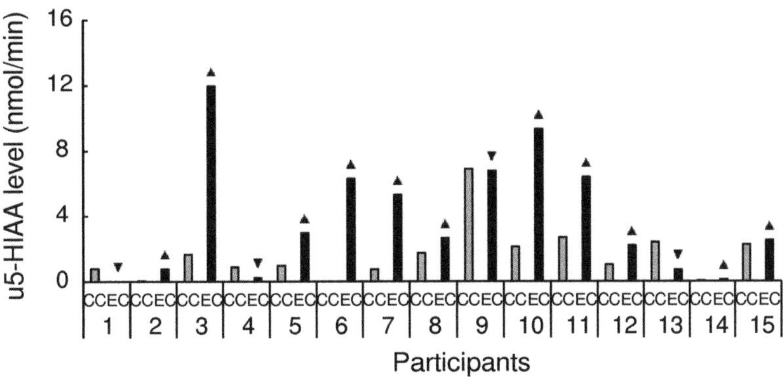

Figure 5. Increased and decreased ratio of urinary 5-hydroxyindoleacetic acid level between conditions for each participant. Despite the large interindividual variance, 11 out of 15 participants had an increased ratio of 5-hydroxyindoleacetic acid level during exercise compared with the control condition. u5-HIAA, urinary 5-hydroxyindoleacetic acid; CC, control condition; EC, exercise condition; ▲, increase ratio in EC compared with CC; ▼, decrease ratio in EC compared with CC.

The cognitive data are shown in Figure 7. There were significant differences between the two conditions in the congruent (Figure 7A: $t_{(14)} = 2.45$, $p = 0.028$, $d = 0.75$) and incongruent word tasks (Figure 7B: $t_{(14)} = 2.39$, $p = 0.032$, $d = 0.30$). As Figures 8 and 9 show, most individuals increased their Stroop performance after rhythmic exercise. These t-values were higher than the adjusted significant level threshold ($t_{(14)} = 2.25$) obtained in the permutation test. These results indicate that participants showed higher attention performance after the rhythmic exercise of long-rope jumping than after the CC.

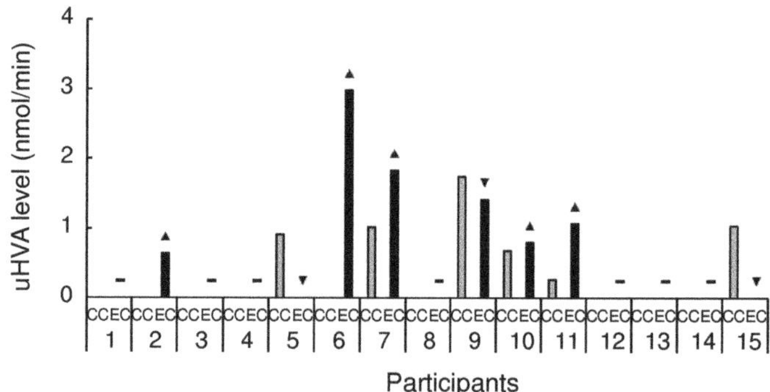

Figure 6. Increased and decreased ratio of urinary homovanillic acid level between conditions for each participant. Although 5 participants had an increased ratio of homovanillic acid level in urine during exercise compared with the control condition, in most participants homovanillic acid was not detected in any condition. Homovanillic acid in urine may not be used as a biomarker of dopamine in the central nervous system. uHVA, urinary homovanillic acid; CC, control condition; EC, exercise condition; ▲, increase ratio in EC compared with CC; ▼, decrease ratio in EC compared with CC; −, no change.

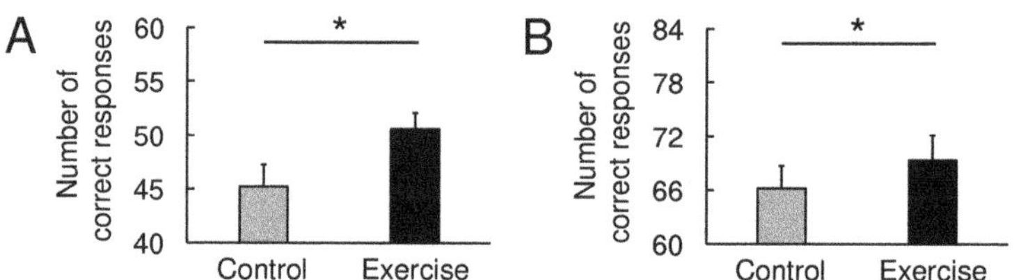

Figure 7. Effect of rhythmic rope jumping exercise on Stroop performance. (**A**) In the congruent word task, rhythmic exercise increased the rate of correct responses compared with the control condition. (**B**) In the incongruent word task, rhythmic exercise increased the rate of correct responses compared with the control condition. Parameters are indicated as mean (*SE*). * $p < 0.05$. *SE*, standard error.

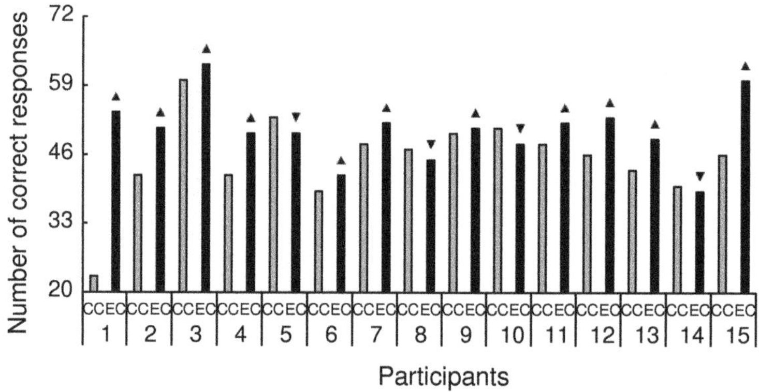

Figure 8. Increased and decreased ratio of Stroop congruent performance between conditions for each participant. Despite the large interindividual variance, 11 out of 15 participants had an increased ratio of Stroop performance during exercise compared with the control condition. CC, control condition; EC, exercise condition; ▲, increase ratio in EC compared with CC; ▼, decrease ratio in EC compared with CC.

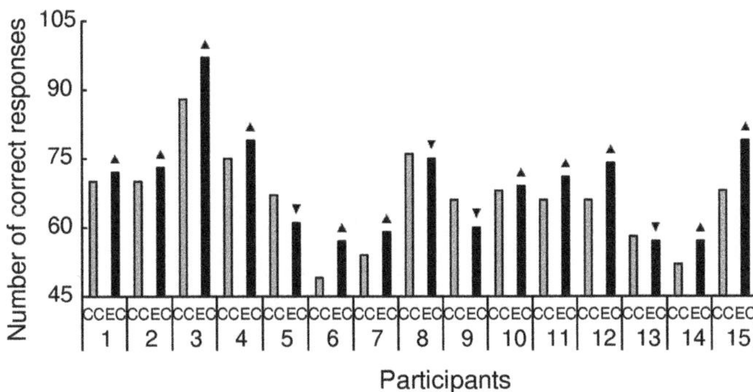

Figure 9. Increased and decreased ratio of Stroop incongruent performance between conditions for each participant. Despite the large interindividual variance, 11 out of 15 participants had an increased ratio of Stroop performance during exercise compared with the control condition. CC, control condition; EC, exercise condition; ▲, increase ratio in EC compared with CC; ▼, decrease ratio in EC compared with CC.

We also investigated the correlation between s/uMHPG levels and u5-HIAA levels and Stroop performance (Table 2). The u5-HIAA levels were positively correlated with the s/uMHPG levels in the rhythmic EC (sMHPG: $r = 0.80$, $p < 0.001$, uMHPG: $r = 0.58$, $p = 0.022$). The absolute Pearson's r values were higher than the adjusted significance level threshold ($|r| = 0.54$) obtained in the permutation test. In contrast, such a correlation with sMHPG levels was not significant in the CC (sMHPG: $r = 0.32$, $p = 0.241$, uMHPG: $r = 0.10$, $p = 0.721$). This shows that higher 5-HIAA levels are linked to higher levels of MHPG.

Table 2. Correlations between monoamine metabolites and attention.

Pair of Variables	Control	Exercise
(A) Correlation with u5-HIAA		
sMHPG	0.32	0.80 ***
uMHPG	0.10	0.58 *
(B) Correlation with Stroop performance		
Congruent word task		
sMHPG	0.26	0.61 *
uMHPG	0.44	0.03
u5-HIAA	0.35	0.38
Incongruent word task		
sMHPG	0.27	0.50
uMHPG	0.59 *	−0.30
u5-HIAA	−0.25	0.38

* $p < 0.05$, *** $p < 0.001$. u5-HIAA, urinary 5-hydroxyindloeacetic acid; sMHPG, salivary 3-methoxy-4-hydroxyphenylglycol; uMHPG, urinary 3-methoxy-4-hydroxyphenylglycol.

In subsequent correlation analyses including sex, age, and education as the control variables (Table 2), sMHPG levels were positively correlated with correct responses in the congruent word task of the Stroop test in the EC ($r = 0.61$, $p = 0.036$). The absolute Pearson's r value was higher than the adjusted significance level threshold ($|r| = 0.54$) obtained in the permutation test. In contrast, such a correlation was not found for the CC ($r = 0.26$, $p = 0.417$). These results indicate that enhancement of the MHPG level 30 min after completion of the rhythmic exercise of long-rope jumping is linked to higher attention performance.

4. Discussion

This study aimed to investigate the association between the rhythmic exercise of long-rope jumping and central monoaminergic and cognitive functions. The main findings revealed that after the rhythmic rope jumping exercise, participants showed (1) a significant reduction of anxiety scores; (2) increased s/uMHPG and u5-HIAA levels; and (3) improved Stroop performance. Moreover, the higher release of 5-HIAA was associated with higher MHPG levels in the EC. In addition, the MHPG levels that were increased by the rhythmic rope jumping exercise were associated with higher attention performance on the Stroop task. Our results indicate that long-rope jumping may affect cognitive function by activating the norepinephrinergic and serotonergic systems and their interactions.

Consistent with previous findings [34,35], this study revealed that the rhythmic rope jumping exercise significantly reduced anxiety scores compared to the CC. The improvement in anxiety score might be explained by the decrease of hypothalamic corticotropin-releasing factor and activation of the central monoaminergic system (e.g., norepinephrine and serotonin) [36], both of which are based on the antianxiety and antidepressant effects of exercise. Psychologically, exercise is also associated with a self-efficacy improvement through progressive positive feedback, such as fitness gains [36]. The multimodal nature of exercise may contribute to improving anxiety in our study.

In contrast to previous studies using aerobic exercise showing significant improvement in depressive symptoms of high-depressive participants [37,38], we did not find similar improvement in the POMS depression score of healthy participants. One possible reason is the initial level of depression (e.g., high-depressive participants vs. healthy participants). Another possible reason is that continued rope jumping exercise may strongly reduce depression levels.

Concerning monoaminergic functions, s/uMHPG and u5-HIAA levels and their interaction were found to be increased by rhythmic rope jumping exercise. These findings can be interpreted as central norepinephrinergic and serotonergic activation due to long-rope jumping. Previous studies have reported that exhaustive exercise is associated with higher serotonin synthesis [9,39] and lower norepinephrine levels in the brain [40], indicating that monoamine changes are implicated in the central fatigue mechanism. However, one study reported that after supplementation with 2-μM L-tryptophan (serotonin precursor), nerve terminals took up the serotonin over a 60-min period, rapidly metabolizing it to 5-HIAA to return the concentration of serotonin to its original level after 90 min [41]. Another study emphasized that exercise performance is not influenced by fluoxetine (selective serotonin reuptake inhibitor) [42]. These aspects of serotonin could contribute to increased brain plasticity [43,44] but not induce central fatigue [9,41,42]. In particular, the increased u5-HIAA levels due to the rhythmic exercise of long-rope jumping may help facilitate neuroplasticity. Alternatively, the heightened u5-HIAA levels may be related to cognitive demand and motor plan in long-rope jumping. The serotonergic system projects from raphe nuclei to the precuneus and the hippocampus [45]. The precuneus is involved in attention shift and timing function [46–48] and plays an important role in visuospatial imagery for body movement control [49–51]. Malouin et al. reported activation of the precuneus in imagery tasks of walking with obstacles through a virtual environment [52], suggesting the involvement of the precuneus in efficient predictive adaptation of postural control, motor coordination, spatial orientation, and reaction to moving objects/persons. Moreover, the hippocampus has been implicated in the processing capacity of spatial information as well as attention [53,54]. In addition, neuroimaging and neurophysiological studies have showed that serotonergic modulation influences motor planning and sensory perception (e.g., rhythm and timing) [55,56]. The accumulation of training in motor skills involving high cognitive demand may strongly influence the precuneus and hippocampus associated with the serotonergic system, presumably because visuospatial processing, coordinating movements to maintain balance and rhythm, and attention shift are essential for long-rope umping.

Interestingly, although increased dopamine availability in the brain has an exercise performance-enhancing effect [57], we could not detect uHVA excretion in most individuals, nor were there significant differences between the two conditions. Dopaminergic neurons are restricted to the nigrostriatal pathway [43], suggesting that dopamine content is much lower in other regions, except for the striatum [58]. In contrast, serotonergic and norepinephrinergic neurons are localized in the whole brain [58]. Therefore, higher u5-HIAA and uMHPG excretion would be expected as these metabolites are transported more from the whole brain via the OAT3, but not HVA excretion. Thus, uHVA excretion may not function well as the central dopamine biomarker in our study.

Further analyses revealed that higher sMHPG levels were associated with better Stroop performance in the rhythmic rope jumping exercise, indicating a rhythmic exercise-specific link between processes controlling attention and the norepinephrinergic system. This result suggests that long-rope jumping may facilitate attentional performance by central norepinephrinergic activation.

The norepinephrinergic pathway projects from the locus coeruleus to the whole brain [59]. A pharmacological study reported that after clonidine microinjection in rats, the reduction of locus coeruleus-norepinephrinergic system activity reduces prefrontal-dependent visuospatial attention performance [60], indicating a role of the norepinephrinergic system in attention associated with locus coeruleus-prefrontal circuit. In addition, aerobic exercise is associated with the activation of these regions [61,62]. Although we did not evaluate these brain functions, these aspects of the rhythmic exercise of long-rope jumping may enhance the connectivity between the locus coeruleus and the prefrontal cortex, associated with the ascending norepinephrinergic pathway. The increased peripheral MHPG levels and their association with attention performance in our study suggest that long-rope jumping may enhance central norepinephrine release for improved attention.

Finally, we provided the first evidence of a positive correlation between uMHPG levels and incongruency processing, as measured by the Stroop task in healthy participants in the CC. Since Stroop incongruent condition requires more attention and control of competitive responses [63], the higher uMHPG levels may be fast to respond to the Stroop incongruent effect. Although baseline sMHPG levels were associated with effort performance on the Uchida-Kraepelin test [22], baseline uMHPG may predict effortful attention on the Stroop interference. Therefore, baseline uMHPG could serve as a non-invasive biological marker for detecting central norepinephrinergic activity and a useful predictive marker for arousal and attention in healthy participants.

Our study has several limitations. As small sample size limits statistical power, the results in our study might be considered as preliminary. However, our study size was comparable to those in some biochemical sports trials [64,65] and, as indicated by the interindividual differences in Figures 3–6, 8 and 9, increased s/uMHPG and u5-HIAA levels and improved Stroop performance were observed in EC participants, suggesting that the changes in central norepinephrinergic-serotonergic systems and attention performance were caused by the rhythmic rope jumping exercise. Moreover, while the effects of long-rope jumping have been demonstrated, it is still unclear which aspect of the exercise is effective. For example, CC participants did not undergo any intervention, which limits the interpretation of our findings. That is, social interaction, as well as the rope jumping exercise, may have partially influenced the EC participants. However, it is unlikely that our exercise effects are merely due to an increase in social interaction because the two conditions did not differ in social orientation scores. Furthermore, the present study was unable to recruit a sufficient number of participants for other aerobic exercises of similar intensity (e.g., walking and dance) to further clarify the effects of rhythmic rope jumping on monoamine levels and cognitive functions. Therefore, the specific effects on aerobic demand or more complex cognitive demands of the motor planning involved in rope jumping remain unclear and should be investigated in future studies. Finally, there were few female participants in our sample. Therefore, the present results may not be generalizable to both sexes. Future research should focus on these issues to confirm or reject our findings.

5. Conclusions

We found that the rhythmic exercise of long-rope jumping enhances cognitive performance and central monoaminergic system function through central norepinephrinergic activation. Biochemically, the increased s/uMHPG was associated with higher u5-HIAA levels in the rhythmic rope jumping exercise. Moreover, the behavioral results indicated exercise-specific improved attention performance, suggesting that long-rope jumping may affect the central norepinephrinergic system for attention improvement. The main findings of the present study shed new light on how long-rope jumping possibly influences the central monoaminergic system.

Author Contributions: Conceptualization, M.Y. and T.Y.; methodology, M.Y. and T.Y.; validation, M.Y. and T.Y.; software, M.Y.; formal analysis, M.Y.; investigation, M.Y. and T.Y.; data curation, M.Y. and T.Y.; writing—original draft preparation, M.Y.; writing—review and editing, M.Y. and T.Y.; visualization, M.Y. and T.Y.; supervision, M.Y.; project administration, M.Y.; funding acquisition, M.Y. All authors have read and agreed to the published version of the manuscript.

Funding: This work was supported by the Japan Society for the Promotion of Science, Grants-in-Aid for Scientific Research (KAKENHI) (grant numbers 14J12587, 20K14255) to Masatoshi Yamashita. The APC was funded by KAKENHI (grant number 20K14255) to Masatoshi Yamashita.

Institutional Review Board Statement: The study was conducted according to the guidelines of the Declaration of Helsinki, and the protocol was approved by the Psychological Research Ethics Committee of Tezukayama University (approval number 27–15).

Informed Consent Statement: Informed consent was obtained from all subjects involved in the study.

Data Availability Statement: The raw data that support the findings of this study are available from the corresponding author upon reasonable request.

Acknowledgments: We are grateful to Yasuna Nakao, Ikue Kuwahara, Mako Miyaoka, Jin Kikuchi, Akira Sakashita, Madoka Enouchi and Masakazu Morinaga for recruiting the students and supervising the long-rope jumping and Stroop tasks. We thank Takahiro Soshi for supporting the statistical analyses. This study was conducted in the fitness room of Tezukayama University.

Conflicts of Interest: The authors declare no conflict of interest.

References

1. Bharath, L.P.; Choi, W.W.; Cho, J.-M.; Skobodzinski, A.A.; Wong, A.; Sweeney, T.E.; Park, S.-Y. Combined resistance and aerobic exercise training reduces insulin resistance and central adiposity in adolescent girls who are obese: Randomized clinical trial. *Eur. J. Appl. Physiol.* **2018**, *118*, 1653–1660. [CrossRef]
2. Erickson, K.I.; Voss, M.W.; Prakash, R.S.; Basak, C.; Szabo, A.; Chaddock, L.; Kim, J.; Heo, S.; Alves, H.; White, S.M.; et al. Exercise training increases size of hippocampus and improves memory. *Proc. Natl. Acad. Sci. USA* **2011**, *108*, 3017–3022. [CrossRef]
3. Pitreli, J.; O'shea, P. SPORTS PERFORMANCE SERIES: Rope Jumping: The biomechanics, techniques of and application to athletic conditioning. *Natl. Strength Cond. Assoc. J.* **1986**, *8*, 5–13. [CrossRef]
4. Toplak, M.E.; Dockstader, C.; Tannock, R. Temporal information processing in ADHD: Findings to date and new methods. *J. Neurosci. Methods* **2006**, *151*, 15–29. [CrossRef]
5. Coleman, M.; Offen, K.; Markant, J. Exercise Similarly Facilitates Men and Women's Selective Attention Task Response Times but Differentially Affects Memory Task Performance. *Front. Psychol.* **2018**, *9*, 1405. [CrossRef]
6. Ha, A.S.; Ng, J.Y.Y. Rope skipping increases bone mineral density at calcanei of pubertal girls in Hong Kong: A quasi-experimental investigation. *PLoS ONE* **2017**, *12*, e0189085. [CrossRef]
7. Sung, K.-D.; Pekas, E.J.; Scott, S.D.; Son, W.-M.; Park, S.-Y. The effects of a 12-week jump rope exercise program on abdominal adiposity, vasoactive substances, inflammation, and vascular function in adolescent girls with prehypertension. *Eur. J. Appl. Physiol.* **2019**, *119*, 577–585. [CrossRef]
8. Dahl, M.J.; Mather, M.; Sander, M.C.; Werkle-Bergner, M. Noradrenergic Responsiveness Supports Selective Attention across the Adult Lifespan. *J. Neurosci.* **2020**, *40*, 4372–4390. [CrossRef] [PubMed]
9. Yamashita, M. Potential Role of Neuroactive Tryptophan Metabolites in Central Fatigue: Establishment of the Fatigue Circuit. *Int. J. Tryptophan Res.* **2020**, *13*, 1178646920936279. [CrossRef] [PubMed]

10. Bouret, S.; Sara, S.J. Network reset: A simplified overarching theory of locus coeruleus noradrenaline function. *Trends Neurosci.* **2005**, *28*, 574–582. [CrossRef] [PubMed]

11. Corbetta, M.; Patel, G.; Shulman, G.L. The reorienting system of the human brain: From environment to theory of mind. *Neuron* **2008**, *58*, 306–324. [CrossRef]

12. Vernaleken, I.; Buchholz, H.-G.; Kumakura, Y.; Siessmeier, T.; Stoeter, P.; Bartenstein, P.; Cumming, P.; Gründer, G. 'Prefrontal' cognitive performance of healthy subjects positively correlates with cerebral FDOPA influx: An exploratory [18F]-fluoro-L-DOPA-PET investigation. *Hum. Brain Mapp.* **2007**, *28*, 931–939. [CrossRef]

13. Madsen, K.; Erritzoe, D.; Mortensen, E.L.; Gade, A.; Madsen, J.; Baaré, W.; Knudsen, G.M.; Hasselbalch, S.G. Cognitive function is related to fronto-striatal serotonin transporter levels – a brain PET study in young healthy subjects. *Psychopharmacology* **2011**, *213*, 573–581. [CrossRef]

14. Morinaga, M.; Shimizu, T.; Yamashita, M.; Yamamoto, T. The measurement of the quantity of urinary 5-hydroxyindoleacetic acid excretion as the noninvasive marker of the 5-HT content in the brain: About brain-urine correlation after the 5,7-dihydroxytryptamine microinjection. *Jan. J. Cogn. Neurosci.* **2017**, *19*, 95–101, (In Japanese, English abstract).

15. Yano, S.; Moseley, K.; Fu, X.; Azen, C. Evaluation of Tetrahydrobiopterin Therapy with Large Neutral Amino Acid Supplementation in Phenylketonuria: Effects on Potential Peripheral Biomarkers, Melatonin and Dopamine, for Brain Monoamine Neurotransmitters. *PLoS ONE* **2016**, *11*, e0160892. [CrossRef]

16. Sheline, Y.I.; Miller, K.; Bardgett, M.E.; Csernansky, J.G. Higher cerebrospinal fluid MHPG in subjects with dementia of the Alzheimer type. Relationship with cognitive dysfunction. *Am. J. Geriatr. Psychiatry* **1998**, *6*, 155–161. [CrossRef] [PubMed]

17. Takase, M.; Kimura, H.; Kanahara, N.; Nakata, Y.; Iyo, M. Plasma monoamines change under dopamine supersensitivity psychosis in patients with schizophrenia: A comparison with first-episode psychosis. *J. Psychopharmacol.* **2020**, *34*, 540–547. [CrossRef]

18. Pohl, R.; Ettedgui, E.; Bridges, M.; Lycaki, H.; Jimerson, D.; Kopin, I.; Rainey, J.M. Plasma MHPG levels in lactate and isoproterenol anxiety state. *Biol. Psychiatry* **1987**, *22*, 1127–1136. [CrossRef]

19. Uhde, T.W.; Joffe, R.T.; Jimerson, D.C.; Post, R.M. Normal urinary free cortisol and plasma MHPG in panic disorder: Clinical and theoretical implications. *Biol. Psychiatry* **1988**, *23*, 575–585. [CrossRef]

20. Reuster, T.; Rilke, O.; Oehler, J. High correlation between salivary MHPG and CSF MHPG. *Psychopharmacology* **2002**, *162*, 415–418. [CrossRef]

21. Yang, R.K.; Yehuda, R.; Holland, D.D.; Knott, P.J. Relationship between 3-methoxy-4-hydroxypheylglycol and homovanillic acid in saliva and plasma of healthy volunteers. *Biol. Psychiatry* **1997**, *42*, 821–826. [CrossRef]

22. Li, G.Y.; Ueki, H.; Kawashima, T.; Sugataka, K.; Muraoka, T.; Yamada, S. Involvement of the noradrenergic system in per-formance on a continuous task requiring effortful attention. *Neuropsychobiology* **2004**, *50*, 336–340. [CrossRef] [PubMed]

23. Li, G.Y.; Watanabe, I.; Kunitake, Y.; Sugataka, K.; Muraoka, T.; Kojima, N.; Kawashima, T.; Yamada, S. Relationship between saliva level of 3-methoxy-4-hydroxyphenylglycol and mental health in the elderly general population. *Psychiatry Clin. Neurosci.* **2008**, *62*, 562–567. [CrossRef] [PubMed]

24. Maas, J.W. Relationships between central nervous system noradrenergic function and plasma and urinary concentrations of norepinephrine metabolites. *Adv. Biochem. Psychopharmacol.* **1984**, *39*, 45–55. [PubMed]

25. Kurita, M.; Nishino, S.; Numata, Y.; Okubo, Y.; Sato, T. The noradrenaline metabolite MHPG is a candidate biomarker between the depressive, remission, and manic states in bipolar disorder I: Two long-term naturalistic case reports. *Neuropsychiatr. Dis. Treat.* **2015**, *11*, 353–358. [CrossRef] [PubMed]

26. Ohtsuki, S.; Hori, S.; Terasaki, T. Molecular mechanisms of drug influx and efflux transport at the blood-brain barrier. *Folia Pharmacol. Jpn.* **2003**, *122*, 55–64. [CrossRef]

27. Mori, S.; Takanaga, H.; Ohtsuki, S.; Deguchi, T.; Kang, Y.S.; Hosoya, K.; Terasaki, T. Rat organic anion transporter 3 (rOAT3) is responsible for brain-to-blood efflux of homovanillic acid at the abluminal membrane of brain capillary endothelial cells. *J. Cereb. Blood Flow Metab.* **2003**, *23*, 432–440. [CrossRef]

28. Constantini, N.W.; Dubnov, G.; Lebrun, C.M. The Menstrual Cycle and Sport Performance. *Clin. Sports Med.* **2005**, *24*, e51–e82. [CrossRef] [PubMed]

29. Hansson, S.R.; Bottalico, B.; Noskova, V.; Casslén, B. Monoamine transporters in human endometrium and decidua. *Hum. Reprod. Update* **2008**, *15*, 249–260. [CrossRef]

30. Janse de Jonge, X.A. Effects of the menstrual cycle on exercise performance. *Sports Med.* **2003**, *33*, 833–851. [CrossRef]

31. Ghimire, N.; Paudel, B.H.; Khadka, R.; Singh, P.N. Reaction time in Stroop test in Nepalese Medical Students. *J. Clin. Diagn. Res.* **2014**, *8*, BC14–BC16. [CrossRef] [PubMed]

32. Camargo, A.; Azuaje, F.; Wang, H.; Zheng, H. Permutation – based statistical tests for multiple hypotheses. *Source Code Biol. Med.* **2008**, *3*, 15. [CrossRef] [PubMed]

33. Nichols, T.E.; Holmes, A.P. Nonparametric permutation tests for functional neuroimaging: A primer with examples. *Hum. Brain Mapp.* **2002**, *15*, 1–25. [CrossRef] [PubMed]

34. Bonhauser, M.; Fernandez, G.; Püschel, K.; Yañez, F.; Montero, J.; Thompson, B.; Coronado, G. Improving physical fitness and emotional well-being in adolescents of low socioeconomic status in Chile: Results of a school-based controlled trial. *Health Promot. Int.* **2005**, *20*, 113–122. [CrossRef]

35. De Moor, M.; Beem, A.; Stubbe, J.; Boomsma, D.; de Geus, E. Regular exercise, anxiety, depression and personality: A population-based study. *Prev. Med.* **2006**, *42*, 273–279. [CrossRef]

36. Anderson, E.; Shivakumar, G. Effects of exercise and physical activity on anxiety. *Front. Psychiatry* **2013**, *4*, 27. [CrossRef]
37. Saeed, S.A.; Cunningham, K.; Bloch, R.M. Depression and Anxiety Disorders: Benefits of Exercise, Yoga, and Meditation. *Am. Fam. Physician* **2019**, *99*, 620–627.
38. Meyer, J.D.; Koltyn, K.F.; Stegner, A.J.; Kim, J.-S.; Cook, D.B. Influence of Exercise Intensity for Improving Depressed Mood in Depression: A Dose-Response Study. *Behav. Ther.* **2016**, *47*, 527–537. [CrossRef] [PubMed]
39. Acworth, I.; Nicholass, J.; Morgan, B.; Newsholme, E. Effect of sustained exercise on concentrations of plasma aromatic and branched-chain amino acids and brain amines. *Biochem. Biophys. Res. Commun.* **1986**, *137*, 149–153. [CrossRef]
40. Foley, T.E.; Fleshner, M. Neuroplasticity of dopamine circuits after exercise: Implications for central fatigue. *Neuromol. Med.* **2008**, *10*, 67–80. [CrossRef]
41. Yamamoto, T.; Azechi, H.; Board, M. Essential role of excessive tryptophan and its neurometabolites in fatigue. *Can. J. Neurol. Sci.* **2012**, *39*, 40–47. [CrossRef] [PubMed]
42. Meeusen, R.; Piacentini, M.F.; Magnus, L.; Eynde, S.V.D.; De Meirleir, K. Exercise performance is not influenced by a 5-HT reuptake inhibitor. *Int. J. Sports Med.* **2001**, *22*, 329–336. [CrossRef] [PubMed]
43. Lin, T.-W.; Kuo, Y.-M. Exercise benefits brain function: The monoamine connection. *Brain Sci.* **2013**, *3*, 39–53. [CrossRef] [PubMed]
44. Del Angel-Meza, A.; Ramírez-Cortés, L.; Olvera-Cortés, M.E.; Pérez-Vega, M.I.; González-Burgos, I. A tryptophan-deficient corn-based diet induces plastic responses in cerebellar cortex cells of rat offspring. *Int. J. Dev. Neurosci.* **2001**, *19*, 447–453. [CrossRef]
45. Vanicek, T.; Kutzelnigg, A.; Philippe, C.; Sigurdardottir, H.L.; James, G.M.; Hahn, A.; Kranz, G.; Höflich, A.; Kautzky, A.; Traub-Weidinger, T.; et al. Altered interregional molecular associations of the serotonin transporter in attention deficit/hyperactivity disorder assessed with PET. *Hum. Brain Mapp.* **2017**, *38*, 792–802. [CrossRef]
46. Nagahama, Y.; Okada, T.; Katsumi, Y.; Hayashi, T.; Yamauchi, H.; Sawamoto, N.; Toma, K.; Nakamura, K.; Hanakawa, T.; Konishi, J.; et al. Transient neural activity in the medial superior frontal gyrus and precuneus time locked with attention shift between object features. *Neuroimage* **1999**, *10*, 193–199. [CrossRef]
47. Naghavi, H.R.; Nyberg, L. Common fronto-parietal activity in attention, memory, and consciousness: Shard demands on integration? *Conscious. Cogn.* **2005**, *14*, 390–425. [CrossRef]
48. Hart, H.; Radua, J.; Mataix-Cols, D.; Rubia, K. Meta-analysis of fMRI studies of timing in attention-deficit hyperactivity disorder (ADHD). *Neurosci. Biobehav. Rev.* **2012**, *36*, 2248–2256. [CrossRef]
49. Ogiso, T.; Kobayashi, K.; Sugishita, M. The precuneus in motor imagery: A magnetoencephalographic study. *Neuroreport* **2000**, *11*, 1345–1349. [CrossRef]
50. Suchan, B.; Yaguez, L.; Wunderlich, G.; Canavan, A.G.; Herzog, H.; Tellmann, L.; Homberg, V.; Seitz, R.J. Hemispheric dissociation of visual-pattern processing and visual rotation. *Behav. Brain Res.* **2002**, *136*, 533–544. [CrossRef]
51. Tian, Q.; Resnick, S.M.; Davatzikos, C.; Erus, G.; Simonsick, E.M.; Studenski, S.A.; Ferrucci, L. A prospective study of focal brain atrophy, mobility and fitness. *J. Intern. Med.* **2019**, *286*, 88–100. [CrossRef]
52. Malouin, F.; Richards, C.L.; Jackson, P.; Dumas, F.; Doyon, J. Brain activations during motor imagery of locomotor-related tasks: A PET study. *Hum. Brain Mapp.* **2003**, *19*, 47–62. [CrossRef]
53. Goldfarb, E.V.; Chun, M.M.; Phelps, E.A. Memory-Guided Attention: Independent Contributions of the Hippocampus and Striatum. *Neuron* **2016**, *89*, 317–324. [CrossRef]
54. Kaplan, R.; Horner, A.J.; Bandettini, P.A.; Doeller, C.F.; Burgess, N. Human hippocampal processing of environmental novelty during spatial navigation. *Hippocampus* **2014**, *24*, 740–750. [CrossRef]
55. Biskup, C.S.; Helmbold, K.; Baurmann, D.; Klasen, M.; Gaber, T.J.; Bubenzer-Busch, S.; Königschulte, W.; Fink, G.R.; Zepf, F.D. Resting state default mode network connectivity in children and adolescents with ADHD after acute tryptophan depletion. *Acta Psychiatr. Scand.* **2016**, *134*, 161–171. [CrossRef] [PubMed]
56. Flaive, A.; Fougère, M.; Van Der Zouwen, C.I.; Ryczko, D. Serotonergic Modulation of Locomotor Activity From Basal Vertebrates to Mammals. *Front. Neural Circuits* **2020**, *14*, 590299. [CrossRef]
57. Balthazar, C.H.; Leite, L.H.; Rodrigues, A.G.; Coimbra, C.C. Performance-enhancing and thermoregulatory effects of intracerebroventricular dopamine in running rats. *Pharmacol. Biochem. Behav.* **2009**, *93*, 465–469. [CrossRef] [PubMed]
58. Yamashita, M.; Yamamoto, T. Tryptophan circuit in fatigue: From blood to brain and cognition. *Brain Res.* **2017**, *1675*, 116–126. [CrossRef] [PubMed]
59. Moore, R.Y.; E Bloom, F. Central catecholamine neuron systems: Anatomy and physiology of the dopamine systems. *Annu. Rev. Neurosci.* **1978**, *1*, 129–169. [CrossRef]
60. Mair, R.D.; Zhang, Y.; Bailey, K.R.; Toupin, M.M.; Mair, R.G. Effects of clonidine in the locus coeruleus on prefrontal- and hippocampal-dependent measures of attention and memory in the rat. *Psychopharmacology* **2005**, *181*, 280–288. [CrossRef]
61. Kujach, S.; Byun, K.; Hyodo, K.; Suwabe, K.; Fukuie, T.; Laskowski, R.; Dan, I.; Soya, H. A transferable high-intensity inter-mittent exercise improves executive performance in associated with dorsolateral prefrontal activation in young adults. *Neu-roimage* **2018**, *169*, 117–125. [CrossRef]
62. Murray, P.S.; Groves, J.L.; Pettett, B.J.; Britton, S.L.; Koch, L.G.; Dishman, R.K.; Holmes, P.V. Locus coeruleus galanin expres-sion is enhanced after exercise in rats selectively bred for high capacity for aerobic activity. *Peptides* **2010**, *31*, 2264–2268. [CrossRef]
63. Kane, M.J.; Engle, R.W. Working-memory capacity and the control of attention: The contributions of goal neglect, response competition, and task set to Stroop interference. *J. Exp. Psychol. Gen.* **2003**, *132*, 47–70. [CrossRef] [PubMed]

64. Oikawa, S.Y.; Macinnis, M.J.; Tripp, T.R.; Mcglory, C.; Baker, S.K.; Phillips, S.M. Lactalbumin, Not Collagen, Augments Muscle Protein Synthesis with Aerobic Exercise. *Med. Sci. Sports Exerc.* **2020**, *52*, 1394–1403. [CrossRef] [PubMed]
65. Olsen, T.; Sollie, O.; Nurk, E.; Turner, C.; Jernerén, F.; Ivy, J.L.; Vinknes, K.J.; Clauss, M.; Refsum, H.; Jensen, J. Exhaustive Exercise and Post-exercise Protein Plus Carbohydrate Supplementation Affect Plasma and Urine Concentrations of Sulfur Amino Acids, the Ratio of Methionine to Homocysteine and Glutathione in Elite Male Cyclists. *Front. Physiol.* **2020**, *11*, 609335. [CrossRef] [PubMed]

Article

Influence of Aerobic Fitness on White Matter Integrity and Inhibitory Control in Early Adulthood: A 9-Week Exercise Intervention

Hao Zhu [1,2], Lina Zhu [2,3], Xuan Xiong [1,2], Xiaoxiao Dong [1,2], Dandan Chen [1,2], Jingui Wang [1,2], Kelong Cai [1,2], Wei Wang [4] and Aiguo Chen [1,2,*]

1   College of Physical Education, Yangzhou University, Yangzhou 225127, China; MX120180365@yzu.edu.cn (H.Z.); DX120190066@yzu.edu.cn (X.X.); DX120190065@yzu.edu.cn (X.D.); DX120200078@yzu.edu.cn (D.C.); MX120170362@yzu.edu.cn (J.W.); MX120170353@yzu.edu.cn (K.C.)
2   Institute of Sports, Exercise and Brain, Yangzhou University, Yangzhou 225127, China; zhulina827@mail.bnu.edu.cn
3   School of Physical Education and Sports Science, Beijing Normal University, Beijing 100875, China
4   Department of Medical Imaging, The Affiliated Hospital of Yangzhou University, Yangzhou University, Yangzhou 225009, China; 090223@yzu.edu.cn
*   Correspondence: agchen@yzu.edu.cn; Tel.: +86-1395-272-5968

Abstract: Previous cross-sectional studies have related aerobic fitness to inhibitory control and white matter (WM) microstructure in young adults, but there is no longitudinal study to confirm whether these relationships exist. We carried out a longitudinal study comparing aerobic fitness, inhibitory control, and WM integrity across time points, before versus after completing an exercise intervention in young adults (18–20 years old) relative to a control group. The exercise group ($n = 35$) participated in a 9-week exercise protocol, while the control group ($n = 24$) did not receive any regular exercise training. Behavioral data and diffusion tensor imaging (DTI) data were collected prior to and following the intervention. After the exercise intervention, aerobic fitness and inhibitory control performance were significantly improved for the exercise group, but not for the control group. Analyses of variance (ANOVA) of the DTI data demonstrated significantly increased fractional anisotropy (FA) in the right corticospinal tract and significantly decreased FA in the left superior fronto-occipital fasciculus in the exercise group after the intervention versus before. The enhanced aerobic fitness induced by exercise was associated with better inhibitory control performance in the incongruent condition and lower FA in the Left superior fronto-occipital fasciculus (SFOF). Regression analysis of a mediation model did not support Left SFOF FA as a mediator of the relationship between improvements in aerobic fitness and inhibitory control. The present data provide new evidence of the relationship between exercise-induced changes in aerobic fitness, WM integrity, and inhibitory control in early adulthood. Longer-duration intervention studies with larger study cohorts are needed to confirm and further explore the findings obtained in this study.

Keywords: diffusion tensor imaging; fractional anisotropy; inhibitory control; aerobic fitness; early adulthood

Citation: Zhu, H.; Zhu, L.; Xiong, X.; Dong, X.; Chen, D.; Wang, J.; Cai, K.; Wang, W.; Chen, A. Influence of Aerobic Fitness on White Matter Integrity and Inhibitory Control in Early Adulthood: A 9-Week Exercise Intervention. *Brain Sci.* **2021**, *11*, 1080. https://doi.org/10.3390/brainsci11081080

Academic Editors: Filipe Manuel Clemente, Ana Filipa Silva and Stéphane Perrey

Received: 2 June 2021
Accepted: 16 August 2021
Published: 18 August 2021

Publisher's Note: MDPI stays neutral with regard to jurisdictional claims in published maps and institutional affiliations.

1. Introduction

Extensive research involving subjects across various age groups has demonstrated that aerobic fitness can benefit several aspects of cognitive function [1–3], including attention [4], inhibitory control [5], memory [6], and executive function [7]. Inhibitory control, which is regarded as a core component of executive function, refers to one's ability to control his or her attention, behavior, thoughts, and emotions and to overcome his or her strong internal tendencies and external temptations so that he or she is able to initiate and stay engaged in appropriate or necessary tasks [8]. Poorly developed inhibitory control can compromise cognitive, emotional, and social functioning [8]. Thus, a better understanding of the relationship between aerobic fitness and inhibitory control will provide theoretical insights applicable to promoting cognitive development in young people and preventing cognitive decline in the elderly.

To elucidate the relationship between aerobic fitness and inhibitory control, it will be important to clarify how particular brain areas, processes, and inter-regional connections are affected by aerobic fitness and involved in inhibitory control. The right inferior frontal gyrus has emerged as a key region for mediating inhibitory control [9]. The relatively few studies in the literature examining the relationship between fitness and inhibitory control have focused on children or the elderly, and the findings have been ambiguous. A cross-sectional study employing white matter (WM) tractography based on diffusion tensor imaging (DTI) data demonstrated a positive correlation between fitness level and fractional anisotropy (FA) in a cohort of 9- and 10-year-old children [10]. A cross-sectional study involving elderly subjects related fitness to spatial working memory performance, and obtained data suggesting that FA may play a mediating role between aerobic fitness and cognition [11]. Conversely, Herting et al. found no relationship between aerobic fitness and FA in male adolescents [12]. These inconsistent results suggest that the relationship between aerobic fitness, WM integrity, and cognition may vary across different stages of life.

Although exercise can improve aerobic fitness [13], there is limited information regarding how exercise interventions that enhance aerobic fitness affect WM integrity and cognition. Notably, an intervention study involving 70 participants showed that changes in aerobic fitness induced by 1 year of fitness training were associated with FA changes in prefrontal and temporal cortices as well as with improved short-term memory performance [3]. A recent path analysis study showed that FA could play a mediating role between aerobic fitness and cognition [2]. Although those data were derived from cross-sectional research, the model provides a sound basis from which to examine WM changes associated with the effects of improved fitness on cognition. In general, cross-sectional studies focusing on the relationship between aerobic fitness and WM integrity have lacked cognitive measurements. Moreover, the lack of intervention studies precludes us from making causal inferences about the relationships observed in cross-sectional studies.

Adolescence to early adulthood is a unique developmental period characterized by immature brain processing and vulnerability to psychopathology emergence [14,15]. Although cognitive and WM development changes during this period have slowed relative to earlier developmental phases, ample plasticity remains [16,17]. To the best of our knowledge, the influence of aerobic fitness on WM integrity and behavioral inhibitory control in early adulthood has not been examined in an intervention study. The aim of this study was to examine the influence of aerobic fitness on WM integrity and inhibitory control in young adults, thus addressing this knowledge gap, and to attempt to explore the interaction between the three variables caused by exercise interventions. We hypothesized that after 9-week aerobic exercise intervention, enhanced aerobic fitness would be associated with better inhibitory control performance and alterations in WM integrity based on the supposition that WM integrity may be an important indirect pathway by which aerobic fitness affects inhibitory control. Empirical support for our hypothesis would support the view that aerobic fitness is an important life factor for healthy brain and intellectual development and support the use of DTI-derived FA data for illuminating neural mechanisms underlying exercise effects on brain function.

2. Materials and Methods

2.1. Participants

A cohort of 84 college freshmen (35 males and 49 females) were recruited from a university in cities of Yangzhou in Jiangsu Province, China. The inclusion criteria were age of 18–20 years old, right-handedness, and normal vision without color blindness. The exclusion criteria were drug abuse, any genetic disease, general intelligence problems, or any medical condition that limits physical activity or that could affect the research results. The study protocol was approved by the ethics and human protection committees of Yangzhou University Affiliated Hospital (2017-YKL045-01). All participants provided written informed consent after receiving a detailed explanation of the experimental procedure. All research procedures were in accordance with the latest version of the Declaration of Helsinki.

2.2. Aerobic Fitness Assessment

Peak aerobic oxygen uptake (VO_2peak) was used as an index of aerobic fitness level [18]. VO_2peak was assessed while participants exercised on a stress-test stationary bicycle EGT 1000 (ELMED, Zimmer Elektromedizin GmbH in Neu-Ulm, Germany). Before fitness testing, resting heart rate (HR) was determined and then each subject undertook a 3~5-min warm-up to prevent injury. Testing commenced when the subject had returned to his or her resting HR. The testing program started with a 50 W load (a starting load of 50 W was selected to prevent a risk to subjects due to an excessively high initial load) at a cycling speed of 55–60 rotations/min. Maintaining this speed, the load was increased by 50 W every 3 min until the subject reached exhaustion. The subject was considered to have achieved maximum VO_2 when one of the following criteria was met: (1) no compensatory increase in oxygen intake upon load increase; (2) respiratory quotient exceeded 1.1; (3) HR exceeded 180 beats/min; (4) a speed of 55–60 rotations/min could not be maintained despite repeated encouragement. After the test, the data were recorded. The subjects were instructed to rest for 5 min before being allowed to leave. During this period, a measure of perceived exertion was attained by using the participants' RPE scale of perceived exertion and they were observed to ensure that they were not having an abnormal reaction.

2.3. Exercise Intervention

Participants were randomized into exercise and control groups. Those in the exercise group agreed to take part in exercise training four times a week for 9 weeks. The daily exercise program consisted of a 10-min warm-up, 40 min of moderate-intensity endurance training, and a 10-min cool-down/relaxation period. The exercise program design and implementation were developed and monitored by two trained exercise coaches. We employed a moderate-intensity aerobic load (60–69% of the maximum HR, where maximum HR = 220 − age in years), as defined by the American College of Sports Medicine (2006) [19]. Exercise intensity was assessed with HR monitoring (BHT GOFIT; Beijing, China) the monitors were pre-tested on five subjects before the intervention to ensure the program was appropriate. A return visit was applied to the subjects in the control group. If a subject participated in physical exercise of moderate intensity or above during the intervention period (more than or equal to twice a week), this subject was not included in the final statistical analysis.

2.4. Inhibitory Control Testing

Inhibitory control was assessed with a modified Eriksen Flanker task [20]. Briefly, a series of English letters appeared on the screen under two conditions: congruent (FFFFF and LLLLL); and incongruent (LLFLL and FFLFF). Participants were asked to identify the middle letter as quickly as possible by pressing the F key or L key. The two conditions were equally represented and randomly presented. The formal test was composed of two blocks, and each block contained 48 trials, in which the duration of fixation was 500 ms, the duration of letter presentation was 1000 ms, and the stimulation interval was 2000 ms. There were 12 practice sessions before the formal test. A shorter response time (RT) represented better inhibitory control.

2.5. Imaging Acquisition

Images were acquired on a 3-T magnetic resonance imaging (MRI) scanner (GE Discovery MR750W) at the Affiliated Hospital of Yangzhou University. None of the subjects had participated in high-intensity physical exercise in the 48 hours prior to the scan. Participants were told to stay relaxed and move as little as possible during the scan. Diffusion images were acquired with an echo planar imaging sequence (acquisition matrix size = 112 × 112, 70 interleaved slices, voxel size = 2 × 2 × 2 mm^3, field of view 224 × 224 mm^2, repetition time = 16,500 ms, echo time = 96.2 ms, flip angle = 90°, three B0 images, 30 diffusion weighted images, and a b value of 1000 s/mm^2).

2.6. Image Analysis

PANDA (Pipeline for Analysing braiN Diffusion imAges) (http://www.nitrc.org/projects/panda/), which has automated processing flows, was used to analyze diffusion images [21], including data preprocessing and calculating mean DTI parameter values for whole WM fibers. The preprocessing parameters were: local diffusion homogeneity = 7 voxels; normalizing resolution = 2 mm; and smoothing kernel = 6 mm. The following specific data processes were applied: format conversion, mask generation, image clipping, eddy current and head motion correction, parameter calculation, spatial registration and Gaussian smoothing. Then, using the atlas-based analysis, we normalized FA data in Montreal Neurological Institute space and calculated regional DTI parameters by averaging values within each region of the ICBM DTI-81 atlas [22]. PANDA generates twelve Excel files after completing the automatic processing of data under the folder 'AllAtlasResults', which contains the regional average values for diffusion metrics images with a voxel size of 1 × 1 × 1 mm^3 in the standard space. The values in Excel were then copied to Statistical Package for the Social Sciences (SPSS) for statistics.

2.7. Experimental Procedure

A two-factor mixed experiment was carried out with a 2 (groups: exercise and control) × 2 (time: before and after intervention) design. Group was a between-subject factor, while time was a within-subject factor. Data were collected for three periods: pre-intervention test, during the exercise intervention, and post-intervention test. All subjects completed MRI scanning and cognitive testing before the exercise intervention in the pre-test period. One week after completing the pre-intervention test, the exercise group commenced the aforementioned 9-week exercise intervention, while the control group lived normally. After the intervention, all subjects underwent a second MRI scan and cognitive test.

2.8. Statistical Analyses

All statistical analysis was implemented in SPSS 22.0 (IBM, Armonk, NY, USA). Mean behavioral test values are reported with standard deviations (SDs). Independent sample t tests and χ^2 tests were used to compare demographic variables across the two groups. Normal distribution test was performed before analysis of covariance. ANOVAs was used to determine whether the two groups differed with respect to change in aerobic fitness before versus after the exercise intervention, with post hoc paired sample t-tests to detect intra-group differences. Repeated-measures analyses of variance (ANOVAs) were performed to analyze effects of the intervention on WM integrity and inhibitory control, with post hoc simple effect analyses. Pearson correlation coefficients (r values) were calculated among changes in aerobic fitness, FA, and inhibitory control. p values < 0.05 were considered significant. Causal step regression [23] was used to test whether the influence of aerobic fitness on inhibitory control was mediated by FA. Variables involved in correlation and regression processes were converted into z-scores.

3. Results

3.1. Participants' Characteristics

Of the 84 initially enrolled subjects, 59 completed the full study, including 24 freshmen in the control group and 35 freshmen in the exercise group. The demographic data of the exercise group and the control group are presented in Table 1. Due to the loss of data in the control group, there was a significant inter-group difference ($p < 0.05$) in aerobic fitness at baseline. Therefore, we treated aerobic fitness at baseline as a covariate to exclude the influence of baseline differences on the results.

Table 1. Participant demographics (M ± SD).

Variable	Control Group	Exercise Group	p
N	24	35	
male/female	9/15	15/20	0.68
age	18.5 ± 0.78	18.66 ± 0.48	0.39
BMI	22.46 ± 4.28	20.8 ± 2.22	0.09
aerobic fitness at baseline	24.07 ± 6.08	17.96 ± 4.41	0.00
HR during treatment		136.64 ± 7.77	

3.2. Aerobic Fitness

Treating group as an independent variable, change of aerobic fitness as a dependent variable, and aerobic fitness at baseline as a covariate, an analysis of covariance detected a significant difference in aerobic fitness change between the two groups ($F_{1,56} = 37.20$, $p < 0.05$, partial $\eta^2 = 0.571$). Paired sample *t*-tests showed that the exercise group showed a significant improvement in aerobic fitness from before (17.96 ± 4.41) to after (30.53 ± 6.45) the exercise intervention (t = −12.03, $p < 0.05$), whereas the control group did not (pre-intervention: 24.07 ± 6.08; post-intervention: 24.49 ± 6.57; t = −0.42, $p > 0.05$).

3.3. Inhibitory Control

Descriptive data for RT and accuracy are presented in Table 2. Two-way Repeated Measures ANOVA, with baseline aerobic fitness as a covariate, revealed a significant group × time interaction in RT in the incongruent condition of the inhibitory control task ($F_{1,56} = 25.63$, $p < 0.05$, partial $\eta^2 = 0.314$), but not in the congruent condition ($F_{1,56} = 0.193$, $p > 0.05$, partial $\eta^2 = 0.003$). Simple effect analysis revealed that RT was significantly reduced in the exercise group ($p < 0.05$) and significantly increased in the control group ($p < 0.05$) from the pre-intervention to the post-intervention test. We did not observe a significant group × time interaction in accuracy in the inhibitory control task in either the incongruent ($F_{1,56} = 0.121$, $p > 0.05$, partial $\eta^2 = 0.002$) or the congruent ($F_{1,56} = 0.817$, $p > 0.05$, partial $\eta^2 = 0.014$) condition.

Table 2. Mean RTs and accuracy rates (M ± SD) in the Eriksen Flanker task by group.

Variable	Control Group (N = 24)		Exercise Group (N = 35)	
	Pre-Intervention	Post-Intervention	Pre-Intervention	Post-Intervention
RT, ms				
Congruent	486.24 ± 50.11	451.78 ± 61.34	501.39 ± 99.28	473.78 ± 48.40
Incongruent *	520.85 ± 41.96	560.96 ± 63.15	547.49 ± 48.36	507.09 ± 38.53
Accuracy, %				
Congruent	95.57 ± 4.04	95.23 ± 5.05	95.77 ± 3.70	95.99 ± 3.15
Incongruent	93.84 ± 4.49	95.14 ± 5.20	94.32 ± 3.71	95.84 ± 3.23

* There is a significant group × time interaction ($p < 0.05$) under the incongruent condition.

3.4. WM Structure

Treating aerobic fitness at baseline as a covariate, ANOVAs of FA revealed significant group by time interactions for the right corticospinal tract (Right CST) ($F_{1,56} = 4.033$, $p < 0.05$, partial $\eta^2 = 0.067$) and left superior fronto-occipital fasciculus (Left SFOF) ($F_{1,56} = 5.682$, $p < 0.05$, partial $\eta^2 = 0.092$) (Figure 1). Simple effect analysis indicated that the FA of the right corticospinal tract presented an upward trend in the exercise group, but a downward trend in the control group. Meanwhile, FA of the Left SFOF showed a downward trend and an upward trend in the exercise group and control group, respectively. The specific FA values of the brain regions with significant interactions are listed in Table 3.

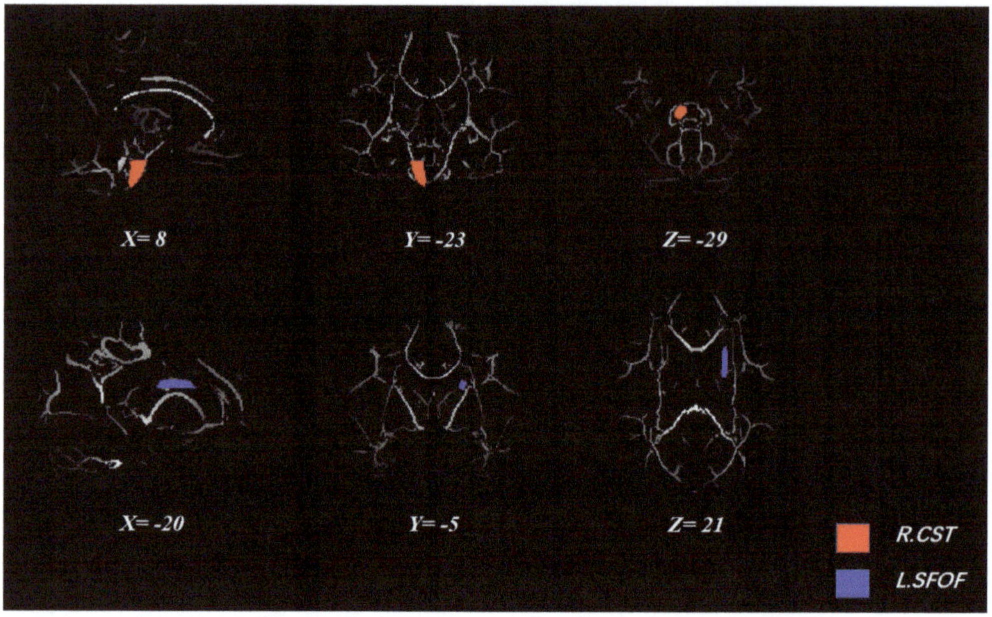

Figure 1. Brain regions with significant interaction effects on FA ($p < 0.05$). Abbreviations: R.CST, right corticospinal tract; L.SFOF, left superior fronto-occipital fasciculus.

Table 3. FA values of the brain regions with interaction (M ± SD).

Brain Region	Control Group ($N = 24$)		Exercise Group ($N = 35$)	
	Pre-Intervention	Post-Intervention	Pre-Intervention	Post-Intervention
Right CST	$0.561 \pm 3.20 \times 10^{-2}$	$0.551 \pm 3.07 \times 10^{-2}$	$0.557 \pm 3.69 \times 10^{-2}$	$0.562 \pm 3.65 \times 10^{-2}$
Left SFOF	$0.460 \pm 3.11 \times 10^{-2}$	$0.470 \pm 3.19 \times 10^{-2}$	$0.472 \pm 3.20 \times 10^{-2}$	$0.465 \pm 2.47 \times 10^{-2}$

Note: Right CST, right corticospinal tract; Left SFOF, left superior fronto-occipital fasciculus.

3.5. Changes in Aerobic Fitness, FA, and Inhibitory Control

Aerobic fitness correlated inversely with FA of the L.SFOF (r = −0.26, 95% CI (−0.473, 0.032), $p < 0.05$), with increases in aerobic fitness being accompanied by L.SFOF FA decreases, but did not correlate significantly with FA of the right corticospinal tract (r = 0.06, $p > 0.05$). Additionally, aerobic fitness correlated inversely with RT in the inhibitory control task (r = −0.45, 95% CI (−0.657, −0.213), $p < 0.05$), indicating that a higher aerobic fitness level was predictive of better inhibitory control performance. Regression analysis of the mediation model wherein aerobic fitness change and FA change of left superior fronto-occipital fasciculus were treated as independent variables, while inhibitory control change was treated as a dependent variable, yielded a non-significant ($p > 0.05$) regression coefficient (−0.024), thereby invalidating the FA mediation model.

4. Discussion

The present study demonstrated that a 9-week aerobic exercise intervention, which was confirmed to improve aerobic fitness, could induce changes in WM integrity and inhibitory control in young adults. Specifically, the exercise group exhibited improved inhibitory control together with increased FA in right corticospinal tract and decreased FA in left superior fronto-occipital fasciculus. The exercise-induced changes in aerobic fitness correlated with observed effects on inhibitory control and FA of the Left SFOF. However, regression analysis indicated that the altered FA of the Left SFOF, examined as a potential mediator, could not account for the relationship between aerobic fitness and inhibitory control.

4.1. Behavior

Three prior meta-analyses focusing on different age groups [24–26] showed that long-term participation in aerobic exercise improved aerobic fitness in both healthy participants and patients with mental illness [27] and cancer [28]. Our findings reinforce the notion that exercise interventions improve aerobic fitness, specifically during early adulthood.

The view that exercise promotes inhibitory control has been supported by behavioral, physiological, and imaging studies [29–31]. The exercise paradigm used in the present study yielded RT improvement selectively in the incongruent condition of the Eriksen Flanker task, without a concomitant significant improvement in accuracy. Prior studies in children and in the elderly showed RT improvements in both incongruent and congruent conditions. The more limited improvement in our study could be related to the ages of the subjects. That is, in early adulthood, which is not long after puberty, brain morphology and core cognitive abilities are more firmly established than in children, whereas elderly participants are more likely to have acquired some mild cognitive deficits that may benefit from exercise [32]. Furthermore, only the relatively more difficult of the conditions in the Eriksen Flanker task, which is a relatively simple task, may have been sufficiently sensitive to reflect a mild improvement in inhibitory control. In general, our behavioral data confirm that exercise can improve aerobic fitness and inhibitory control in early adulthood.

4.2. WM Integrity

WM, which is composed of glial cells and myelinated neurons, matures mostly before early adulthood, and myelin formation contributes to efficient neurotransmission throughout the brain, which in turn improves cognitive function [12,33]. The most commonly examined index of WM integrity is FA, with larger FA values representing greater diffusion of water molecules in the axial direction than that in the radial direction within analyzed WM tracts. Overweight children who participated in an exercise intervention were found to have a significant increase in bilateral uncinate fasciculus FA after completing an 8-month exercise program [34]. A recent study showed that a 12-week badminton intervention increased FA in the posterior limb of the internal capsule and upper corona radiate in participants in their mid-twenties [35]. The present study showed distinct sites of WM changes relative to the sites of WM changes observed in these prior studies. Notwithstanding, the increased FA of the right corticospinal tract observed in our study is consistent with the findings of a prior cross-sectional study [12]. However, our finding of reduced FA of the left superior fronto-occipital fasciculus after the exercise intervention was unexpected. It might be related to the age of our subjects (18 to 20 years old). FA in tracts of the frontal lobe peaks in late adolescence and 50% of corticospinal tract voxels reach their FA peak at about 20 years old, followed by downward trends [36]. Thus, because the FA of the left superior fronto-occipital fasciculus in our subjects would be at or near peak developmental values, it may difficult to augment with an intervention. Alternatively, a meta-analysis suggested that open-skill exercises may be more effective than closed-skill exercises for mobilizing cognitive load and promoting inhibitory control improvement [37]. Badminton is an open-skill exercise wherein one needs to respond quickly to a dynamic and unpredictable environment. Here, we employed a closed-skill exercise paradigm that was relatively consistent and well controlled. Hence, WM integrity may be differentially influenced under different exercise intervention treatments. The presently employed 9-week exercise intervention reshaped the WM integrity in young adults, an age group for which data were lacking. Longer-term studies are needed to clarify how exercise and fitness improvement affect WM.

4.3. The Relationship between Improved Aerobic Fitness and Inhibitory Control Cannot Be Explained by FA Alterations

Pair-wise correlation analysis demonstrated a significant correlation between increased aerobic fitness and decreased RT in the Eriksen Flanker task. This finding fits with prior research showing that individuals with high aerobic fitness perform better than less fit individuals on various cognitive tasks. With respect to mechanism, authors of prior studies have suggested that this relationship may be related to effects on brain functional connectivity [38], prefrontal cortex thickness [39], and event-related potentials [5]. The present work adds evidence of WM integrity changes to the potential mechanisms that may contribute to aerobic fitness effects on cognitive performance.

Both FA of the right corticospinal tract and FA of the left superior fronto-occipital fasciculus were found to have significant interaction effects in this study, with FA of the left superior fronto-occipital fasciculus having a significant negative correlation with aerobic fitness, while FA of the right corticospinal tract had only a non-significant trend toward a positive correlation with aerobic fitness. By contrast, previous cross-sectional studies have shown significant positive correlations between aerobic fitness and FA in adolescents with a mean age of 14.3 years ([40] and in (older) young adults with a mean age of 28.8 years [2]). The present study provides data for an intermediate group (mean age, 18.6 years), thus supplementing the literature. The relationship between the changes in aerobic fitness and changes in FA induced by an exercise intervention has not been examined in very early adulthood previously. Meanwhile, our novel finding of a negative correlation between aerobic fitness and FA of the left superior fronto-occipital fasciculus might be explained by a reduction of FA in association with gaining fluency with an automated movement [41–44]. Furthermore, we found no correlation between changes in FA and RT reduction in the Eriksen Flanker task. This negative finding may be due to our relatively small sample size or short intervention duration. Recent studies have been inconsistent, with one study showing a positive correlation between FA and inhibitory control [45], and another showing that higher FA was associated with worse inhibitory control performance [46]. Future studies with more rigorous experimental designs are needed to clarify whether there is specificity regarding exercise interventions that lead to associations between FA and aerobic fitness.

Our supposition that FA of the left superior fronto-occipital fasciculus may play a mediating role in the influence of aerobic fitness on inhibitory control was not confirmed, which may be due to the short duration of intervention or the specific characteristics of the structural development phase of the brain in early adulthood. Early adulthood is a highly plastic cognitive developmental period when people transition from secondary school to social and economic independence, and thus is of particular interest for educational decision-makers and researchers. Notably, WM developmental properties during this phase of life are distinct from those during childhood, when there is more rapid WM development, and also distinct from those during old age, when there are declines in WM integrity [47,48].

4.4. Strengths and Limitations

A major advantage of our study is that an exercise intervention was employed as a breakthrough point to explore the relationship between increased aerobic fitness, changes in WM integrity, and inhibitory control improvement in early adulthood, extending the results of previous cross-sectional studies. The conclusions of this study should be considered in light of several limitations. First, aerobic fitness at baseline differed between the exercise and control groups due to loss of subjects. Although aerobic fitness at baseline was treated as a statistical covariate, it still could have impacted the accuracy of the results. In addition, the loss of subjects leads to the reduction of sample size, which can be avoided as far as possible in future studies to improve the reliability of the study. Second, there are additional indicators of WM integrity that were not analyzed in our study, including mean diffusivity, radial diffusivity, and axial diffusivity. Further exploration of these indicators could further clarify the relationships among aerobic fitness, WM integrity, and cognition. Third, prior studies involving children [49] and elderly participants [3] adopted longer term interventions (8 months and 12 months, respectively) than that used in this study. It is possible that the 9-week intervention examined here was not sufficiently long in duration to reveal a significant positive relationship between aerobic fitness and FA of the right corticospinal tract. Therefore, the effects of higher exercise doses (duration, intensity, and frequency) should be examined in the future. Finally, we only compared data across two time points (pre-intervention and post-intervention). Because exercise-induced FA changes may be nonlinear, future research should collect data at multiple time points during the intervention process.

5. Conclusions

The present results showed that a 9-week exercise intervention that improves aerobic fitness resulted in an improvement in inhibitory control performance and changes in WM integrity in very young adults. Enhanced aerobic fitness correlated positively with an inhibitory control performance parameter and correlated negatively with FA of the left superior fronto-occipital fasciculus. However, a mediator role of altered FA in the relationship between improved aerobic fitness and improved inhibitory control could not be demonstrated. This work provides a reference for future research exploring the impact of aerobic fitness on cognition from the perspective of WM integrity.

Author Contributions: Conceptualization, H.Z. and L.Z.; methodology, H.Z. and K.C.; software, X.X. and K.C.; validation, W.W., J.W. and D.C.; formal analysis, H.Z. and K.C.; investigation, X.D.; resources, X.X. and X.D.; data curation, H.Z. and L.Z.; writing—original draft preparation, H.Z.; writing—review and editing, A.C.; visualization, H.Z.; supervision, A.C.; project administration, A.C.; funding acquisition, A.C. All authors have read and agreed to the published version of the manuscript.

Funding: This research was supported by grants from the National Natural Science Foundation of China (31771243) to Aiguo Chen.

Institutional Review Board Statement: The study protocol was approved by the ethics and human protection committees of Yangzhou University Affiliated Hospital (2017-YKL045-01). All participants provided written consent after receiving a de-tailed explanation of the experimental procedure. All research procedures were in line with the latest version of the Declaration of Helsinki.

Conflicts of Interest: The authors declare no conflict of interest.

References

1. Hillman, C.H.; Buck, S.M.; Themanson, J.R.; Pontifex, M.B.; Castelli, D.M. Aerobic fitness and cognitive development: Event-related brain potential and task performance indices of executive control in preadolescent children. *Dev. Psychol.* **2009**, *45*, 114–129. [CrossRef]
2. Opel, N.; Martin, S.; Meinert, S.; Redlich, R.; Enneking, V.; Richter, M.; Goltermann, J.; Johnen, A.; Dannlowski, U.; Repple, J. White matter microstructure mediates the association between physical fitness and cognition in healthy, young adults. *Sci. Rep.* **2019**, *9*, 12885. [CrossRef]
3. Voss, M.W.; Heo, S.; Prakash, R.S.; Erickson, K.I.; Alves, H.; Chaddock, L.; Szabo, A.N.; Mailey, E.L.; Wójcicki, T.R.; White, S.M.; et al. The influence of aerobic fitness on cerebral white matter integrity and cognitive function in older adults: Results of a one-year exercise intervention. *Hum. Brain Mapp.* **2013**, *34*, 2972–2985. [CrossRef]
4. Erickson, K.I.; Hillman, C.H.; Kramer, A.F. Physical activity, brain, and cognition. *Curr. Opin. Behav. Sci.* **2015**, *4*, 27–32. [CrossRef]
5. Stroth, S.; Kubesch, S.; Dieterle, K.; Ruchsow, M.; Heim, R.; Kiefer, M. Physical fitness, but not acute exercise modulates event-related potential indices for executive control in healthy adolescents. *Brain Res.* **2009**, *1269*, 114–124. [CrossRef]
6. Erickson, K.I.; Voss, M.W.; Prakash, R.S.; Basak, C.; Szabo, A.; Chaddock, L.; Kim, J.S.; Heo, S.; Alves, H.; White, S.M.; et al. Exercise training increases size of hippocampus and improves memory. *Proc. Natl. Acad. Sci. USA* **2011**, *108*, 3017–3022. [CrossRef]
7. Hillman, C.H.; Erickson, K.I.; Kramer, A.F. Be smart, exercise your heart: Exercise effects on brain and cognition. *Nat. Rev. Neurosci.* **2008**, *9*, 58–65. [CrossRef] [PubMed]

8. Diamond, A. Executive functions. *Annu. Rev. Psychol.* **2013**, *64*, 135–168. [CrossRef]
9. Hampshire, A.; Chamberlain, S.R.; Monti, M.M.; Duncan, J.; Owen, A.M. The role of the right inferior frontal gyrus: Inhibition and attentional control. *Neuroimage* **2010**, *50*, 1313–1319. [CrossRef]
10. Chaddock, L.; Erickson, K.I.; Prakash, R.S.; VanPatter, M.; Voss, M.W.; Pontifex, M.B.; Raine, L.B.; Hillman, C.H.; Kramer, A.F. Basal ganglia volume is associated with aerobic fitness in preadolescent children. *Dev. Neurosci.* **2010**, *32*, 249–256. [CrossRef]
11. Oberlin, L.E.; Verstynen, T.D.; Burzynska, A.Z.; Voss, M.W.; Prakash, R.S.; Chaddock-Heyman, L.; Wong, C.; Fanning, J.; Awick, E.; Gothe, N.; et al. White matter microstructure mediates the relationship between cardiorespiratory fitness and spatial working memory in older adults. *Neuroimage* **2016**, *131*, 91–101. [CrossRef]
12. Herting, M.M.; Colby, J.B.; Sowell, E.R.; Nagel, B.J. White matter connectivity and aerobic fitness in male adolescents. *Dev. Cogn. Neurosci.* **2014**, *7*, 65–75. [CrossRef]
13. Caspersen, C.J.; Powell, K.E.; Christenson, G.M. Physical activity, exercise, and physical fitness: Definitions and distinctions for health-related research. *Public Health Rep.* **1985**, *100*, 126–131. [CrossRef]
14. Chau, D.T.; Roth, R.M.; Green, A.I. The neural circuitry of reward and its relevance to psychiatric disorders. *Curr. Psychiatry Rep.* **2004**, *6*, 391–399. [CrossRef] [PubMed]
15. Paus, T.; Keshavan, M.; Giedd, J.N. Why do many psychiatric disorders emerge during adolescence? *Nat. Rev. Neurosci.* **2008**, *9*, 947–957. [CrossRef] [PubMed]
16. Luna, B.; Garver, K.E.; Urban, T.A.; Lazar, N.A.; Sweeney, J.A. Maturation of cognitive processes from late childhood to adulthood. *Child Dev.* **2004**, *75*, 1357–1372. [CrossRef]
17. Asato, M.R.; Terwilliger, R.; Woo, J.; Luna, B. White matter development in adolescence: A DTI study. *Cereb. Cortex* **2010**, *20*, 2122–2131. [CrossRef] [PubMed]
18. Armstrong, N.; Tomkinson, G.; Ekelund, U. Aerobic fitness and its relationship to sport, exercise training and habitual physical activity during youth. *Br. J. Sports Med.* **2011**, *45*, 849–858. [CrossRef]
19. Pescatello, L.S. *ACSM's Guidelines for Exercise Testing and Prescription*, 9th ed.; Wolters Kluwer/Lippincott Williams & Wilkins Health: Philadelphia, PA, USA, 2014; ISBN 9781609136055.
20. Chen, A.-G.; Yan, J.; Yin, H.-C.; Pan, C.-Y.; Chang, Y.-K. Effects of acute aerobic exercise on multiple aspects of executive function in preadolescent children. *Psychol. Sport Exerc.* **2014**, *15*, 627–636. [CrossRef]
21. Cui, Z.; Zhong, S.; Xu, P.; He, Y.; Gong, G. PANDA: A pipeline toolbox for analyzing brain diffusion images. *Front. Hum. Neurosci.* **2013**, *7*, 42. [CrossRef]
22. Mori, S.; Oishi, K.; Jiang, H.; Jiang, L.; Li, X.; Akhter, K.; Hua, K.; Faria, A.V.; Mahmood, A.; Woods, R.; et al. Stereotaxic white matter atlas based on diffusion tensor imaging in an ICBM template. *Neuroimage* **2008**, *40*, 570–582. [CrossRef]
23. Baron, R.M.; Kenny, D.A. The moderator–mediator variable distinction in social psychological research: Conceptual, strategic, and statistical considerations. *J. Personal. Soc. Psychol.* **1986**, *51*, 1173–1182. [CrossRef]
24. García-Hermoso, A.; Alonso-Martinez, A.M.; Ramírez-Vélez, R.; Izquierdo, M. Effects of exercise intervention on health-related physical fitness and blood pressure in preschool children: A systematic review and meta-analysis of randomized controlled trials. *Sports Med.* **2020**, *50*, 187–203. [CrossRef] [PubMed]
25. De Carvalho Souza Vieira, M.; Boing, L.; Leitão, A.E.; Vieira, G.; Coutinho de Azevedo Guimarães, A. Effect of physical exercise on the cardiorespiratory fitness of men-A systematic review and meta-analysis. *Maturitas* **2018**, *115*, 23–30. [CrossRef]
26. Huang, G.; Gibson, C.A.; Tran, Z.V.; Osness, W.H. Controlled endurance exercise training and VO2max changes in older adults: A meta-analysis. *Prev. Cardiol.* **2005**, *8*, 217–225. [CrossRef] [PubMed]
27. Krogh, J.; Speyer, H.; Nørgaard, H.C.B.; Moltke, A.; Nordentoft, M. Can exercise increase fitness and reduce weight in patients with schizophrenia and depression? *Front. Psychiatry* **2014**, *5*, 89. [CrossRef] [PubMed]
28. Sweegers, M.G.; Altenburg, T.M.; Brug, J.; May, A.M.; van Vulpen, J.K.; Aaronson, N.K.; Arbane, G.; Bohus, M.; Courneya, K.S.; Daley, A.J.; et al. Effects and moderators of exercise on muscle strength, muscle function and aerobic fitness in patients with cancer: A meta-analysis of individual patient data. *Br. J. Sports Med.* **2019**, *53*, 812. [CrossRef]
29. Chang, Y.-K.; Chen, F.-T.; Kuan, G.; Wei, G.-X.; Chu, C.-H.; Yan, J.; Chen, A.-G.; Hung, T.-M. Effects of acute exercise duration on the inhibition aspect of executive function in late middle-aged adults. *Front. Aging Neurosci.* **2019**, *11*, 227. [CrossRef] [PubMed]
30. Hwang, J.; Kim, K.; Brothers, R.M.; Castelli, D.M.; Gonzalez-Lima, F. Association between aerobic fitness and cerebrovascular function with neurocognitive functions in healthy, young adults. *Exp. Brain Res.* **2018**, *236*, 1421–1430. [CrossRef]
31. Hillman, C.H.; Pontifex, M.B.; Castelli, D.M.; Khan, N.A.; Raine, L.B.; Scudder, M.R.; Drollette, E.S.; Moore, R.D.; Wu, C.-T.; Kamijo, K. Effects of the FITKids randomized controlled trial on executive control and brain function. *Pediatrics* **2014**, *134*, e1063–e1071. [CrossRef] [PubMed]
32. Huttenlocher, P.R. Morphometric study of human cerebral cortex development. *Neuropsychologia* **1990**, *28*, 517–527. [CrossRef]
33. Casey, B.J.; Jones, R.M.; Hare, T.A. The adolescent brain. *Ann. N. Y. Acad. Sci.* **2008**, *1124*, 111–126. [CrossRef] [PubMed]
34. Schaeffer, D.J.; Krafft, C.E.; Schwarz, N.F.; Chi, L.; Rodrigue, A.L.; Pierce, J.E.; Allison, J.D.; Yanasak, N.E.; Liu, T.; Davis, C.L.; et al. An 8-month exercise intervention alters frontotemporal white matter integrity in overweight children. *Psychophysiology* **2014**, *51*, 728–733. [CrossRef] [PubMed]
35. Bai, X.; Shao, M.; Liu, T.; Yin, J.; Jin, H. Altered structural plasticity in early adulthood after badminton training. *Acta Psychol. Sin.* **2020**, *52*, 173–183. [CrossRef]

36. Westlye, L.T.; Walhovd, K.B.; Dale, A.M.; Bjørnerud, A.; Due-Tønnessen, P.; Engvig, A.; Grydeland, H.; Tamnes, C.K.; Ostby, Y.; Fjell, A.M. Life-span changes of the human brain white matter: Diffusion tensor imaging (DTI) and volumetry. *Cereb. Cortex* **2010**, *20*, 2055–2068. [CrossRef]
37. Zhu, H.; Chen, A.; Guo, W.; Zhu, F.; Wang, B. Which type of exercise is more beneficial for cognitive function? A meta-analysis of the effects of open-skill exercise versus closed-skill exercise among children, adults, and elderly populations. *Appl. Sci.* **2020**, *10*, 2737. [CrossRef]
38. Talukdar, T.; Nikolaidis, A.; Zwilling, C.E.; Paul, E.J.; Hillman, C.H.; Cohen, N.J.; Kramer, A.F.; Barbey, A.K. Aerobic fitness explains individual differences in the functional brain connectome of healthy young adults. *Cereb. Cortex* **2018**, *28*, 3600–3609. [CrossRef]
39. Weinstein, A.M.; Voss, M.W.; Prakash, R.S.; Chaddock, L.; Szabo, A.; White, S.M.; Wojcicki, T.R.; Mailey, E.; McAuley, E.; Kramer, A.F.; et al. The association between aerobic fitness and executive function is mediated by prefrontal cortex volume. *Brain Behav. Immun.* **2012**, *26*, 811–819. [CrossRef]
40. Ruotsalainen, I.; Gorbach, T.; Perkola, J.; Renvall, V.; Syväoja, H.J.; Tammelin, T.H.; Karvanen, J.; Parviainen, T. Physical activity, aerobic fitness, and brain white matter: Their role for executive functions in adolescence. *Dev. Cogn. Neurosci.* **2020**, *42*, 100765. [CrossRef]
41. Schmithorst, V.J.; Wilke, M. Differences in white matter architecture between musicians and non-musicians: A diffusion tensor imaging study. *Neurosci. Lett.* **2002**, *321*, 57–60. [CrossRef]
42. Jäncke, L.; Koeneke, S.; Hoppe, A.; Rominger, C.; Hänggi, J. The architecture of the golfer's brain. *PLoS ONE* **2009**, *4*, e4785. [CrossRef]
43. Piervincenzi, C.; Ben-Soussan, T.D.; Mauro, F.; Mallio, C.A.; Errante, Y.; Quattrocchi, C.C.; Carducci, F. White matter microstructural changes following quadrato motor training: A longitudinal study. *Front. Hum. Neurosci.* **2017**, *11*, 590. [CrossRef] [PubMed]
44. Mayinger, M.C.; Merchant-Borna, K.; Hufschmidt, J.; Muehlmann, M.; Weir, I.R.; Rauchmann, B.-S.; Shenton, M.E.; Koerte, I.K.; Bazarian, J.J. White matter alterations in college football players: A longitudinal diffusion tensor imaging study. *Brain Imaging Behav.* **2018**, *12*, 44–53. [CrossRef]
45. Seghete, K.L.M.; Herting, M.M.; Nagel, B.J. White matter microstructure correlates of inhibition and task-switching in adolescents. *Brain Res.* **2013**, *1527*, 15–28. [CrossRef] [PubMed]
46. Treit, S.; Chen, Z.; Rasmussen, C.; Beaulieu, C. White matter correlates of cognitive inhibition during development: A diffusion tensor imaging study. *Neuroscience* **2014**, *276*, 87–97. [CrossRef] [PubMed]
47. Johnson, M.H. Development of human brain functions. *Biol. Psychiatry* **2003**, *54*, 1312–1316. [CrossRef]
48. Stillman, C.M.; Erickson, K.I. Physical activity as a model for health neuroscience. *Ann. N. Y. Acad. Sci.* **2018**, *1428*, 103–111. [CrossRef]
49. Krafft, C.E.; Schaeffer, D.J.; Schwarz, N.F.; Chi, L.; Weinberger, A.L.; Pierce, J.E.; Rodrigue, A.L.; Allison, J.D.; Yanasak, N.E.; Liu, T.; et al. Improved frontoparietal white matter integrity in overweight children is associated with attendance at an after-school exercise program. *Dev. Neurosci.* **2014**, *36*, 1–9. [CrossRef]

Article

Pupillometry Reveals the Role of Arousal in a Postexercise Benefit to Executive Function

Naila Ayala [1,2,3,*] and Matthew Heath [1,2]

1 Department of Kinesiology, School of Kinesiology, University of Western Ontario, London, ON N6G 3K7, Canada; mheath2@uwo.ca
2 Graduate Program in Neuroscience, University of Western Ontario, London, ON N6G 3K7, Canada
3 Department of Kinesiology, University of Waterloo, Waterloo, ON N2L 3G1, Canada
* Correspondence: nayala@uwaterloo.ca

Abstract: A single bout of aerobic exercise improves executive function; however, the mechanism(s) underlying this improvement remains unclear. Here, we employed a 20-min bout of aerobic exercise, and at pre- and immediate post-exercise sessions examined executive function via pro- (i.e., saccade to veridical target location) and anti-saccade (i.e., saccade mirror symmetrical to a target) performance and pupillometry metrics. Notably, tonic and phasic pupillometry responses in oculomotor control provided a framework to determine the degree that arousal and/or executive resource recruitment influence behavior. Results demonstrated a pre- to post-exercise decrease in pro- and anti-saccade reaction times ($p = 0.01$) concurrent with a decrease and increase in tonic baseline pupil size and task-evoked pupil dilations, respectively (ps < 0.03). Such results demonstrate that an exercise-induced improvement in saccade performance is related to an executive-mediated "shift" in physiological and/or psychological arousal, supported by the locus coeruleus norepinephrine system to optimize task engagement.

Keywords: antisaccade; cognition; exercise; oculomotor; task-evoked pupil dilation; vision

Citation: Ayala, N.; Heath, M. Pupillometry Reveals the Role of Arousal in a Postexercise Benefit to Executive Function. *Brain Sci.* **2021**, *11*, 1048. https://doi.org/10.3390/brainsci11081048

Academic Editors: Filipe Manuel Clemente and Ana Filipa Silva

Received: 6 July 2021
Accepted: 5 August 2021
Published: 7 August 2021

Publisher's Note: MDPI stays neutral with regard to jurisdictional claims in published maps and institutional affiliations.

1. Introduction

Top-down executive control includes response suppression, working memory, and cognitive flexibility, and each component is essential for daily living [1]. A single bout of aerobic and/or resistance training improves executive function [2,3] and is a benefit that persists for up to 60 min [4]. A prominent mechanism associated with the benefit is an exercise-mediated change in arousal [5]. Notably, arousal is a multidimensional construct (i.e., physiological, cognitive/psychological, and affective components) [5], and little research has examined the effect of exercise on the distinct factors clustered within the term. A crucial component of cognitive/psychological arousal is the locus coeruleus norepinephrine (LC-NE) system, which is a collection of noradrenergic neurons within the brainstem that have an essential role in modulating the neural system's level of alertness and the brain's attentional state [6]. Therefore, the present investigation sought to determine whether a single bout of aerobic exercise influences LC-NE activity and how a putative change may influence a post exercise benefit to executive function.

The LC-NE system influences sensory processing and cognition through the regulation of attention [7]. Accordingly, behavioral and neuropsychological measures that reflect the LC-NE system provide a framework to understand the mechanism(s) supporting the post exercise improvement in executive function. The present work used concurrent behavioral and pupillometry metrics of pro- (i.e., saccade to veridical target location) and anti-saccades (i.e., saccade mirror symmetrical to a target stimulus) to examine pre- and immediate post-exercise executive function. The basis for this was three-fold. First, behavioral and electrophysiological studies have shown that antisaccades are mediated via an extensive frontoparietal network [8] that shows task-dependent changes in activity following single and chronic bouts of exercise [9,10]. Second, antisaccades provide the necessary resolution to detect subtle post exercise improvements in executive function (i.e., decreased reaction times (RT)) across a range of exercise intensities and durations in healthy young and older adults, as well as those at risk for cognitive decline [11,12]. Third, the pupil size changes observed in the antisaccade task provide a proxy for task-dependent changes in executive function (i.e., inhibitory control) and cognitive/psychological arousal [7,13,14]. Indeed, the locus coeruleus (LC) receives direct—and indirect—input from the prefrontal cortex and the insula, which in turn influence the efferent gain throughout the cortical and subcortical regions, serving sensorimotor and cognitive processing [15,16]. These projections influence a biphasic pupil response that are indicative of arousal and executive resource recruitment: a tonic (baseline pupil size) and a phasic (task-evoked pupil dilation: TEPD) response [7,17]. During tasks that require a focusing of attention, neurons in the LC exhibit moderate levels of tonic activation, which enable phasic bursts of activity to occur that support the execution of a response to task-related events [7]. Arousal modulates baseline pupil size in a manner resembling the classic Yerkes–Dodson inverted-U relationship, and this influences TEPD and cognitive performance in a stereotypical manner [17]. Previous work has demonstrated that increasing tonic activation decreases phasic bursts, increases distractibility, and decreases performance (i.e., an increase in RT and/or response errors), whereas suppressing tonic activation to moderate levels increases phasic bursts, reduces distractibility, and improves performance (i.e., a decrease in RT and/or response errors) [7,17]. Consequently, if tonic activation falls below this moderate level, then task-relevant information is not processed and reflects a low level of alertness in the neural system [6,7].

Here, we examined antisaccade performance and pupillometry metrics prior to and following a 20-min single-bout of aerobic activity via a cycle ergometer at 80% of the participants' predicted maximum heart rate (HR_{max}: 220 minus age in years). In terms of research predictions, if exercise enhances LC-NE system attentional modulation via a reduction in cognitive/psychological arousal, then decreased post exercise antisaccade RTs should be paired with a pre- to post-exercise decrease in baseline pupil size (i.e., suppression of tonic activity to moderate levels) and a concomitant increase in TEPD (i.e., increased task-evoked phasic bursts). Such a pattern of results would evince improved executive control via a narrowing of selective attention and serve to optimize the processing of task-relevant information. In contrast, if a post exercise improvement in antisaccade planning is executive-specific and not related to a modulation of attentional control via cognitive/psychological arousal, then decreased post exercise antisaccade RTs should be paired with a post exercise increase in antisaccade TEPDs and no change in baseline pupil size.

2. Materials and Methods

2.1. Participants

Sixteen (eight females in the age range of 20–26 years) members of the University of Western Ontario community participated in this study. All were self-declared right-hand dominant, had normal or corrected-to-normal vision, and no current or previous history of neuropsychiatric or neurological impairment. Participants obtained a full score on the Physical Activity Readiness Questionnaire (PAR-Q) and were "recreationally active" as determined by the Godin Leisure-Time Exercise Questionnaire (GLTEQ; mean = 63, SD = 16, min = 32, and max = 88). Participants refrained from caffeine and tobacco use 8 h prior to participation. Participants signed a consent form approved by the Health Sciences Research Ethics Board, University of Western Ontario, and this research was conducted according to the Declaration of Helsinki.

2.2. Exercise Intervention

The exercise intervention involved a 20-min bout of aerobic exercise via a cycle ergometer (Monark 818E Ergometer, Monark Exercise AB, Vonsbro, Sweden) at 80% of the participants' HR_{max}. Prior to and after the intervention, a 2.5 min warm-up and cool down, respectively, were performed at 50% of HR_{max} (Heath and Shukla, 2020). Heart rate was continuously monitored during the intervention (Polar Wearlink and Coded Transmitter, Polar Electro Inc., Lack Success, NY, USA), and the experimenter or participant adjusted ergometer resistance to maintain a work rate in the prescribed intensity.

2.3. Oculomotor Task

Prior to and following the exercise session, participants sat on a height adjustable chair in front of a table with their head placed in a head−chin rest. A 30-in LCD monitor (60 Hz, 8 ms response rate, 1280 × 960 pixels; Dell 3007WFP, Round Rock, TX, USA) was located at the participants' midline and 550 mm from the front edge of the tabletop and was used to present visual stimuli. Gaze position and pupil size of the left eye were sampled at 1000 Hz (EyeLink 1000 Plus; SR Research Ltd., Ottawa, ON, Canada). Stimulus presentation and data acquisition were controlled via MATLAB (R2018b, TheMathWorks, Natick, MA, USA) and the Psychophysics Toolbox extensions (v. 3.0) [18], including the EyeLink Toolbox [19]. Prior to data collection, a nine-point calibration was performed and followed by a validation (i.e., $<1°$ of error).

Visual stimuli were presented on a high-contrast black background ($1\ cd/m^2$) and included a centrally presented red or green fixation cross ($1°$). The color of the fixation was equiluminant ($42\ cd/m^2$) and instructed the nature of the required response (i.e., prosaccade = green and antisaccade = red). Open white circles served as targets ($2.7°$ diameter: $132\ cd/m^2$) and were $13.5°$ (i.e., proximal) and $16.5°$ (i.e., distal) left and right of the fixation, respectively, and in the same horizontal axis. The different eccentricities were used to prevent participants from adopting stereotyped responses. A trial began with the appearance of the fixation for 1000 ms, after which it was extinguished and a target appeared 200 ms thereafter (i.e., gap paradigm). Targets were presented for 50 ms and this brief presentation—in part –served to equate pro- and anti-saccades for the absence of extraretinal feedback [20]. Target onset cued participants to pro- (i.e., saccade to veridical target location) or anti-saccade (i.e., saccade mirror symmetrical to target location) "quickly and accurately". Pro- and anti-saccades, as well as stimulus location (i.e., left and right of fixation at proximal and distal eccentricities), in each assessment were pseudorandomized within a block of 80 trials. The intertrial interval was set to 2.5 s to ensure the pupil diameter returned to baseline prior to the next trial [21].

Following the pre exercise oculomotor assessment, participants immediately commenced the exercise intervention, whereas the post exercise assessment began when participants' heart rates were less than 100 beats per minute (i.e., <5 min following the cool-down). Each oculomotor assessment required less than 10 min to complete.

2.4. Data Reduction, Dependent Variables and Statistical Analysis

Gaze position data were filtered offline via a dual-pass Butterworth filter employing a low-pass cut-off frequency of 15 Hz. Filtered displacement data were used to calculate the instantaneous velocities via a five-point central-finite difference algorithm. Acceleration data were similarly obtained from the velocity. Saccade onset was determined when velocity and acceleration exceeded $30°/s$ and $8000°/s$, respectively. Saccade offset was marked by a velocity of less than $30°/s$ for 42 consecutive frames (i.e., 42 ms). Trials with missing data (i.e., loss of signal $>25\%$ of fixation period), RT less than 85 ms, and/or an amplitude less than $2°$ or greater than $26°$ were excluded from the data analysis ($<10\%$ of trials).

Pupil data were filtered offline via a 10 Hz low-pass filter. Trials missing more than 40% of data or an eye position deviation more than $2°$ from the fixation during the initial fixation period (i.e., 0–1200 ms after fixation cross onset) were excluded from the analyses. A blink correction algorithm involving linear interpolation beginning 50 ms before the blink and ending 150 ms after the blink was used to avoid task-uncorrelated high-frequency changes in pupil size [22]. A pupil size greater than 2.5 standard deviations from a participant's mean were also removed (<15% of trials). Notably, for all participants, at least 76% of trials were available for the statistical analyses. At least 20 trials remained for each condition from each participant. Because video-based tracking systems can distort pupil size following changes in gaze location, this measure was restricted to epochs involving central fixation and prior to saccade initiation (i.e., when gaze was located at the center of the screen). In line with previous work, [14,23,24] pupil size was determined in three epochs prior to saccade initiation (i.e., when gaze was located at the center of the screen): (1) the start of the visual fixation (FIX$_{st}$; 100–300 ms after fixation onset), (2) maximal pupil constriction (CON$_{max}$; 650–750 ms after fixation onset), and (3) end of gap (GAP$_{end}$; 150–200 ms following gap onset; Figure 1).

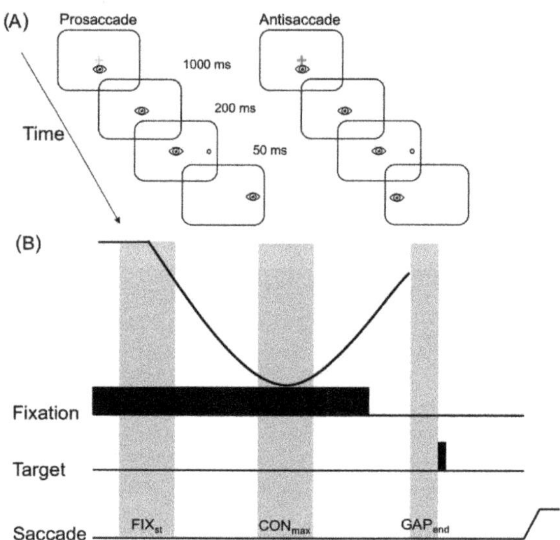

Figure 1. Panel A shows the timeline of visual and motor events for pro- and anti-saccades. A single target eccentricity is depicted to the right of the fixation; however, in the current study, two eccentricities were employed and the targets are presented left and right of the fixation. Panel B presents the epochs for the pupil analysis: fixation start (FIX$_{st}$; 100–300 ms after fixation onset), maximal pupil constriction (CON$_{max}$; 650–750 ms after fixation onset), and gap end (GAP$_{end}$; 150–200 ms after gap onset). The solid black line in Panel B depicts the time course change in the absolute pupil diameter.

Dependent variables included the reaction time (RT; time from response cueing to saccade onset), saccade duration (time from saccade onset to saccade offset), percentage of directional errors (i.e., the percentage of trials involving a prosaccade instead of instructed antisaccade and vice versa), baseline pupil diameter (average pupil diameter during FIX$_{st}$), and task evoked pupil dilation (TEPD; GAP$_{end}$ minus CON$_{max}$). Dependent variables were analyzed via two (assessment: pre- and post-exercise) by two (task: pro- and anti-saccade) fully repeated measures ANOVA ($p < 0.05$). An alpha level of 0.05 was used for statistical significance, and simple-effects (i.e., paired-samples t-tests) were employed to decompose the main effects and interactions.

3. Results

The RT results yielded a main effect for the assessment ($F(1,15) = 8.70$, $p = 0.01$, and $\eta_p^2 = 0.37$) and task ($F(1,15) = 34.9$, $p < 0.01$, and $\eta_p^2 = 0.7$). The prosaccade RTs (260 ms, SD = 42) were shorter than the antisaccades (302 ms, SD = 36; Figure 2A), and the difference scores in Figure 2B show that RTs for the pro- and anti-saccades decreased from the pre- to post-exercise assessments. Directional errors also demonstrated a main effect of task ($F(1,15) = 6.07$, $p = 0.026$, and $\eta_p^2 = 0.29$), such that the prosaccades produced fewer directional errors (4%, SD = 4) than the antisaccades (9%, SD = 6; Figure 2C). Difference scores in Figure 2D show that this result did not vary from pre- to post-exercise assessments: ($F(1,15) = 0.06$, $p = 0.804$, and $\eta_p^2 < 0.01$). In terms of the saccade duration, the grand mean was 55 ms (SD = 0.6) and no reliable main effects or interactions were observed (all $F(1,15) < 0.47$, $ps > 0.5$, and all $\eta_p^2 < 0.03$).

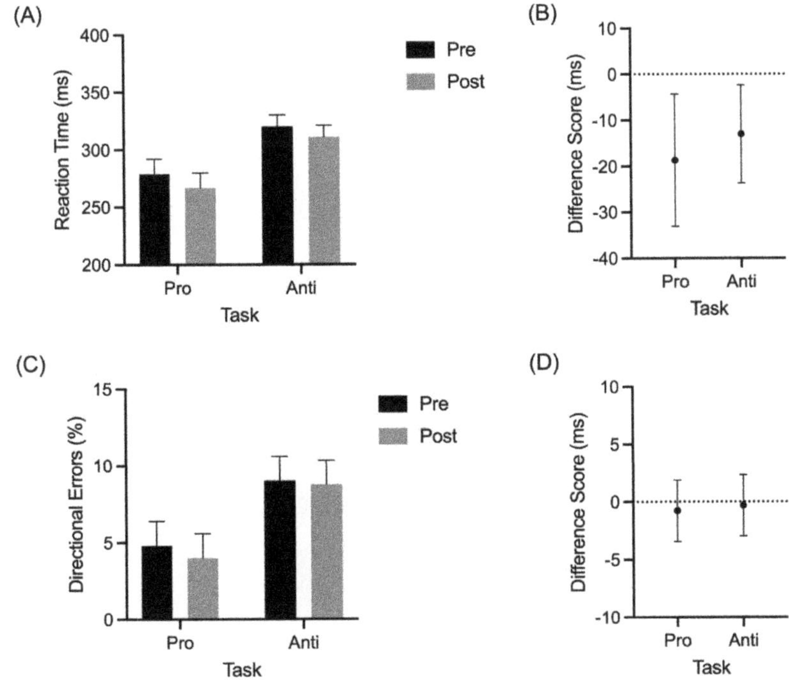

Figure 2. Group mean reaction time (RT) (**A**) and directional errors (**C**) for pro- and antisaccades at pre- and postexercise assessments. The offset panels show an RT (**B**) and directional error (**D**) difference scores (postexercise minus pre-exercise) with error bars representing 95% between-participant confidence intervals (**B**). An absence of overlap between the error bars and zero (i.e., horizontal dashed line) indicates a reliable difference inclusive to a test of the null hypothesis.

Baseline pupil diameter produced a main effect of assessment ($F(1,15) = 11.95$, $p = 0.004$, and $\eta_p^2 = 0.44$), whereas TEPD demonstrated the main effects for the assessment ($F(1,15) = 5.97$, $p = 0.027$, and $\eta_p^2 = 0.29$) and task ($F(1,15) = 4.58$, $p = 0.049$, and $\eta_p^2 = 0.23$). Figure 3C,E show that baseline pupil diameter and TEPD decreased and increased, respectively, from pre- to postexercise assessments. As expected, the baseline pupil diameter did not significantly differ between the pro- and antisaccades (Figure 3A), while the TEPDs for the prosaccades were less than TEPDs for the antisaccades (Figure 3B).

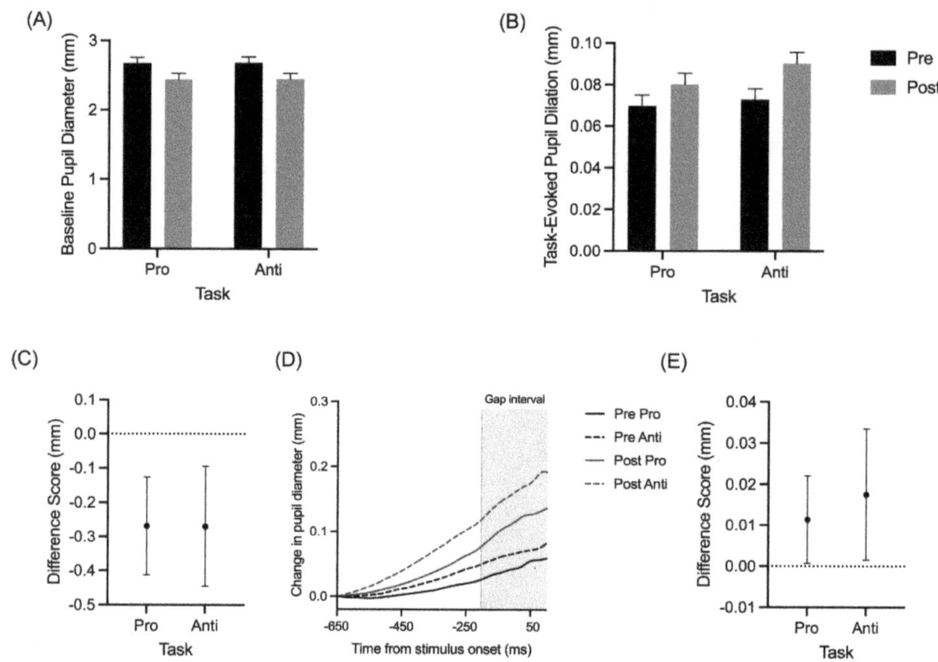

Figure 3. Group mean absolute pupil diameter (**A**) and task-evoked pupil dilation (**B**) for pro- and anti-saccades at pre- and postexercise assessments. Group mean difference scores (postexercise minus pre-exercise) for pupil diameter (**C**) and task-evoked pupil dilation (**E**). Baseline corrected pro- (solid lines) and antisaccade (dashed lines) pupil size changes by time traces for an exemplar participant during pre- (black lines) and postexercise (grey lines) assessments (**D**). Error bars represent 95% between-participant confidence intervals.

4. Discussion

The pro- and antisaccade RTs decreased from pre- to postexercise assessments. In terms of the antisaccade findings, the decrease in RT was independent of any change in saccade duration or directional errors—a result evincing that improved planning times were unrelated to a speed−accuracy trade-off. These results support previous work by our group [12,25], and provide convergent evidence that a single bout of aerobic exercise improves executive function. It is, however, important to recognize that prosaccade RTs also decreased from pre- to postexercise—a finding not observed in previous work. In reconciling this discrepancy, we note that previous work examined pro- and anti-saccades in separate blocks, whereas the current work randomly interleaved task-type on a trial-by-trial basis—a necessary manipulation to prevent TEPD attenuation due to task predictability [26]. As such, the current paradigm not only required the executive demand of response suppression for antisaccade trials, but also the executive component of cognitive flexibility (i.e., task-switching) across antisaccades and prosaccades [13,23,27]. Furthermore, and given that cognitive flexibility and task-switching efficiency elicits a robust postexercise benefit [4,28], it is possible that the postexercise decrease in prosaccade RTs observed here reflects a global benefit to executive function.

Wang and colleagues [14] established that interleaved pro- and anti-saccades produce equivalent baseline pupil sizes; however, the latter task-type was associated with larger TEPDs. The increased TEPDs for antisaccades, in combination with single-cell recording work in non-human primates [29,30], has been taken to evince that TEPDs are a direct neural proxy for the increased executive demands of the antisaccade task (i.e., response suppression). In support of Wang et al. [14], we found that pre- and postexercise TEPDs were larger for antisaccades than prosaccades. Of course, the goal of the present work was to extend Wang et al.'s results in determining whether an exercise intervention modulates pro- and antisaccade TEPDs. To that end, our results demonstrate that baseline pupil size and TEPDs decreased and increased, respectively, from pre- to post-exercise. In accounting for this, Aston-Jones and Cohen [7] proposed that the modulation of activity in the LC-NE system underlies an optimal range of arousal (i.e., tonic activity) that serves to enhance neural gain (i.e., phasic activity) in executive-related cortical structures. Specifically, the activation pattern of the LC-NE system exhibits a causal relationship with behavioral performance and attention. Indeed, microinjection experiments in non-human primates have demonstrated that increasing the tonic activation of the LC-NE system via injection of a muscarinic cholinergic agonist increases distractibility, reduces phasic responsiveness, and decreases performance [7]. In contrast, suppressing tonic activation to moderate levels via the injection of an adrenoreceptor agonist decreases distractibility, increases phasic responsiveness, and increases performance. Bouret and Sara [6] proposed that these moderate levels of tonic activity in the LC entrains other neural systems to reduce responsiveness to irrelevant stimuli, thus preventing distractions, with the task-related phasic bursts of activity serving to selectively facilitate goal-directed behaviors by providing a brief attentional filter. Accordingly, we propose that the postexercise decrease in pro- and anti-saccade RTs, combined with the decrease in preparatory phase tonic pupil size, reflect an optimal modulation of the LC-NE system. Importantly, this modulation is proposed to underly enhanced attentional control via the processing of task-relevant information and increased phasic recruitment of executive control networks supporting saccade generation.

An alternate account for the observed pupillometry findings is that the increased TEPDs across both pro- and anti-saccades reflect a broader improvement to general cognition. Our paradigm employed an interleaved pro- and antisaccade condition that introduced the executive component of cognitive flexibility. Our paradigm employed an interleaved pro- and antisaccade condition that introduced the executive component of cognitive flexibility. In line with the RT findings, the current prosaccade TEPD results provide evidence of a postexercise benefit to cognitive flexibility and task-switching efficiency. In support of this, previous work by our lab involving a blocked pro- and antisaccade paradigm with the same target eccentricities demonstrated a selective postexercise antisaccade benefit [13,23,27]. Additionally, another study involving concussed individuals demonstrated evidence of suppressed pro- and antisaccade RTs in a similar interleaved saccade paradigm that were proposed to reflect a concussion-related dysfunction to inhibitory control and cognitive flexibility [23]. Taken together, this evidence supports the stance that the current findings reflect a global improvement to executive function rather than a broader improvement to general cognition.

We note that our work contradicts the postexercise pupillometry results reported in a similar study by McGowan and colleagues [31]. This work examined the behavioral, electrophysiological, and pupillometric changes in response to the Eriksen flanker task before and after a 20-min bout of either aerobic exercise (i.e., 70% of the age-predicted maximum heart rate) or an active-control condition (i.e., walking on a treadmill at the lowest speed (0.5 mph) and incline (0 settings)). The discrepant findings may be accounted for by several between-experiment differences in methodology. First, the computation of TEPD in the current study was based on a baseline correction with CON_{max} (i.e., maximal pupil contraction)—a necessary procedure to control for pupil size variability in response to the pupillary light reflex associated with stimulus presentation [14,32]. In contrast, McGowan et al. employed a baseline correction for the initial size of the pupil prior to stimulus presentation. Although McGowan et al.'s procedure provides a normalization of pupil size change with respect to stimulus/task onset, it does not account for the additional variability that an individual's pupillary light reflex contributes to pupil size once a TEPD response emerges [14,32]. Second, to maintain an accurate measure of the pupil size, the selected epochs for pupil analysis in the current study were either during the central fixation period or before saccade initiation; that is, when the eye position was located at the center of the screen and no other motor responses were being completed. In contrast, McGowan et al. selected a pupil epoch that included a manual motor response associated with the task, and it is unknown if the participants' eye positions were fixated and directed to the center of the screen during this epoch. Finally, McGowan and colleagues employed the letter versions of the Eriksen flanker task, whereas the current study employed an interleaved pro- and antisaccade task. The benefit of the saccade paradigm used here is that it provides a directed measure of executive function without introducing concurrent non-executive task processes such as receptive language (i.e., letter identification), and does not result in larger manual-motor movements from impacting pupil responses and measurement [22].

Study Limitations

We recognize that our findings are limited by at least three methodological factors. First, the current study did not employ a non-exercise control condition. As a result, we cannot directly assert that the post exercise changes in saccade performance and pupillometry metrics are specific to exercise or underlie a practice-related improvement in the current task. With that being said, our lab has repeatedly shown that a non-exercise control condition does not exhibit a practice-related improvement in pro- or antisaccade performance measures when performed in separate blocks [11,12] or randomly interleaved trials [4,23,28]. Specifically, three studies by our group [12,25,28], employing null hypothesis testing in conditions involving exercise (same exercise intensity used in the current study) and control (rest) conditions, reported that antisaccade RTs reliably decreased from to pre- to postexercise (all ps < 0.001; all d_z > 1.10), whereas no reliable change was associated with the pre- to post-rest assessments (all ps > 0.50, all d_z < 0.14). Moreover, here, we computed supplementary two one-sided test (TOST) statistics from our group's previous work, and the results showed that pre- to post-rest antisaccade RTs for Samani and Heath (t(24) = 1.79, *p* = 0.043) [12], Heath and Shukla (t(17) = 2.07, *p* = 0.027) [28], and Tari et al. (t(14) = 1.88, *p* < 0.044) [25] were all within an equivalence boundary. Accordingly, null and equivalence tests from previous work support the direct assertion that antisaccade performance metrics do not relate to a practice-related improvement; rather, the results indicate that improved antisaccade RTs are specific to an exercise intervention. As such, we believe that the current findings, in combination with the extant literature, support the view that the oculomotor changes reported here reflect the exercise intervention. Second, the participants were young and reported a healthy lifestyle as determined via the PAR-Q and GLTEQ. It is therefore unknown whether populations outside this age range and fitness level would demonstrate a comparable postexercise improvement in arousal and executive function. Lastly, the postexercise assessment was completed within 15-min of the exercise intervention. Therefore, it is unclear whether the executive "boost" associated with LC-NE system modulation persisted beyond 15 min. In a follow-up study, we will examine pro- and antisaccade performance and pupillometry metrics across a range of post exercise intervals (i.e., immediate, for 30, 45, and 60 min) to determine the time frame through which a single bout of exercise modulates LC-NE activity and supports a postexercise benefit to executive function.

5. Conclusions

The present findings demonstrate that a 20-min single-bout of moderate-intensity aerobic exercise improved task-related arousal and preparatory activity in pro- and antisaccades. Specifically, the results demonstrated a pre- to postexercise decrease in pro- and antisaccade RTs, decreased tonic baseline pupil size, and increased phasic TEPDs. The results demonstrate that exercise-related saccade performance improvements are associated with a decrease in cognitive/psychological arousal—closer to an optimal level for task engagement—and augmented phasic recruitment of executive control resources. Accordingly, our findings provide evidence to suggest that the modulation of the LC-NE system is a mechanism underlying exercise-induced enhancements in cognition.

Author Contributions: Conceptualization, M.H. and N.A.; methodology, M.H. and N.A.; validation, N.A.; formal analysis, N.A.; investigation, N.A.; writing- original draft preparation, N.A.; writing-review and editing, M.H. and N.A.; visualization, M.H. and N.A.; supervision, M.H.; funding acquisition, M.H. All authors have read and agreed to the published version of the manuscript.

Funding: This work was supported by a Discovery Grant from the Natural Sciences and Engineering Research Council (NSERC) of Canada (M.H.), as well as Faculty Scholar (M.H.) and Major Academic Development Fund (M.H.) awards from the University of Western Ontario.

Institutional Review Board Statement: The study was conducted according to the guidelines of the Declaration of Helsinki, and approved by the Health Sciences Research Ethics Board of the University of Western Ontario (112323).

Informed Consent Statement: Informed consent was obtained from all subjects involved in the study.

Data Availability Statement: Data are available from the corresponding author upon request. The data are not publicly available due to ethical restrictions.

Conflicts of Interest: The authors declare that the research was conducted in the absence of any commercial or financial relationships that could be construed as a potential conflict of interest.

References

1. Miyake, A.; Friedman, N.P.; Emerson, M.J.; Witzki, A.H.; Howerter, A.; Wager, T.D. The unity and diversity of executive functions and their contributions to complex "frontal lobe" tasks: A latent variable analysis. *Cogn. Psychol.* **2000**, *41*, 49–100. [CrossRef]
2. Chang, Y.K.; Labban, J.D.; Gapin, J.I.; Etnier, J.L. The effects of acute exercise on cognitive performance: A meta-analysis. *Brain Res.* **2012**, *1453*, 87–101. [CrossRef] [PubMed]
3. Lambourne, K.; Tomporowski, P. The effect of exercise-induced arousal on cognitive task performance: A meta-regression analysis. *Brain Res.* **2010**, *1341*, 12–24. [CrossRef] [PubMed]
4. Shukla, D.; Heath, M. A single bout of exercise provides a persistent benefit to cognitive flexibility. *Res. Q. Exerc. Sport.* **2021**, in press.
5. Pontifex, M.B.; McGowan, A.L.; Chandler, M.C.; Gwizdala, K.L.; Parks, A.C.; Fenn, K.; Kamijo, K. A primer on investigating the after effects of acute bouts of physical activity on cognition. *Psychol. Sport Exerc.* **2019**, *40*, 1–22. [CrossRef]
6. Sara, S.J.; Bouret, S. Orienting and reorienting: The locus coeruleus mediates cognition through arousal. *Neuron.* **2012**, *76*, 130–141. [CrossRef] [PubMed]
7. Aston-Jones, G.; Cohen, J.D. An integrative theory of locus coeruleus-norepinephrine function: Adaptive gain and optimal performance. *Annu. Rev. Neurosci.* **2005**, *28*, 403–450. [CrossRef] [PubMed]
8. Munoz, D.P.; Everling, S. Look away: The anti-saccade task and the voluntary control of eye movement. *Nat. Rev. Neurosci.* **2004**, *5*, 218–228. [CrossRef] [PubMed]
9. Colcombe, S.J.; Kramer, A.F.; Erickson, K.I.; Scalf, P.; McAuley, E.; Cohen, N.J.; Webb, A.; Jerome, G.J.; Marquez, D.X.; Elavsky, S. Cardiovascular fitness, cortical plasticity, and aging. *Proc. Natl. Acad. Sci. USA* **2004**, *101*, 3316–3321. [CrossRef]
10. Hillman, C.H.; Snook, E.M.; Jerome, G.J. Acute cardiovascular exercise and executive control function. *Int. J. Psychophysiol.* **2003**, *48*, 307–314. [CrossRef]
11. Petrella, A.F.; Belfry, G.; Heath, M. Older adults elicit a single-bout post-exercise executive benefit across a continuum of aerobically supported metabolic intensities. *Brain Res.* **2019**, *1712*, 197–206. [CrossRef]
12. Samani, A.; Heath, M. Executive-related oculomotor control is improved following a 10-min single-bout of aerobic exercise: Evidence from the antisaccade task. *Neuropsychologia* **2018**, *108*, 73–81. [CrossRef]
13. Ayala, N.; Niechwiej-Szwedo, E. Effects of blocked vs. interleaved administration mode on saccade preparatory set revealed using pupillometry. *Exp. Brain Res.* **2021**, *239*, 245–255. [CrossRef]
14. Wang, C.A.; Brien, D.C.; Munoz, D.P. Pupil size reveals preparatory processes in the generation of pro-saccades and anti-saccades. *Eur. J. Neurosci.* **2015**, *41*, 1102–1110. [CrossRef]
15. Ballinger, E.C.; Ananth, M.; Talmage, D.A.; Role, L.W. Basal forebrain cholinergic circuits and signaling in cognition and cognitive decline. *Neuron* **2016**, *91*, 1199–1218. [CrossRef] [PubMed]
16. Hasselmo, M.E.; Sarter, M. Modes and models of forebrain cholinergic neuromodulation of cognition. *Neuropsychopharmacology* **2011**, *36*, 52–73. [CrossRef] [PubMed]
17. Gilzenrat, M.S.; Nieuwenhuis, S.; Jepma, M.; Cohen, J.D. Pupil diameter tracks changes in control state predicted by the adaptive gain theory of locus coeruleus function. *Cogn. Affect Behav. Neurosci.* **2010**, *10*, 252–269. [CrossRef]
18. Brainard, D.H. The psychophysics toolbox. *Spat. Vis.* **1997**, *10*, 433–436. [CrossRef]
19. Cornelissen, F.W.; Peters, E.M.; Palmer, J. The Eyelink Toolbox: Eye tracking with MATLAB and the Psychophysics Toolbox. *Behav. Res. Methods.* **2002**, *34*, 613–617. [CrossRef] [PubMed]
20. Heath, M.; Weiler, J.; Marriott, K.; Welsh, T. The antisaccade task: Dissociating stimulus and response influences online saccade control. *J. Vis.* **2011**, *11*, 549. [CrossRef]

21. Eckstein, M.K.; Guerra-Carrillo, B.; Singley, A.T.M.; Bunge, S.A. Beyond eye gaze: What else can eyetracking reveal about cognition and cognitive development? *Dev. Cogn. Neurosci.* **2017**, *25*, 69–91. [CrossRef]

22. Winn, M.B.; Wendt, D.; Koelewijn, T.; Kuchinsky, S.E. Best practices and advice for using pupillometry to measure listening effort: An introduction for those who want to get started. *Trends Hear.* **2018**, *22*, 2331216518800869. [CrossRef]

23. Ayala, N.; Heath, M. Executive Dysfunction Following a Sport-Related Concussion is Independent of Task-Based Symptom Burden. *J. Neurotrauma.* **2020**, *37*, 2558–2568. [CrossRef] [PubMed]

24. Wang, C.A.; McInnis, H.; Brien, D.C.; Pari, G.; Munoz, D.P. Disruption of pupil size modulation correlates with voluntary motor preparation deficits in Parkinson's disease. *Neuropsychologia* **2016**, *80*, 176–184. [CrossRef]

25. Tari, B.; Vanhie, J.J.; Belfry, G.R.; Shoemaker, J.K.; Heath, M. Increased cerebral blood flow supports a single-bout postexercise benefit to executive function: Evidence from hypercapnia. *J. Neurophysiol.* **2020**, *124*, 930–940. [CrossRef]

26. Zekveld, A.A.; Koelewijn, T.; Kramer, S.E. The pupil dilation response to auditory stimuli: Current state of knowledge. *Trends Heart* **2018**, *22*, 2331216518777174. [PubMed]

27. Weiler, J.; Heath, M. Task-switching in oculomotor control: Unidirectional switch-cost when alternating between pro-and antisaccades. *Neurosci. Lett.* **2012**, *530*, 150–154. [CrossRef]

28. Heath, M.; Shukla, D. A single bout of aerobic exercise provides an immediate "boost" to cognitive flexibility. *Front. Psychol.* **2020**, *11*, 1106. [CrossRef] [PubMed]

29. Lehmann, S.J.; Corneil, B.D. Transient pupil dilation after subsaccadic microstimulation of primate frontal eye fields. *J. Neurosci.* **2016**, *36*, 3765–3776. [CrossRef]

30. Wang, C.A.; Boehnke, S.E.; White, B.J.; Munoz, D.P. Microstimulation of the monkey superior colliculus induces pupil dilation without evoking saccades. *J. Neurosci.* **2012**, *32*, 3629–3636. [CrossRef] [PubMed]

31. McGowan, A.L.; Chandler, M.C.; Brascamp, J.W.; Pontifex, M.B. Pupillometric indices of locus-coeruleus activation are not modulated following single bouts of exercise. *Int. J. Psychophysiol.* **2019**, *140*, 41–52. [CrossRef] [PubMed]

32. Mathôt, S. Pupillometry: Psychology, physiology, and function. *J. Cogn.* **2018**, *1*, 1–23. [CrossRef] [PubMed]

Article

Regular Vigorous-Intensity Physical Activity and Walking Are Associated with Divergent but not Convergent Thinking in Japanese Young Adults

Chong Chen [1,*], Yasuhiro Mochizuki [2], Kosuke Hagiwara [1], Masako Hirotsu [1] and Shin Nakagawa [1]

[1] Division of Neuropsychiatry, Department of Neuroscience, Yamaguchi University Graduate School of Medicine, Ube, Yamaguchi 755-8505, Japan; khagi@yamaguchi-u.ac.jp (K.H.); hirotsu@yamaguchi-u.ac.jp (M.H.); snakaga@yamaguchi-u.ac.jp (S.N.)

[2] RIKEN Center for Brain Science, Wako, Saitama 351-0198, Japan; mochizuki@brain.riken.jp

* Correspondence: cchen@yamaguchi-u.ac.jp

Citation: Chen, C.; Mochizuki, Y.; Hagiwara, K.; Hirotsu, M.; Nakagawa, S. Regular Vigorous-Intensity Physical Activity and Walking Are Associated with Divergent but not Convergent Thinking in Japanese Young Adults. *Brain Sci.* **2021**, *11*, 1046. https://doi.org/10.3390/brainsci11081046

Academic Editors: Filipe Manuel Clemente and Ana Filipa Silva

Received: 27 July 2021
Accepted: 5 August 2021
Published: 6 August 2021

Abstract: The beneficial effects of regular physical activity (PA) on cognitive functions have received much attention. Recent research suggests that regular PA may also enhance creative thinking, an indispensable cognitive factor for invention and innovation. However, at what intensity regular PA brings the most benefits to creative thinking remains uninvestigated. Furthermore, whether the levels of regular PA affect the acute PA effects on creative thinking is also unclear. In the present study, using a previous dataset that investigated the effects of an acute bout of aerobic exercise on creative thinking in healthy Japanese young adults (22.98 ± 1.95 years old) in the year 2020, we tested the association between different intensities of regular PA (i.e., vigorous, moderate, and walking) and creative thinking with the cross-sectional baseline data using multiple linear regression. We also investigated whether regular PA levels were associated with the acute aerobic exercise intervention effects on creative thinking. The results showed that cross-sectionally, the regular PAs were differentially associated with divergent but not convergent thinking. Specifically, whereas the amount of vigorous-intensity PA was positively associated with fluency and flexibility, the amount of walking was positively associated with novelty on the alternate uses test (AUT) measuring divergent thinking. Importantly, the explained variances of fluency, flexibility, and novelty were 20.3% ($p = 0.040$), 18.8% ($p = 0.055$), and 20.1% ($p = 0.043$), respectively. None of the regular PAs predicted convergent thinking (i.e., an insight problem-solving task), nor were they associated with the acute aerobic exercise intervention effects on divergent and convergent thinking. These findings suggest that engaging in regular vigorous-intensity PA and walking may be useful strategies to enhance different aspects of divergent thinking in daily life.

Keywords: international physical activity questionnaires; exercise; creativity; divergent thinking; alternate uses test; convergent thinking

1. Introduction

The World Health Organization (WHO) recommends that adults aged 18–61 years should conduct at least 150 min of moderate-intensity physical activity (PA), 75 min of vigorous-intensity PA, or an equivalent combination of both moderate- and vigorous-intensity PA [1]. In line with this recommendation, a growing body of research has confirmed that engaging in regular PA enhances a wide range of cognitive functions, including attention, executive functions, memory storage and retrieval, and so on (for reviews and meta-analyses, [2–5]). Potential neurobiological mechanisms of such benefits include the activation of the prefrontal cortex and the release of lactate, cortisol, neurotrophins, and neurotransmitters (see the above reviews for details). However, little is known about the effects of PA on creative thinking.

Creative thinking is generally believed to comprise two fundamental cognitive processes: divergent and convergent thinking [6–8]. Divergent thinking involves generating multiple, novel solutions and is commonly assessed by the alternate uses test (AUT, [9]) or the Torrance Tests of Creative Thinking [10]. In the AUT, for instance, subjects are asked to write down as many as possible unique, original uses of common objects such as "brick" and "umbrella". The number of generated uses (known as fluency), the number of different conceptually categories of the uses (known as flexibility), and the rareness of the uses (known as originality) are frequently used as indicators of divergent thinking. In contrast, convergent thinking involves approaching a single correct solution [6] and is often assessed by the remote associates test (RAT, [11]), the compound remote associates test (CRA, [12]), or insight problem-solving quizzes (e.g., matchstick arithmetic problems, [13]). In the CRA, subjects are asked to think of a single word that is associated with each of three given words, for instance, "time", "hair", and "stretch" (the answer is "long"). In the matchstick arithmetic problem, several sticks form a false arithmetic equation and subjects are asked to move a single stick to make the equation correct. It has been suggested that divergent and convergent thinking differ in such a way that the latter relies more on top-down, executive control [14]. Since both divergent and convergent thinking are indispensable for invention and innovation [6–8], the development of effective strategies to enhance creative thinking may have great social significance.

We have recently conducted a systematic review of studies on acute PA and creative thinking and concluded that acute PA has the potential to enhance divergent and convergent thinking (see Introduction and Supplementary Materials of Aga et al. (2021) [15]). For instance, a recent study showed that compared to stay seated, a 4-min walk on a treadmill or outdoors increased the originality of divergent thinking, without affecting performance on CRA that measured convergent thinking [16]. However, acute PA at too low an intensity may not reliably enhance divergent and convergent thinking, while at too high an intensity, it may impair divergent and convergent thinking. We further conducted a randomized controlled trial and showed that a 15-min physical test program improved divergent thinking (in terms of flexibility on the AUT) independent of post-exercise mood [15]. In contrast, the program affected convergent thinking (matchstick arithmetic problems) in a post-exercise mood-dependent way: it tended to enhance convergent thinking in subjects reporting high vigor but impair convergent thinking in those reporting low vigor [15].

However, to the best of our knowledge, the effects of regular or chronic PA on creative thinking has not been systematically reviewed so far. Whereas the effects of acute PA are transient, those of regular PA are long lasting and may be more valuable for long-term enhancement of creative thinking. We, therefore, searched (1) interventional studies that investigated the effects of multiple sessions of PA interventions on creative thinking and (2) observational studies that investigated the association between the levels of regular PA and creative thinking (see the literature search strategy described in the footnote of Table S1). As a result, we identified six interventional studies ([17–22], see Table S1) and four observational studies ([23–26], see Table S2).

To summarize the findings, the identified interventional studies consistently reported significant enhancing effects of PA programs lasting 6–12 weeks (weekly 2–5 sessions) on one or several measures of divergent thinking, such as fluency (i.e., the number of generated uses) on AUT and figural fluency on TTCT [17–22]. However, two studies that evaluated convergent thinking with matchstick puzzles failed to find any effects of similar PA programs [17,18]. Similarly, the identified cross-sectional studies consistently found positive associations between self-reported or actigraph-monitored regular PA and measures of divergent thinking in all or a subgroup of subjects [23,24,26] but not convergent thinking (i.e., insight problem-solving quizzes, [25]).

Although the results of our systematic review indicate beneficial effects of regular PA on creative thinking (especially for divergent thinking), the influence of PA intensity or, in other words, at what intensity regular PA brings the most benefits remains unclear. None of the identified interventional studies specified the intensity of their PA programs. Two of the cross-sectional studies did differentiate the intensity of regular PA, but the results were inconclusive. Rominger et al. (2020) [24] reported a significant correlation between total, no-to-light, and moderate but not vigorous and very vigorous everyday bodily movement (EBM) and divergent thinking. However, given the frequently observed correlation among different intensities of regular PA [25], the independent effects of different intensities of regular PA on divergent thinking were unknown in this study. Nakagawa et al. (2020) [25] included measures of walking and moderate- and vigorous-intensity PA in a regression model to predict convergent thinking evaluated with insight problem-solving quizzes but failed to find any significant effects. The lack of evidence on the relation between regular PA intensity and creative thinking is unfortunate because intensity is a critical and frequently discussed issue in PA prescription and planning for enhancing cognitive and mental health [25,27].

Furthermore, none of the studies we reviewed have examined the interaction effect between regular and acute PA on creative thinking. In the field of PA and mood, studies have reported that subjects with high levels of regular PA may show more enhanced mood in response to a new, acute bout of PA [28,29]. However, there were also studies suggesting that this difference only occurs when conducting high- but not low-intensity acute PA [30,31] or that the psychological effects of acute PA are independent of regular PA altogether [32]. Clarifying the interaction effect between regular and acute PA on creative thinking may provide insights into the mechanism by which regular PA enhances creative thinking, and if the interaction effect does exist, it will be worthy to be carefully considered during PA prescription and planning.

Therefore, in the present study, we conducted a secondary analysis of a previously published study [15] that investigated the effects of an acute bout of aerobic exercise on divergent and convergent thinking. Specifically, we tested three hypotheses. First, the cross-sectional associations between regular PA and divergent and convergent thinking are different such that regular PA is only associated with divergent thinking. Second, different intensities of regular PA may have distinct associations with divergent thinking. Third, regular PA may affect the acute aerobic exercise intervention effects on creative thinking. To our knowledge, this is the first study that evaluated the independent effects of different intensities of regular PA on divergent thinking and the first study that investigated whether regular and acute PA interact to affect divergent and convergent thinking.

2. Materials and Methods

2.1. Participants and Procedure

The study was approved by the Institutional Review Board of Yamaguchi University Hospital and preregistered on the University hospital Medical Information Network Clinical Trial Registry (UMIN-CTR, register ID: UMIN000041122). The study was carried out following the latest version of the Declaration of Helsinki, and all subjects agreed to participate in the study and provided written informed consent. The characteristics of the participants and procedure of the intervention have been described in [15]. In brief, the study was a randomized controlled trial using a between-subjects pre-test–post-test comparison design. Based on a priori power analysis, forty healthy subjects (all undergraduate students, 11 females, 29 males, age: 22.98 ± 1.95 years) were recruited via posters placed on campus and department homepage and through word-of-mouth during the period of July–October 2020. The inclusion criteria were being 20–29 years old at the time of the visit and the exclusion criteria were (1) reporting any history of diseases that greatly affect cardiopulmonary functions, such as chronic heart failure, (2) currently suffering from any mental illness or being scheduled to receive any medical examinations due to suspicion of mental illness, (3) being a member of our department who receives the personnel evaluation directly by the principal investigator of this study, and (4) being judged to be unsuitable for this study (e.g., bodyweight exceeding the applicable weight of the exercise bike). No participant was excluded due to meeting any of the exclusion criteria.

As shown in Figure 1, subjects first filled out a form assessing their age, sex, education level, regular PA, and so on. Subjects then performed tests of creative thinking at baseline, after which they were randomized to receive either the acute aerobic exercise intervention or control intervention. Immediately after the intervention, subjects rated their mood in terms of pleasure, relaxation, and vigor and, then, conducted tests of creative thinking again.

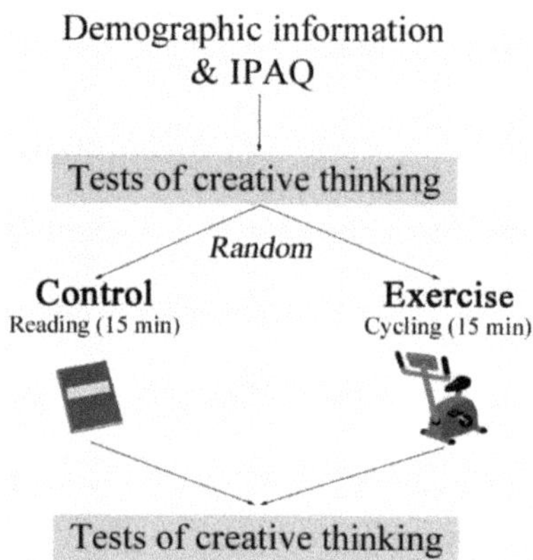

Figure 1. Schematic illustration of the procedure. Adapted from [15].

2.2. Measures of Regular PA

The international physical activity questionnaire long-form (IPAQ, [33]) was employed here. The IPAQ has been translated into many languages including Japanese, and its reliability and validity have been established by Craig et al. (2003) [33]. The IPAQ asks subjects to indicate the duration and frequency of different intensities (i.e., walking, moderate, and vigorous) of PA within four different domains (i.e., occupational, transport, household, and leisure related) during the past seven days. To investigate the independent effects of different intensities of regular PA, we calculated the total amount of PA (minutes/week) for each intensity, respectively. Furthermore, we calculated the total weekly PA metabolic equivalents (MET-minutes, hereafter referred to as total PA), by weighting the amount of PA of each intensity (minutes/week) with a MET energy expenditure estimate assigned to each intensity (8 for vigorous, 4 for moderate, and 3.3 for walking).

2.3. Measures of Creative Thinking

Divergent thinking was measured with the AUT [9]. In this test, subjects were presented with three common objects and asked to write down as many as possible uncommon, original, and unique uses of those objects within 4 min on an A4 size blank paper. We calculated the fluency, flexibility, and originality for each subject [34,35]. Fluency was the number of generated uses, flexibility was the number of different conceptual categories the uses are from, and originality was the rareness of the generated uses here defined as the number of conceptual categories. After sufficient training [36] under the supervision of the corresponding author, a primary coder scored all responses and a secondary coder scored responses of a randomly selected object; the two coders (both were undergraduate research assistants enrolled in the medical department) reached a substantial or almost perfect agreement, indicated by Cohen's $\kappa = 0.936$ for flexibility and Cohen's $\kappa = 0.706$ for originality.

Convergent thinking was evaluated with the matchstick arithmetic problems developed by Knoblich et al. [13]. At pre-test, six problems were presented, and subjects had 12 min to solve these problems. The unsolved problems were presented again at post-text, together with another set of creative problem-solving puzzles that require visuospatial and logical reasoning [37]. To measure the intervention effects, we created two measures of convergent thinking for data analysis: one was a creative problem-solving (CPS) score consisting of matchstick arithmetic problems at pre-test and creative problem-solving puzzles at post-test; the other was a matchstick re-test score obtained at post-test only and calculated as the proportion of correctly solved problems that they failed at pre-test (subjects that correctly solved all problems at pre-test were removed from this analysis).

2.4. Intervention

For the acute aerobic exercise intervention, we used an automated physical test program built in an exercise bike (Wellbike BE-260, Fukuda Denshi, Tokyo, Japan). The program lasted 15 min and consisted of 10 min of physical test and 5 min of cooldown. This was essentially a graded exercise program designed for convenient physical testing. Under this program, subjects sat quietly on the bike during the first minute to measure their resting pulse, after which they started pedaling at a pace of 50 rpm. The workload increased at 4 and 7 min after the start of the program according to the pulse rate of the subjects at the moment. This program was chosen as the exercise intervention because its workload was predicted to fall between normal cycling at moderate intensity and intense cycling with maximal effort, two intensities tested by Colzato et al. (2013) [38]. Exercise at this intensity was believed to improve subjects' mood and be more likely to enhance creative thinking [15]. Subjects' heart rate was monitored with an Apple Watch Series 4 (Apple Inc., Cupertino, CA, USA). The mean heart rate was 118.25 (standard deviation or SD 11.93) bpm, the maximal heart rate was 158.65 (SD 13.00) bpm, and the heart rate in the last minute of the program was 102.31 (SD 11.93) bpm. The details of the exercise intervention are available in [15]. For the control intervention, we asked subjects to read mood-neutral materials on the association between PA and brain functions at a self-selected pace. The mean, maximal, and last-minute heart rate under the control intervention were 75.10 (SD 11.27), 89.70 (SD 12.18), and 75.53 (SD 12.40) bpm, respectively. Both interventions were conducted by an undergraduate research assistant under the supervision of the corresponding author.

2.5. Statistical Analysis

The statistical analysis was conducted with IBM SPSS Statistics 26.0. To investigate the association between regular PA and creative thinking at baseline, we combined subjects from the exercise and control groups together (n = 40). The normality of the data was checked using the Shapiro–Wilk test. Due to non-normal distribution of total PA, Spearman correlation analysis was used to examine the association between total PA and creative thinking at baseline as well as the acute aerobic exercise intervention effects. Confidence intervals (CI) for the Spearman correlation coefficients were calculated based on [39]. Multiple linear regression was used to evaluate the independent effects of different intensities of regular PA on creative thinking at baseline and the acute intervention effects. The normal P-P plot of regression standardized residual was confirmed and shown in Figures S1 and S2. We did not detect any obvious multicollinearity (i.e., variance inflation factors all <5) or homoscedasticity issue with the multiple linear regression. Given that we had a small set of predictors (i.e., three) and that we were not clear which was the best predictor, we used the standard "Enter" method for the multiple linear regression. A significance level of $p < 0.05$ was used.

3. Results

3.1. Regular PA and Divergent and Convergent Thinking at Baseline

Spearman correlation analysis indicated that total PA was marginally associated with fluency (rho = 0.293, 95% CI= [−0.027, 0.559], p = 0.067) and flexibility (rho = 0.283, 95% CI = [−0.038, 0.551], p = 0.077) but not novelty (rho = 0.237, 95% CI = [−0.085, 0.514], p = 0.142) on AUT of divergent thinking (Figure 2). Total PA was not associated with the matchstick pre-test measure of convergent thinking (rho = 0.091, 95% CI= [−0.228, 0.392], p = 0.575).

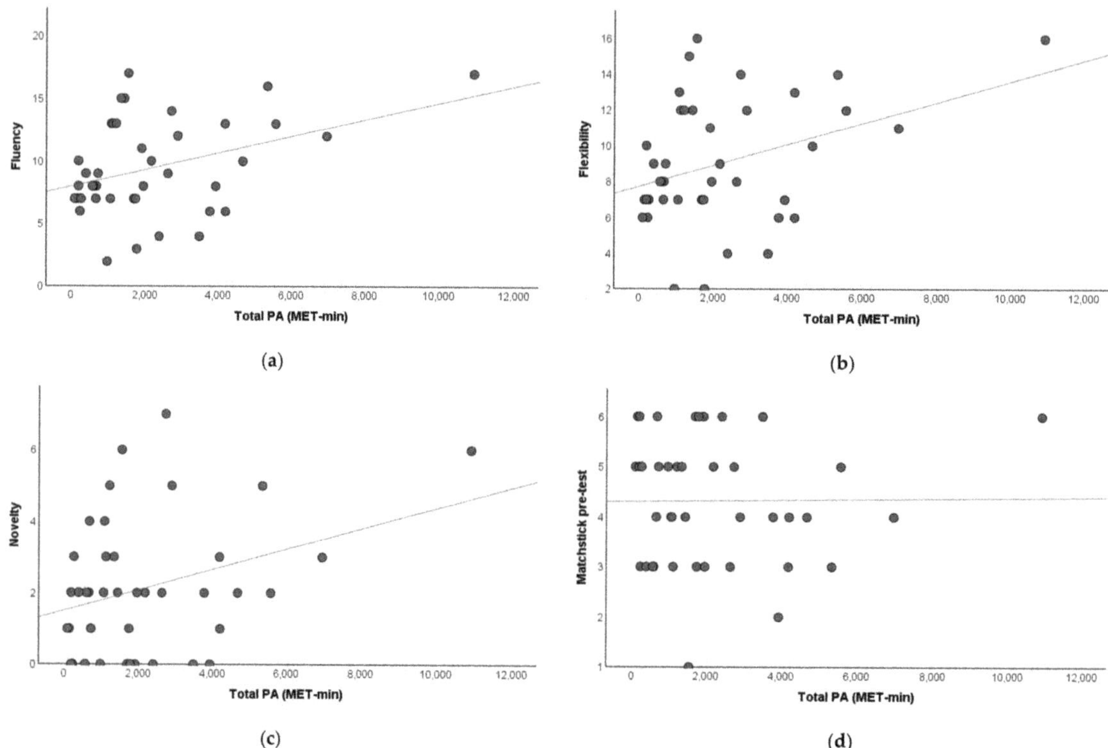

Figure 2. Scatterplot of the association between total PA and creative thinking at baseline. (**a**) Fluency. (**b**) Flexibility. (**c**) Novelty. (**d**) Matchstick pre-test.

To investigate the independent effects of different intensities of regular PA on creative thinking at baseline, we incorporated all three intensities of regular PA as independent variables to predict creative thinking using multiple linear regression models (Table 1). The results showed that whereas the amount of vigorous-intensity PA positively predicted fluency ($p = 0.016$) and flexibility ($p = 0.032$), walking positively predicted novelty ($p = 0.016$) on AUT measure of divergent thinking. Notably, the models explained 18.8–20.3% variances of the AUT measures. A partial regression plot of the associations is shown in Figure 3. In contrast, the model incorporating all three intensities of regular PA did not predict convergent thinking ($p = 0.976$).

Table 1. Multiple linear regression results using intensity-specific regular PA to predict creative thinking at baseline.

Dependent Variables		Divergent Thinking (AUT)			Convergent Thinking
		Fluency	Flexibility	Novelty	Matchstick Pre-Test
Independent variables	Vigorous	**0.382 ***	**0.342 ***	0.253	0.062
	Moderate	0.031	0.046	0.080	−0.028
	Walking	0.262	0.283	**0.378 ***	−0.039
Model statistics	$F_{(3,36)}$	3.062	2.774	3.013	0.070
	R^2	0.203	0.188	0.201	0.006
	p	**0.040 ***	**0.055 +**	**0.043 ***	0.976

Note: Standardized coefficients are presented here. The unstandardized coefficients and their 95% confidence intervals are reported in Table S3. Vigorous refers to vigorous-intensity PA; moderate refers to moderate-intensity PA. * $p < 0.05$; + $p < 0.06$.

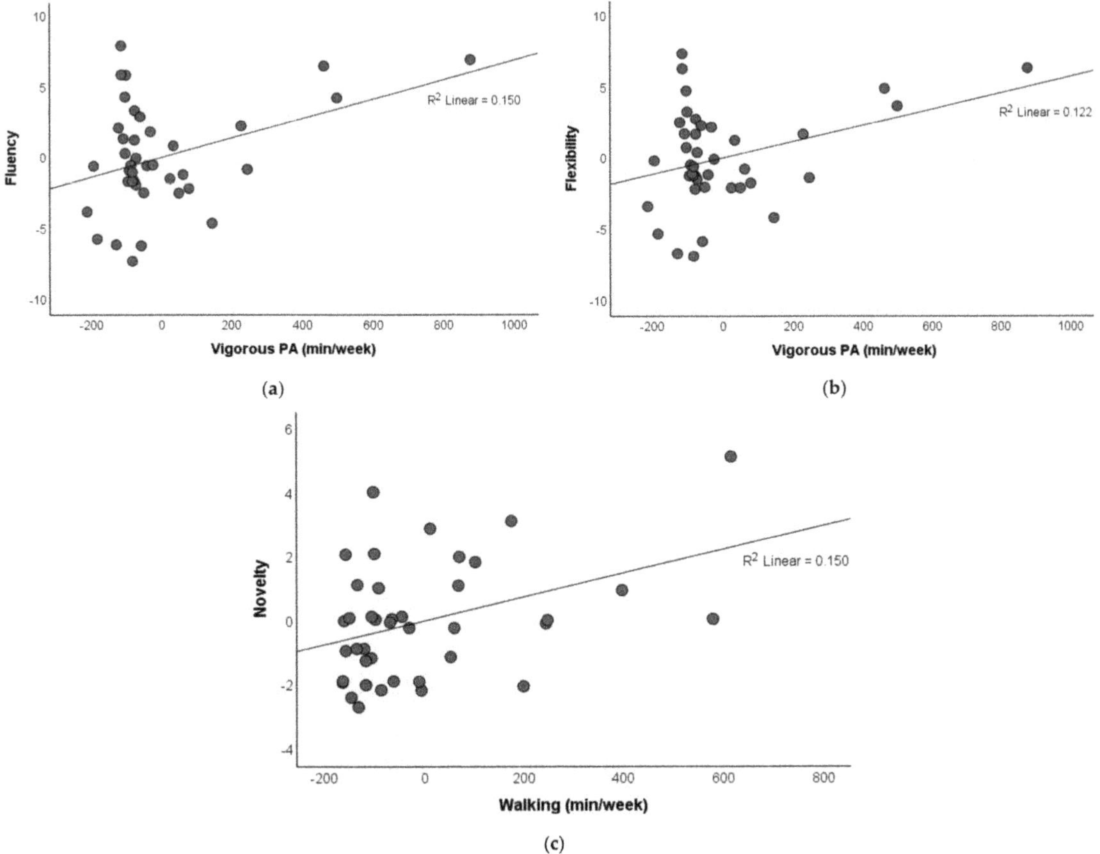

Figure 3. Partial regression plot of the association between different intensities of regular PA and creative thinking at baseline. (**a**) Vigorous-intensity PA and fluency. (**b**) Vigorous-intensity PA and flexibility. (**c**) Walking and novelty.

3.2. Regular PA and Acute Aerobic Exercise Intervention Effects

Spearman correlation analysis indicated that total PA was not associated with any of the acute intervention effects on divergent or convergent thinking in either the exercise (all $p > 0.09$) or control (all $p > 0.50$) group.

We next included all three intensities of regular PA as independent variables to predict the acute intervention effects using multiple linear regression models (Table 2). The results showed that the amount of different intensities of regular PA did not predict the change in divergent or convergent thinking in response to the acute aerobic exercise intervention (all $p > 0.30$). However, in the control group, the amount of moderate-intensity PA positively predicted the matchstick retest measure of convergent thinking in response to the control intervention ($p = 0.024$, Figure 4a). This was not due to a low score at pre-test, as moderate-intensity PA did not predict matchstick pre-test scores ($p = 0.856$, Figure 4b).

Table 2. Multiple linear regression results using intensity-specific regular PA to predict the intervention effects on creative thinking.

Dependent Variables		Divergent Thinking (AUT)			Convergent Thinking	
		△Fluency	△Flexibility	△Novelty	△CPS	Matchstick Retest
Exercise group						
Independent variables	Vigorous	−0.214	−0.152	−0.191	0.177	0.297
	Moderate	−0.071	−0.106	−0.069	0.3369	0.033
	Walking	−0.078	−0.038	−0.166	0.051	−0.329
	$F_{(3,16)}$	0.310	0.215	0.330	1.042	1.302
	R^2	0.055	0.039	0.058	0.163	0.246
	p	0.818	0.885	0.804	0.401	0.319
Control group						
Independent variables	Vigorous	−0.157	−0.263	−0.168	−0.228	−0.170
	Moderate	0.301	0.204	0.011	−0.029	**0.566** *
	Walking	−0.144	0.207	−0.305	−0.265	−0.421
Model statistics	$F_{(3,16)}$	0.703	0.789	0.697	0.725	3.554
	R^2	0.116	0.129	0.116	0.120	0.492
	p	0.564	0.517	0.567	0.552	**0.051 +**

Note: Standardized coefficients are presented here. The unstandardized coefficients and their 95% confidence intervals are reported in Table S4. Vigorous refers to vigorous-intensity PA; moderate refers to moderate-intensity PA. For matchstick retest, $F_{(3,12)}$ for exercise, $F_{(3,11)}$ for control. * $p < 0.05$; + $p < 0.06$.

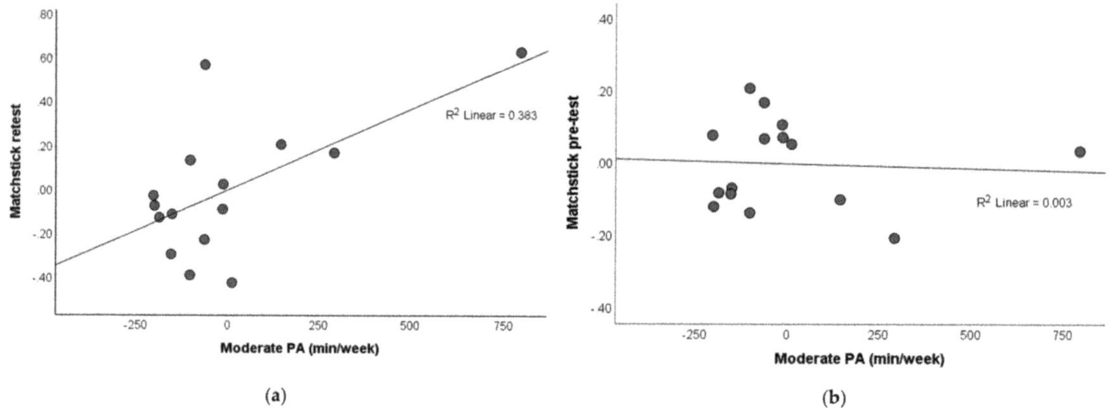

Figure 4. Partial regression plot of the association between moderate-intensity PA and convergent thinking in control group. (a) Matchstick retest after the intervention. (b) Matchstick pre-test at baseline.

4. Discussion

We found that total PA in terms of weekly MET-minutes was marginally correlated with fluency and flexibility but not novelty of divergent thinking nor convergent thinking. These correlations (rho = 0.293 for fluency, rho = 0.283 for flexibility) are considered small in magnitude [40]. More importantly, our multiple linear regression incorporating different intensities of regular PA showed that whereas the amount of vigorous-intensity PA (minutes/week) positively predicted fluency and flexibility, that of walking positively predicted novelty of divergent thinking. In contrast, none of them predicted convergent thinking. These results suggest that the effects of regular PA on divergent thinking are intensity-specific. Thus, in the case of divergent thinking, it is perhaps not the total amount of PA that matters, rather, the intensity of the PA is more important. To enhance divergent thinking, people should conduct more vigorous-intensity PA as well as walking.

Vigorous-intensity PA refers to activities that require hard physical effort and cause large increases in breathing, for instance, heavy lifting, fast bicycling, and intense aerobics. We have previously proposed that higher intensities of PA may bring more benefits to the brain because they generally cause more extensive, enhanced, and long-lasting neurobiological changes [25]. For instance, it has been reported that a single bout of vigorous- but not low-intensity cycling increased the peripheral level of brain-derived neurotrophic factor (BDNF) in young adults [41]. Peripheral BDNF can pass the brain–blood barrier and supports the production, growth, differentiation, and survival of neurons and thereby enhances a large variety of cognitive functions [42]. In rats, higher speed of running has been reported to increase the extracellular concentration of dopamine in the striatum to a greater extent [43]. As an interface between motivation and cognition, striatal dopamine has been associated with cognitive flexibility [44]. However, this proposal is only partially supported in the present study, since vigorous-intensity PA was positively associated with fluency and flexibility but not novelty, while the latter was positively associated with walking, and none of these measures were associated with moderate-intensity PA. The neurobiological and psychological mechanism by which walking enhances novelty of divergent thinking is unclear and remains to be investigated. Nevertheless, this finding is consistent with a recent report that a brief walk on a treadmill or outdoors can boost divergent thinking in terms of novelty on the AUT [16]. In this study, subjects generated more novel responses on a 4-min AUT while walking on a treadmill compared to while staying seated. Furthermore, the brief walk also enhanced subsequent divergent thinking such that those walked on the treadmill for the first AUT also performed better on a second AUT while staying seated compared to those sat for both AUTs. These findings may provide a scientific account of the popular "walking meetings" valued by the academic and managerial fields [45,46]. Although with different tasks to evaluate convergent thinking, both the current study (using matchstick puzzles) and Oppezzo and Schwartz (2014) [16] (using the compound remote associates test) failed to find any significant association between PA and convergent thinking. It remains for future studies to confirm this lack of association and clarify the underlying reasons.

The WHO Guidelines on Physical Activity and Sedentary Behavior has recommended vigorous-intensity and moderate-intensity PA as the primary PA for people of all age groups, including children, adolescents, adults, and older adults [1]. However, in the case of creative thinking, together with recent evidence [16], the current study indicates that walking may be more valuable than vigorous- and moderate-intensity PA, since only walking is associated with novelty of divergent thinking. If these findings are replicated by more robust studies, they will have great implications for cognitive and performance enhancement.

Meanwhile, we found that regular PA was not associated with the acute aerobic exercise intervention effects. The interaction effect of regular and acute PA on mood has been reported and debated (e.g., [28,30,32]). Our current findings are consistent with [32] that the acute effects might be independent of the levels of regular PA. Unexpectedly, however, we found that the amount of moderate-intensity PA was positively associated with the matchstick retest measure of convergent thinking after the control intervention in which subjects quietly read mood-neutral materials while staying seated. This result was not an artifact due to a poorer performance at pre-test. These findings suggest the possibility that when solving insight problems, subjects with a greater amount of moderate-intensity PA might be more likely to benefit from a "second thought" after a physically inactive but not physically active "incubation" period [47]. This speculation, however, remains to be tested by future well-designed experimental studies.

Several limitations of the present study have to be noted. Firstly, our subjects were undergraduate students, which prohibits us from generating our findings to other populations. Secondly, our sample size was relatively small, in particular for the analysis on the interaction between regular and acute PA effects. Thirdly, given our small sample size, we did not probe the potential sex difference in the associations between regular PA and creative thinking. Fourthly, we used only a self-report scale (i.e., the IPAQ) to measure PA. Although the validity of the scale has been established [33], there remains the possibility that the self-reported data are biased by social desirability. Future investigations are required to include measures of social desirability bias or confirm our findings with more accurate, objective measures of PA. We hope the current exploratory study may provide useful insights for future studies with larger sample sizes and more divergent populations to advance our understanding of the associations between regular PA and creative thinking.

5. Conclusions

In a sample of young adults, we found that regular vigorous-intensity PA was positively associated with fluency and flexibility, while regular walking was positively associated with novelty of divergent thinking. None of the different intensities of regular PA were associated with convergent thinking or the acute aerobic exercise intervention effects on divergent and convergent thinking. These findings suggest that engaging in regular vigorous-intensity PA and walking may be employed as useful strategies to enhance different aspects of divergent thinking in daily life. Whereas the WHO recommends vigorous-intensity and moderate-intensity PA as the primary PA for people of all age groups, the current study indicates that in the case of creative thinking, walking may be more valuable than vigorous- and moderate-intensity PA, since only walking is associated with novelty of divergent thinking. If these findings are replicated by more robust studies, they will have great implications for cognitive and performance enhancement.

Supplementary Materials: The following are available online at https://www.mdpi.com/article/10.3390/brainsci11081046/s1, Table S1: Summary of previous studies on the effects of regular physical activity on creative thinking: interventional studies; Table S2: Summary of previous studies on the effects of regular physical activity on creative thinking: observational studies; Figure S1: Normal P-P plot of regression standardized residual of multiple linear regressions shown in Table 1; Figure S2: Normal P-P plot of regression standardized residual of multiple linear regressions shown in Table 2; Table S3: The unstandardized coefficients and their 95% confidence intervals for multiple linear regressions shown in Table 1; Table S4: The unstandardized coefficients and their 95% confidence intervals for multiple linear regressions shown in Table 2.

Author Contributions: Conceptualization, C.C., Y.M., and S.N.; methodology, C.C. and K.H.; formal analysis, C.C. and Y.M.; investigation, C.C., K.H., and M.H.; writing—original draft preparation, C.C.; writing—review and editing, all authors. All authors have read and agreed to the published version of the manuscript.

Funding: This research received no external funding.

Institutional Review Board Statement: The study was conducted according to the guidelines of the Declaration of Helsinki, and approved by the Institutional Review Board of Yamaguchi University Hospital (H2020-065).

Informed Consent Statement: Informed consent was obtained from all subjects involved in the study.

Data Availability Statement: The data that support the findings of this study are available from the corresponding author upon reasonable request.

Acknowledgments: We thank Kohei Aga, Masato Inamura, and Rikuto Yamashita for their assistance with data collection.

Conflicts of Interest: The authors declare no conflict of interest related to this study.

References

1. World Health Organization. *WHO Guidelines on Physical Activity and Sedentary Behaviour*; World Health Organization: Geneva, Switzerland, 2020.
2. Smith, P.J.; Blumenthal, J.A.; Hoffman, B.M.; Cooper, H.; Strauman, T.A.; Welsh-Bohmer, K.; Browndyke, J.N.; Sherwood, A. Aerobic exercise and neurocognitive performance: A meta-analytic review of randomized controlled trials. *Psychosom. Med.* **2010**, *72*, 239. [CrossRef]

3. Guiney, H.; Machado, L. Benefits of regular aerobic exercise for executive functioning in healthy populations. *Psychon. Bull. Rev.* **2013**, *20*, 73–86. [CrossRef]
4. De Greeff, J.W.; Bosker, R.J.; Oosterlaan, J.; Visscher, C.; Hartman, E. Effects of physical activity on executive functions, attention and academic performance in preadolescent children: A meta-analysis. *J. Sci. Med. Sport* **2018**, *21*, 501–507. [CrossRef] [PubMed]
5. Erickson, K.I.; Hillman, C.; Stillman, C.M.; Ballard, R.M.; Bloodgood, B.; Conroy, D.E.; Macko, R.; Marquez, D.X.; Petruzello, S.J.; Powell, K.E. Physical activity, cognition, and brain outcomes: A review of the 2018 physical activity guidelines. *Med. Sci. Sports Exerc.* **2019**, *51*, 1242. [CrossRef]
6. Guilford, J.P. *The Nature of Human Intelligence*; McGraw-Hill: New York, NY, USA, 1967.
7. Runco, M.A. Testing creativity. In *International Encyclopedia of Education*, 3rd ed.; Peterson, P., Baker, E., McGaw, B., Eds.; Elsevier Ltd.: Amsterdam, The Netherlands, 2010; pp. 170–174.
8. Hennessey, B.A.; Amabile, T.M. Creativity. *Annu. Rev. Psychol.* **2010**, *61*, 569–598. [CrossRef] [PubMed]
9. Guilford, J.P. *Alternate Uses*; Form, A., Ed.; Sheridan Supply: Beverly Hills, CA, USA, 1960.
10. Torrance, P.E. *Torrance Tests of Creative Thinking*; Personnel Press: Princeton, NJ, USA, 1966.
11. Mednick, M.T.; Mednick, S.A.; Mednick, E.V. Incubation of creative performance and specific associative priming. *J. Abnorm. Soc. Psychol.* **1964**, *69*, 84–88. [CrossRef] [PubMed]
12. Bowden, E.M.; Jung-Beeman, M. Normative data for compound remote associate problems. *Behav. Res. Methods Instrum. Comput.* **2003**, *35*, 634–639. [CrossRef]
13. Knoblich, G.; Ohlsson, S.; Haider, H.; Rhenius, D. Constraint relaxation and chunk decomposition in insight problem solving. *J. Exp. Psychol. Learn. Mem. Cogn.* **1999**, *25*, 1534. [CrossRef]
14. Colzato, L.S.; Ozturk, A.; Hommel, B. Meditate to create: The impact of focused-attention and open-monitoring training on convergent and divergent thinking. *Front. Psychol.* **2012**, *3*, 116. [CrossRef] [PubMed]
15. Aga, K.; Inamura, M.; Chen, C.; Hagiwara, K.; Yamashita, R.; Hirotsu, M.; Takao, A.; Fujii, Y.; Matsubara, T.; Nakagawa, S. The Effect of Acute Aerobic Exercise on Divergent and Convergent Thinking and Its Influence by Mood. *Brain Sci.* **2021**, *11*, 546. [CrossRef]
16. Oppezzo, M.; Schwartz, D.L. Give your ideas some legs: The positive effect of walking on creative thinking. *J. Exp. Psychol. Learn. Mem. Cogn.* **2014**, *40*, 1142. [CrossRef]
17. Gondola, J.C.; Tuckman, B.W. Effects of a systematic program of exercise on selected measures of creativity. *Percept. Mot. Ski.* **1985**, *60*, 53–54. [CrossRef] [PubMed]
18. Gondola, J.C. The enhancement of creativity through long and short term exercise programs. *J. Soc. Behav. Personal.* **1986**, *1*, 77–82.
19. Tuckman, B.W.; Hinkle, J.S. An experimental study of the physical and psychological effects of aerobic exercise on schoolchildren. *Health Psychol.* **1986**, *5*, 197. [CrossRef] [PubMed]
20. Hinkle, J.S.; Tuckman, B.W.; Sampson, J.P. The psychology, physiology, and creativity of middle school aerobic exercisers. *Elem. Sch. Guid. Couns.* **1993**, *28*, 133–145.
21. Herman-Tofler, L.R.; Tuckman, B.W. The effects of aerobic training on children's creativity, self-perception, and aerobic power. *Child Adolesc. Psychiatr. Clin.* **1998**, *7*, 773–790. [CrossRef]
22. Pedro Ángel, L.R.; Beatriz, B.A.; Jerónimo, A.V.; Antonio, P.V. Effects of a 10-week active recess program in school setting on physical fitness, school aptitudes, creativity and cognitive flexibility in elementary school children. A randomised-controlled trial. *J. Sports Sci.* **2021**, *39*, 1277–1286. [CrossRef]
23. Cavallera, G.M.; Boari, G.; Labbrozzi, D.; Bello, E.D. Morningness-eveningness personality and creative thinking among young people who play recreational sport. *Soc. Behav. Personal.* **2011**, *39*, 503–518. [CrossRef]
24. Rominger, C.; Fink, A.; Weber, B.; Papousek, I.; Schwerdtfeger, A.R. Everyday bodily movement is associated with creativity independently from active positive affect: A Bayesian mediation analysis approach. *Sci. Rep.* **2020**, *10*, 1–9. [CrossRef]
25. Nakagawa, T.; Koan, I.; Chen, C.; Matsubara, T.; Hagiwara, K.; Lei, H.; Hirotsu, M.; Yamagata, H.; Nakagawa, S. Regular moderate-to vigorous-intensity physical activity rather than walking is associated with enhanced cognitive functions and mental health in young adults. *Int. J. Environ. Res. Public Health* **2020**, *17*, 614. [CrossRef]
26. Piya-Amornphan, N.; Santiworakul, A.; Cetthakrikul, S.; Srirug, P. Physical activity and creativity of children and youths. *BMC Pediatrics* **2020**, *20*, 1–7. [CrossRef] [PubMed]
27. Physical Activity Guidelines Advisory Committee. *2018 Physical Activity Guidelines Advisory Committee Scientific Report*; Department of Health and Human Services: Washington, DC, USA, 2018.
28. Hoffman, M.D.; Hoffman, D.R. Exercisers achieve greater acute exercise-induced mood enhancement than nonexercisers. *Arch. Phys. Med. Rehabil.* **2008**, *89*, 358–363. [CrossRef]
29. Hallgren, M.Å.; Moss, N.D.; Gastin, P. Regular exercise participation mediates the affective response to acute bouts of vigorous exercise. *J. Sports Sci. Med.* **2010**, *9*, 629.
30. Parfitt, G.; Markland, D.; Holmes, C. Responses to physical exertion in active and inactive males and females. *J. Sport Exerc. Psychol.* **1994**, *16*, 178–186. [CrossRef]
31. Boutcher, S.H.; McAuley, E.; Courneya, K.S. Positive and negative affective response of trained and untrained subjects during and after aerobic exercise. *Aust. J. Psychol.* **1997**, *49*, 28–32. [CrossRef]
32. Daley, A.J.; Welch, A. Subjective exercise experiences during and after high and low intensity exercise in active and inactive adult females: Some preliminary findings. *J. Sports Med. Phys. Fit.* **2003**, *43*, 220.

33. Craig, C.L.; Marshall, A.L.; Sjöström, M.; Bauman, A.E.; Booth, M.L.; Ainsworth, B.E.; Oja, P. International physical activity questionnaire: 12-country reliability and validity. *Med. Sci. Sports Exerc.* **2003**, *35*, 1381–1395. [CrossRef] [PubMed]
34. Runco, M.A.; Acar, S. Divergent thinking as an indicator of creative potential. *Creat. Res. J.* **2012**, *24*, 66–75. [CrossRef]
35. Reiter-Palmon, R.; Forthmann, B.; Barbot, B. Scoring divergent thinking tests: A review and systematic framework. *Psychol. Aesthet. Creat. Arts* **2019**, *13*, 144. [CrossRef]
36. Czyz-Szypenbej, K.; Medrzycka-Dabrowska, W.; Sak-Dankosky, N. Neurocognitive Testing—Do We Lack in Expertise? *Crit. Care Med.* **2019**, *47*, e531. [CrossRef]
37. Tago, A. *Atama No Taiso Best [Mental Gymnastics: Best]*; Kobunsha: Tokyo, Japan, 2009; pp. 55, 56, 75, 76, 133, 134.
38. Colzato, L.S.; Szapora Ozturk, A.; Pannekoek, J.N.; Hommel, B. The impact of physical exercise on convergent and divergent thinking. *Front. Hum. Neurosci.* **2013**, *7*, 824. [CrossRef] [PubMed]
39. Bonett, D.G.; Wright, T.A. Sample size requirements for estimating Pearson, Kendall and Spearman correlations. *Psychometrika* **2000**, *65*, 23–28. [CrossRef]
40. Zhu, W. Sadly, the earth is still round ($p < 0.05$). *J. Sport Health Sci.* **2012**, *1*, 9–11.
41. Hötting, K.; Schickert, N.; Kaiser, J.; Röder, B.; Schmidt-Kassow, M. The effects of acute physical exercise on memory, peripheral BDNF, and cortisol in young adults. *Neural Plast.* **2016**, *2016*, 12. [CrossRef] [PubMed]
42. Stimpson, N.J.; Davison, G.; Javadi, A.H. Joggin' the noggin: Towards a physiological understanding of exercise-induced cognitive benefits. *Neurosci. Biobehav. Rev.* **2018**, *88*, 177–186. [CrossRef]
43. Hattori, S.; Naoi, M.; Nishino, H. Striatal dopamine turnover during treadmill running in the rat: Relation to the speed of running. *Brain Res. Bull.* **1994**, *35*, 41–49. [CrossRef]
44. Aarts, E.; van Holstein, M.; Cools, R. Striatal dopamine and the interface between motivation and cognition. *Front. Psychol.* **2011**, *2*, 163. [CrossRef]
45. Clayton, R.; Thomas, C.; Smothers, J. How to do Walking Meetings Right. Harvard Business Review. 2015. Available online: https://hbr.org/2015/08/how-to-do-walking-meetings-right (accessed on 26 July 2021).
46. Jonas, W. Walking Meetings: The Future of Safely Collaborating: How to Increase Your Creativity, Productivity, and Wellbeing at Work. Psychology Today. 2021. Available online: https://www.psychologytoday.com/us/blog/how-healing-works/202103/walking-meetings-the-future-safely-collaborating (accessed on 26 July 2021).
47. Ritter, S.M.; Dijksterhuis, A. Creativity—The unconscious foundations of the incubation period. *Front. Hum. Neurosci.* **2014**, *8*, 215. [CrossRef] [PubMed]

Article

Higher Handgrip Strength Is Linked to Better Cognitive Performance in Chinese Adults with Hypertension

Shenghua Lu [1,2], Fabian Herold [3,4], Yanjie Zhang [5,6,*], Yuruo Lei [7,*], Arthur F. Kramer [8,9], Can Jiao [10,11], Qian Yu [10,11], Scott Doig [12], Jinming Li [10,11], Zhe Yan [10,11], Jin Kuang [10,11], Ting Wang [10,11] and Liye Zou [10,11]

1. Hunan Academy of Education Sciences, Changsha 225002, China; lushenghua6666@126.com
2. College of Sports Science, Jishou University, Jishou 416000, China
3. Research Group Neuroprotection, German Center for Neurodegenerative Diseases (DZNE), Leipziger Str. 44, 39120 Magdeburg, Germany; fabian.herold@st.ovgu.de
4. Department of Neurology, Medical Faculty, Otto von Guericke University, Leipziger Str. 44, 39120 Magdeburg, Germany
5. Health and Exercise Science Laboratory, Institute of Sports Science, Seoul National University, Seoul 08826, Korea
6. Physical Education Unit, School of Humanities and Social Science, The Chinese University of Hong Kong-Shenzhen, Shenzhen 518172, China
7. Institute of Urban Governance, Shenzhen University, Shenzhen 518060, China
8. Center for Cognitive and Brain Health, Northeastern University, Boston, MA 02115, USA; a.kramer@northeastern.edu
9. Beckman Institute, University of Illinois at Urbana-Champaign, Champaign, IL 61801, USA
10. Institute of KEEP Collaborative Innovation, Shenzhen University, Shenzhen 518060, China; jiaocan@szu.edu.cn (C.J.); yuqianmiss@163.com (Q.Y.); jinmingli1999@gmail.com (J.L.); zoeyan0610@163.com (Z.Y.); dennyppg89@gmail.com (J.K.); janewang10142021@163.com (T.W.); liyezou123@gmail.com (L.Z.)
11. Exercise Psychophysiology Laboratory, School of Psychology, Shenzhen University, Shenzhen 518060, China
12. Department of Physical Education, Limestone University, Gaffney, SC 29340, USA; srdoig@limestone.edu
*   Correspondence: elite_zhangyj@163.com (Y.Z.); yrlei@szu.edu.cn (Y.L.)

Citation: Lu, S.; Herold, F.; Zhang, Y.; Lei, Y.; Kramer, A.F.; Jiao, C.; Yu, Q.; Doig, S.; Li, J.; Yan, Z.; et al. Higher Handgrip Strength Is Linked to Better Cognitive Performance in Chinese Adults with Hypertension. *Brain Sci.* **2021**, *11*, 985. https://doi.org/10.3390/brainsci11080985

Academic Editors: Filipe Manuel Clemente and Ana Filipa Silva

Received: 29 June 2021
Accepted: 19 July 2021
Published: 25 July 2021

Publisher's Note: MDPI stays neutral with regard to jurisdictional claims in published maps and institutional affiliations.

Abstract: Objective: There is growing evidence that in adults, higher levels of handgrip strength (HGS) are linked to better cognitive performance. However, the relationship between HGS and cognitive performance has not been sufficiently investigated in special cohorts, such as individuals with hypertension who have an intrinsically higher risk of cognitive decline. Thus, the purpose of this study was to examine the relationship between HGS and cognitive performance in adults with hypertension using data from the Global Ageing and Adult Health Survey (SAGE). Methods: A total of 4486 Chinese adults with hypertension from the SAGE were included in this study. Absolute handgrip strength (aHGS in kilograms) was measured using a handheld electronic dynamometer, and cognitive performance was assessed in the domains of short-term memory, delayed memory, and language ability. Multiple linear regression models were fitted to examine the association between relative handgrip strength (rHGS; aHGS divided by body mass index) and measures of cognitive performance. Results: Overall, higher levels of rHGS were associated with higher scores in short-term memory ($\beta = 0.20$) and language ($\beta = 0.63$) compared with the lowest tertiles of rHGS. In male participants, higher HGS was associated with higher scores in short-term memory ($\beta = 0.31$), language ($\beta = 0.64$), and delayed memory ($\beta = 0.22$). There were no associations between rHGS and cognitive performance measures in females. Conclusion: We observed that a higher level of rHGS was associated with better cognitive performance among hypertensive male individuals. Further studies are needed to investigate the neurobiological mechanisms, including sex-specific differences driving the relationship between measures of HGS and cognitive performance in individuals with hypertension.

Keywords: handgrip strength; cognition; hypertension; China

1. Introduction

Hypertension is a major cause of cardiovascular diseases (e.g., stroke), affecting about 40% of the aging population globally [1]. The increasing incidence of hypertension characterized by high blood pressure has become an urgent public health problem, and nearly 60% of Chinese older adults are suffering from hypertension [2]. In this context, it is worth noting that hypertension can increase the risk of cognitive decline since it affects the brain negatively [3–6], although the neurobiological mechanisms driving the relationship between hypertension and cognitive performance are controversially discussed and not fully understood. In a study by Tzourio et al., a non-significant association between hypertension and cognitive performance was observed [7], whereas other studies reported that hypertension (i) was associated with structural brain damage, (ii) was associated with a deficit in cognitive performance, and (iii) could accelerate the development of dementia [3,5,8,9]. Moreover, cognitive impairments can affect individuals' mental health and social functioning and even cause an increase in the economic burden to take care of individual symptoms caused by the disease [10]. Notably, it has also been observed that individuals with hypertension who meet established physical activity guidelines showed superior cognitive performance [11], thus buttressing the claim that regular physical exercises are important to ensure brain health and cognition in adults with hypertension [12]. Collectively, the available evidence shows that hypertension has a detrimental effect on cognitive health, and thus identifying risk and protective factors before a manifestation of cognitive impairment occurs is essential in individuals with hypertension.

Handgrip strength (HGS), an important indicator of physical function, has been linked to various physical health outcomes (e.g., increased risk of falls, depression, and pre-mortality) among older adults [13–16]. Moreover, there is mounting evidence that measures of HGS are associated with cognitive health [17–19]. In particular, previous studies have established a relationship between measures of HGS and cognitive performance in younger adults [20], middle-aged adults [21], and older adults [22,23]. For example, a longitudinal study of older adults indicated that the participants in the highest quartile of rHGS had a 50% lower risk to develop cognitive impairments compared with participants in the lowest quartile of rHGS [24]. In line with this finding, a cross-sectional study of obese females reported that individuals in the lowest tertile of aHGS had 84% higher odds for cognitive impairments compared with obese females in the highest tertile of aHGS [25]. In this context, there is also emerging evidence that biological sex could be an important moderator in the relationship between physical fitness and cognitive performance [26–29]. However, sex-specific effects in the relationships of HGS and cognitive performance have not been extensively investigated, even though there is some evidence that the association between HGS and prevalence of mild cognitive impairments is more pronounced in men [28]. Collectively, these findings imply that individuals with lower measures of HGS may have a higher risk to develop cognitive impairments and that this relationship is moderated by biological sex.

Despite the fact that the above-mentioned studies have shown that measures of HGS can be important markers to identify adults at higher risk for cognitive decline, it has not been sufficiently studied whether this finding can be generalized to an adult population with hypertension. Adults with hypertension are intrinsically at a higher risk to develop cognitive impairments (e.g., due to hypertension-related brain changes). To address this gap in the literature, this study investigated whether rHGS is associated with performance in specific cognitive subdomains in both male and female adults with hypertension. Thus, the current study will add novel evidence to the literature, as it is among the first studies that (i) investigated the relationship between rHGS and performance in specific cognitive domains in hypertensive adults and (ii) examined the moderating role of biological sex in the relationship between rHGS and cognitive performance. Based on the available evidence, we hypothesized that in adults with hypertension, higher levels of rHGS are associated with better cognitive performance.

2. Materials and Methods

2.1. Study Population

This cross-sectional study used publicly available data from the Global Ageing and Adult Health Survey (SAGE: http://www.who.int/healthinfo/sage/en accessed on 29 March 2021). This survey assessed data of a nationally representative sample of Chinese individuals and was carried out between 2007 and 2010 across eight provinces. All procedures of data collection were reported in a previous study in more detail [30]. In brief, well-trained interviewers conducted face-to-face household interview surveys (questionnaire) to collect the data. The survey response rate was 93%. The study was approved by the World Health Organization Ethical Review Committee and the Chinese Ethics Research Review Board of Peking University (IRB00001052-11014 and IRB00001052-11015). Each participant signed informed written consent before she/he participated in the experiment.

Data were extracted on demographic characteristics (e.g., age and sex), handgrip strength, cognitive performance, and diagnosed hypertension from the Chinese cohort of SAGE. All individuals without hypertension were excluded and were not considered in the data analysis. As a result, a total of 4486 individuals with hypertension were included for analysis in this study.

2.2. Hypertension Diagnosis

Participants were defined as hypertensive if they selected "Yes" on "Have you ever been diagnosed with high blood pressure (hypertension)?" or if they had high blood pressure values, which were systolic blood pressure (SBP) $\geq$140 mmHg and/or diastolic blood pressure (DBP) $\geq$90 mm Hg [31].

2.3. Handgrip Strength

Absolute handgrip strength (aHGS) in kilograms (kg) was measured by trained assessors using a Smedley's hand dynamometer (Scandidact, Oldenvej 45, and 3490 Kvistgard, Denmark) [32]. The participants were seated in a comfortable chair with their feet flat on the ground. They were asked to keep their upper arm against their body and to bend their elbow to 90 degrees with their palm facing in (as if shaking hands). The trained assessors instructed the seated participants to hold the dynamometer. The participants were asked to squeeze the hand as hard as they could, and each hand was tested twice. The overall HGS was the average of the best aHGS for each hand. If the participants had any surgery on their hand, arm, or wrist in the past 3 months or arthritis or pain in the hand/wrist/arm, aHGS was not assessed for that hand. As previously described [33], we calculated relative handgrip strength (rHGS) by dividing overall HGS (i.e., the average of the best aHGS) by body mass index (BMI).

2.4. Cognitive Performance

Cognitive performance was assessed using (i) the Digit Span Forward and Backward Test, (ii) the Verbal Fluency Test, and (iii) the Verbal Recall Test.

The Digit Span Forward (DSF) and Digit Span Backward (DSB) tests were used to quantify short-term memory and working memory performance. The participants were required to repeat a set of orally presented items (i.e., numbers) while the level of difficulty was gradually increased by one item for each trial. Both the DSF and the DSB start with a sequence of two numbers (e.g., 9 − 5), and for each span, two trials are performed. For each trial, a new sequence of items was presented. The researcher stopped the test after two consecutive incorrect responses on the same set of items [34]. Performance was scored based on the maximum length of correctly remembered items. The highest achievable scores for the DSF and DSB were 9 points and 8 points, respectively.

The semantic Verbal Fluency Test (VFT) probes language abilities (i.e., lexical retrieval and production). Participants were required to name as many animals as they could within one minute. One point was given for a correctly named animal (nonrepeating) [35].

The Verbal Recall Test (VRT) was used to measure performance in delayed memory. An interviewer read 10 words for participants three times at a standardized pace. After a delay of 10 min, the participants were required to recall as many words as possible from the initially presented 10 words. One point was given for each correctly recalled word [36].

2.5. Independent Variables

Age, sex, years of education, setting (rural or urban), alcohol consumption in the past month, smoking (never, current, or past), chronic physical conditions (stroke and diabetes), and regular level of physical activity (PA; occupational and leisure PA) [37] were used as covariates in the statistical models. Please note that the classifications of stroke and diabetes were based on self-reports of existing diagnoses from medical professionals. In line with a previous study, PA was measured using the Global Physical Activity Questionnaire Version 2 (GPAQ-V2), which consists of 16 questions [38]. The GPAQ-V2, which was developed under the auspices of the World Health Organization (WHO), is an internationally recognized, reliable, and validated questionnaire that assesses levels of regular physical activity [39,40]. According to the PA calculation from a previous study [37], levels of physical activity were classified as $\geq$150 min/week (meeting the recommended guidelines) and <150 min/week (low PA).

2.6. Statistical Analysis

In this study, continuous variables are expressed as means and standard deviation (SD), whereas categorical variables are shown as percentages. On the basis of the known sex difference in aHGS and cognitive performance in later life [22], we analyzed males and females separately (see Tables 1–3). Multivariable linear regression models were used to estimate the associations between rHGS and cognitive performance (Digit Span Forward and Backward Test, Verbal Fluency Test, and Verbal Recall Test). As the interaction terms of rHGS and sex in our linear regression showed a significant difference in all cognitive performance tests ($p < 0.05$), we estimated the associations between rHGS and cognitive performance (Digit Span Forward and Backward Test, Verbal Fluency Test, and Verbal Recall Test) and adjusted the regression models for age, sex, education years, setting, alcohol consumption, smoking, chronic physical conditions, and physical activity. All statistical analyses were conducted using the Stata 15.0 (Stata Corp LP, College Station, Texas). The level of statistical significance for all statistical tests was set at $p < 0.05$ (two-tailed).

Table 1. Overview of demographic characteristics and measures of handgrip strength and cognitive performance.

Variables		Male (3620)	Female (866)
Age (in years)		59.68 ± 0.19	59.39 ± 0.40
Education (in years)		7.54 ± 0.07	6.59 ± 0.18
Setting (in %)	Rural	59.01	52.89
	Urban	40.99	47.11
Alcohol consumption (in %)	Yes	70.80	53.00
	No	29.20	47.00
Smoking (in %)	Never	20.75	86.14
	Current	64.71	10.00
	Past	14.54	3.86
Work physical activity (in %)	≤150 min/week	76.24	83.72
	>150 min/week	23.76	16.28
Leisure physical activity (in %)	≤150 min/week	97.29	97.69
	>150 min/week	2.71	2.31
Stroke (in %)	Yes	3.15	2.08
	No	96.85	97.92
Diabetes (in %)	Yes	4.63	4.52
	No	95.37	95.48
Handgrip strength (kg)		33.74 ± 0.18	22.95 ± 0.28
Digit span forward (score)		7.33 ± 0.03	7.21 ± 0.05
Digit span backward (score)		3.67 ± 0.02	3.06 ± 0.05
Verbal fluency (score)		13.82 ± 0.08	12.77 ± 0.15
Delay recall (score)		5.19 ± 0.04	5.13 ± 0.08

Table 2. Participants' characteristics based on the tertiles of handgrip strength in male and female participants.

Variables	Q3 (1.579–6.347)	Q2 (1.147–1.578)	Q1 (0.224–1.146)	p Value
Male				
Age (n = 3421) (in years)	55.46 ± 0.30	60.49 ± 0.29	65.21 ± 0.37	$p < 0.001$
Handgrip strength (kg)	42.46 ± 0.23	32.68 ± 0.14	21.30 ± 0.22	$p < 0.001$
Digit span forward (n = 3414) (score)	7.61 ± 0.04	7.37 ± 0.04	6.90 ± 0.05	$p < 0.001$
Digit span backward (n = 3406) (score)	3.81 ± 0.04	3.60 ± 0.04	3.50 ± 0.05	$p < 0.001$
Verbal fluency (n = 3414) (score)	14.53 ± 0.14	13.80 ± 0.14	12.84 ± 0.17	$p < 0.001$
Delay recall (n = 3409) (score)	5.58 ± 0.06	5.22 ± 0.06	4.67 ± 0.08	$p < 0.001$
Female				
Age (n = 829) (in years)	53.98 ± 1.53	55.22 ± 0.76	61.20 ± 0.47	$p < 0.001$
Handgrip strength (kg)	37.23 ± 1.00	29.23 ± 0.28	19.29 ± 0.25	$p < 0.001$
Digit span forward (n = 829) (score)	7.41 ± 0.19	7.51 ± 0.09	7.13 ± 0.06	$p = 0.02$
Digit span backward (n = 827) (score)	3.57 ± 0.15	3.25 ± 0.09	2.96 ± 0.06	$p < 0.001$
Verbal fluency (n = 829) (score)	14.05 ± 0.55	13.19 ± 0.32	12.52 ± 0.19	$p = 0.014$
Delay recall (n = 825) (score)	6.07 ± 0.19	5.84 ± 0.15	4.86 ± 0.10	$p < 0.001$

Note: Q1, Low tertile of relative handgrip strength (rHGS); Q2, Moderate tertile of relative handgrip strength (rHGS); Q3, High tertile of relative handgrip strength (rHGS).

Table 3. Regression models of the association between handgrip strength or covariates and the indices of cognitive performance in adults with hypertension.

Variables		Q1	Q2 B (95% Confidence Interval)	Q3 B (95% Confidence Interval)	R2
Digit span forward (n = 3534)	Total	Reference	0.18 (0.07, 0.30) ***	0.20 (0.09, 0.32) ***	0.11
	Male	Reference	0.27 (0.14, 0.40) ***	0.31 (0.18, 0.44) ***	0.12
	Female	Reference	0.11 (−0.14,0.37)	−0.12 (−0.51, 0.28)	0.12
Digit span backward (n = 3530)	Total	Reference	0.01 (−0.09, 0.12)	−0.04 (−0.07, 0.15)	0.16
	Male	Reference	−0.05 (−0.17, 0.08)	−0.05 (−0.18, 0.07)	0.15
	Female	Reference	−0.12 (−0.35, 0.12)	−0.07 (−0.43, 0.30)	0.25
Verbal fluency (n = 3534)	Total	Reference	0.39 (−0.11, 0.79)	0.63 (0.22, 1.05) **	0.09
	Male	Reference	0.46 (−0.01, 0.93)	0.64 (0.16, 1.11) **	0.09
	Female	Reference	0.14 (−0.94, 0.67)	0.55 (−0.71, 1.84)	0.13
Delay recall (n = 3532)	Total	Reference	0.19 (0.01, 0.37) *	0.15 (−0.03, 0.33)	0.13
	Male	Reference	0.22 (0.01, 0.42) *	0.22 (0.01,0.42) *	0.13
	Female	Reference	0.27 (−0.14, 0.70)	0.31 (−0.32,0.95)	0.18

Note: Q1, Low tertile of relative handgrip strength (rHGS); Q2, Moderate tertile of relative handgrip strength (rHGS); Q3, High tertile of relative handgrip strength (rHGS). The adjusted model included age, sex, education years, setting, alcohol consumption, smoking, health condition, work physical activity, and leisure physical activity; *, $p < 0.05$, **, $p < 0.01$; ***, $p < 0.001$.

3. Results

The final sample included in this study consisted of 4486 adults with hypertension. Table 1 provides an overview of the demographic characteristics, rHGS, and cognitive performance. The mean age for the sample was 59.68 ± 0.19 for male participants and 59.39 ± 0.40 for female participants. Furthermore, our sample included more male participants (80.7%) than female participants (19.3%).

As shown in Table 1, we observed significant differences between male and female participants concerning the level of education, setting, alcohol consumption, smoking, level of regular occupational PA, and performance in DSF, DSB, and VFT (all $p < 0.05$).

Table 2 displays the characteristics of participants according to the tertile of HGS for male and female participants. We observed that age, DSF, DSB, VFT, and VRT varied as a function of the tertile of HGS (all $p < 0.05$).

Table 3 displays the adjusted associations between HGS and cognitive function scores in adults with hypertension. Overall, when controlling for age, sex, level of education, setting, alcohol consumption, smoking, level of regular occupational PA, and leisure PA, individuals in the highest tertile of rHGS performed better in DSF and VFT compared with individuals in the lowest tertile of HGS ($p < 0.01$). Independent associations were found between rHGS and performance in DSF and VRT ($p < 0.05$) for male participants in the tertile of moderate and high rHGS. In the highest tertile of rHGS, rHGS was associated with performance on the VFT ($p < 0.01$).

4. Discussion

The current study investigated the associations of rHGS and cognitive performance in a Chinese sample of hypertensive individuals recruited from the SAGE project. Overall, our findings suggest that in adults with hypertension, higher levels of rHGS are associated with better cognitive performance in short-term memory (assessed using the Digit Span Forward and Digit Span Backward tests) and language abilities (assessed using the Verbal Fluency test) after controlling for important covariates (e.g., age, sex). Furthermore, we observed pronounced sex differences, as an association between higher levels of rHGS and better performance on the DSF, VFT, and VRT could be observed in male participants but not in female participants.

Our findings are in line with previous studies showing a relationship between the level of HGS and scores in short-term memory, language, and delayed memory (but not working memory) [18,41]. For example, a scoping review suggested that higher levels of HGS were associated with a lower risk of cognitive impairments in older adults [18]. A prospective study by Alfaro-Acha et al. found that lower aHGS was associated with a deficit in cognitive function in 2160 older Mexican Americans over 7 years [41]. In a cross-sectional study of 449 adults without dementia from Sweden, Praetorius Bjork et al. showed that there are strong relationships between aHGS and short-term memory, delayed memory, and language abilities [42].

Furthermore, the current study substantiates the available evidence regarding the link between measures of HGS and cognitive performance in adults with hypertension. A comparable study observed a link between aHGS and better visuospatial abilities, episodic memory, orientation/attention, and overall cognitive function in a cohort of middle-aged and older adults with hypertension [43]. Thus, our study adds evidence to the current literature that (i) in adults with hypertension, the link between rHGS and cognitive performance does not only comprise visuospatial abilities, episodic memory, orientation/attention, and overall cognitive function but also encompasses short-term memory, delayed memory, and language abilities and (ii) there are pronounced sex-specific differences in the association between rHGS and cognitive performance.

There is some evidence from a cross-sectional study [29] and a meta-analysis of interventional studies [44] that females profit more from physical interventions than males, although this finding is not universal [45]. Even if the mentioned studies are not fully comparable with our study, the direction of the sex-specific differences in our study is somewhat surprising. As suggested by others, this could be caused by sex-specific differences in neurobiological mechanisms driving behavioral performance.

Overall, the findings of the current study suggest that higher levels of rHGS are linked to better cognitive performance, although the neurobiological mechanisms driving this relationship are yet not fully clear and warrant future investigation [46–48]. In this context, several possible mechanisms could explain the positive relationship between measures of HGS and cognitive performance. In particular, it has been proposed that the following levels of analysis need to be considered to understand the effect of physical activity or physical fitness (i.e., level of HGS) on cognitive performance: (i) level 1—molecular and cellular changes, (ii) level 2—functional and structural brain changes, and (iii) level 3—socioemotional changes [47,49].

Our knowledge concerning the relationship of measures of HGS and cognition-relevant changes in level 1 (cellular and molecular changes) is meager. With respect to the association between cardiovascular fitness and/or cardiovascular exercise and cognitive performance, brain-derived neurotrophic factor (BDNF) and other neurotrophic factors have been highlighted to play a crucial role [49–51]. In this regard, there is some evidence that (i) cardiorespiratory fitness level is associated with basal levels of BDNF [52,53], (ii) the lower levels of serum BDNF are associated with a smaller hippocampal volume and poorer memory performance [54], and (iii) changes in BDNF levels in response to long-term aerobic training are associated with changes in executive functions [55] and hippocampal volume [56]. Collectively, these studies suggest that BDNF plays a crucial role in the relationship between cardiorespiratory fitness and cognitive performance. However, comparable studies investigating the relationship between HGS, neurotrophic factors such as BDNF, and cognitive performance are, to the best of our knowledge, currently lacking. Thus, further studies investigating the relationship between HGS and neurotrophic factors (e.g., BNDF) are needed to elucidate the extent to which changes on this level of analysis can explain some variance in the positive relationship between measures of HGS and cognitive performance.

With respect to level 2 (functional and structural brain changes), it has been proposed that HGS and higher-order cognitive functions might share the same neural substrates [7]. Indeed, there is some evidence in the literature suggesting a close relationship between HGS and brain features relevant to higher-order cognitive processes. Concerning functional brain changes, it has been observed that in younger adults, higher levels of normalized HGS are linked to favorable cerebral hemodynamics [47]. However, in this study, normalized HGS was not linked to cognitive performance, nor did this study find convincing evidence that cortical hemodynamics mediate a possible relationship between normalized HGS and cognitive performance [7]. However, as this finding cannot be readily generalized on the basis of our cohort, additional studies are needed to elucidate whether HGS might influence cortical hemodynamics in adults with hypertension. Regarding structural brain changes, there is evidence in the literature linking higher levels of aHGS to greater hippocampal volume [57]. Given that a greater hippocampal volume [58] and an increase in hippocampal volume [56] are linked to better cognitive performance (i.e., spatial memory), it seems plausible to hypothesize that alterations in brain structure at least partly influence the relationship between higher levels of HGS and cognitive performance. This idea is reinforced by the fact that (i) resistance training improves hippocampal integrity in older adults with mild cognitive impairment [59,60] and (ii) the hippocampus mediates the relationship between higher levels of physical fitness (i.e., cardiorespiratory fitness) and spatial memory performance [56,58]. In this context, it could be speculated that the neurobiological processes leading to a higher HGS might be protective against the frequently reported hypertension-related worsening of brain integrity (e.g., faster decline in hippocampal volume and more global brain matter atrophy) and cognitive performance (e.g., more pronounced decline in executive function and memory) [61–64]. However, to the best of our knowledge, there is a lack of cross-sectional and longitudinal studies that assess and analyze the relationship between measures of HGS, brain structure, and cognitive performance. Thus, future investigations that provide empirical evidence to verify or refute the above-mentioned theoretical assumptions while considering sex-related differences in structural brain changes (e.g., those reported by intervention studies [44]) are urgently needed.

Level 3 (socioemotional changes) comprises changes in mood, stress, pain, and sleep. It is beyond the scope and intention of this article to discuss the relationship between HGS and these socioemotional factors. However, sleep has been highlighted as an important mediator in the relationship between physical activity and/or physical fitness and cognitive performance [49,65]. It was observed that unhealthy sleep patterns (e.g., too short or too long sleep duration and insufficient sleep quality) can be linked to weaker HGS and faster decline of HGS in adults [66–68]. Furthermore, empirical evidence suggests that sleep patterns mediate the relationship between regular physical activity and cognitive performance (e.g., inhibition performance) [69,70]. It seems reasonable to hypothesize that changes in socioemotional factors (e.g., sleep) and their influence on the other levels of analysis (e.g., the cellular and molecular level and the functional and structural level) can at least partly explain the positive relationships between measures of HGS and cognitive performance. However, given the lack of research in this direction, future studies are required to test these assumptions empirically.

Limitations

The findings of the current study need to be interpreted in light of some limitations. First, this cross-sectional study does not allow the assessment of causal mechanisms driving the relationship between rHGS and cognitive performance. Second, we used only rHGS as a predictor of physical performance, whereas other measures of physical performance (e.g., gait speed and sit-to-stand performance), which have been used in previous studies investigating the relationship between physical fitness and cognitive performance in hypertensive adults [43,71], were not analyzed because of a lack of available data. Third, the data assessment for the SAGE study was carried out between 2007 to 2010, which might affect the generalizability of the findings to a certain extent. However, given the fact that hypertension is still a major health issue, as it constitutes an important risk factor for heart diseases, stroke, chronic kidney disease, and dementia [72,73], we believe that the findings of the current study are still of interest for the scientific community.

5. Conclusions

In conclusion, the findings of the current study suggest that a higher level of rHGS is related to better performance in specific domains of cognition (i.e., short-term memory, delayed memory, and language abilities) in our sample of Chinese adults with hypertension (especially in male adults). Further studies are needed to investigate the neurobiological mechanisms driving this relationship, including the identification of sex-specific differences.

Author Contributions: Conceptualization, Y.Z. and L.Z.; methodology, S.L. and Y.Z.; software, S.L., C.J. and Y.Z.; validation, Y.Z., F.H., and Y.L.; formal analysis, F.H. and Y.Z.; investigation, F.H., Y.Z., J.L., Z.Y., Y.L., J.K., and T.W.; resources, Y.Z., Q.Y., and L.Z.; data curation, F.H. and Y.Z.; writing—original draft preparation, S.L., F.H., and Y.Z.; writing—review and editing, F.H., Y.Z., A.F.K., C.J., Q.Y., S.D., and L.Z.; project administration, Y.Z., C.J., and L.Z.; funding acquisition, S.L. All authors have read and agreed to the published version of the manuscript.

Funding: This research was funded by the Social Science Foundation of Hunan Province (19YBA286) and the Key Research Project of Hunan Educational Bureau (19A407).

Institutional Review Board Statement: The open available data used in this study was approved by the World Health Organization Ethical Review Committee and the Chinese Ethics Research Review Board of Peking University (IRB00001052-11014 and IRB00001052-11015).

Informed Consent Statement: Informed consent was obtained from all subjects involved in the study.

Data Availability Statement: The publicly archived datasets can be retrieved from http://www.who.int/healthinfo/sage/en accessed on 29 March 2021.

Conflicts of Interest: The authors declare no conflict of interest.

References

1. Leung, A.A.; Daskalopoulou, S.S.; Dasgupta, K.; McBrien, K.; Butalia, S.; Zarnke, K.B.; Nerenberg, K.; Harris, K.C.; Nakhla, M.; Cloutier, L.; et al. Hypertension Canada's 2017 Guidelines for Diagnosis, Risk Assessment, Prevention, and Treatment of Hypertension in Adults. *Can. J. Cardiol.* **2017**, *33*, 557–576. [CrossRef]
2. Lloyd-Sherlock, P.; Beard, J.; Minicuci, N.; Ebrahim, S.; Chatterji, S. Hypertension among older adults in low and middle-income countries: Prevalence, awareness and control. *Int. J. Epidemiol.* **2014**, *43*, 116–128. [CrossRef]
3. Walker, K.A.; Power, M.C.; Gottesman, R.F. Defining the Relationship Between Hypertension, Cognitive Decline, and Dementia: A Review. *Curr. Hypertens. Rep.* **2017**, *19*, 24. [CrossRef]
4. Forte, G.; De Pascalis, V.; Favieri, F.; Casagrande, M. Effects of Blood Pressure on Cognitive Performance: A Systematic Review. *J. Clin. Med.* **2019**, *9*, 34. [CrossRef]
5. Novak, V.; Hajjar, I. The relationship between blood pressure and cognitive function. *Nat. Rev. Cardiol.* **2010**, *7*, 686–698. [CrossRef]
6. Beauchet, O.; Celle, S.; Roche, F.; Bartha, R.; Montero-Odasso, M.; Allali, G.; Annweiler, C. Blood pressure levels and brain volume reduction: A systematic review and meta-analysis. *J. Hypertens.* **2013**, *31*, 1502–1516. [CrossRef] [PubMed]

7. Tzourio, C.; Laurent, S.; Debette, S. Is hypertension associated with an accelerated aging of the brain? *Hypertension* **2014**, *63*, 894–903. [CrossRef]
8. Hughes, T.M.; Sink, K.M. Hypertension and Its Role in Cognitive Function: Current Evidence and Challenges for the Future. *Am. J. Hypertens.* **2016**, *29*, 149–157. [CrossRef] [PubMed]
9. Qiu, C.; Winblad, B.; Fratiglioni, L. The age-dependent relation of blood pressure to cognitive function and dementia. *Lancet Neurol.* **2005**, *4*, 487–499. [CrossRef]
10. Fu, C.; Li, Z.; Mao, Z. Association between social activities and cognitive function among the elderly in china: A cross-sectional study. *Int. J. Environ. Res. Public Health* **2018**, *15*, 231. [CrossRef]
11. Frith, E.; Loprinzi, P.D. Physical activity and cognitive function among older adults with hypertension. *J. Hypertens.* **2017**, *35*, 1271–1275. [CrossRef] [PubMed]
12. Rêgo, M.L.M.; Cabral, D.A.R.; Costa, E.C.; Fontes, E.B. Physical Exercise for Individuals with Hypertension: It Is Time to Emphasize its Benefits on the Brain and Cognition. *Clin. Med. Insights Cardiol.* **2019**, *13*. [CrossRef]
13. García-Hermoso, A.; Ramírez-Vélez, R.; Peterson, M.D.; Lobelo, F.; Cavero-Redondo, I.; Correa-Bautista, J.E.; Martínez-Vizcaíno, V. Handgrip and knee extension strength as predictors of cancer mortality: A systematic review and meta-analysis. *Scand. J. Med. Sci. Sport.* **2018**, *28*, 1852–1858. [CrossRef] [PubMed]
14. Mcgrath, R.; Robinson-Lane, S.G.; Cook, S.; Clark, B.C.; Herrmann, S.; O'connor, M.L.; Hackney, K.J. Handgrip Strength Is Associated with Poorer Cognitive Functioning in Aging Americans. *J. Alzheimer's Dis.* **2019**, *70*, 1187–1196. [CrossRef] [PubMed]
15. McGrath, R.; Vincent, B.M.; Hackney, K.J.; Robinson-Lane, S.G.; Downer, B.; Clark, B.C. The Longitudinal Associations of Handgrip Strength and Cognitive Function in Aging Americans. *J. Am. Med. Dir. Assoc.* **2020**, *21*, 634–639.e1. [CrossRef] [PubMed]
16. McGrath, R.; Cawthon, P.M.; Cesari, M.; Al Snih, S.; Clark, B.C. Handgrip Strength Asymmetry and Weakness Are Associated with Lower Cognitive Function: A Panel Study. *J. Am. Geriatr. Soc.* **2020**, *68*, 2051–2058. [CrossRef] [PubMed]
17. Carson, R.G. Get a grip: Individual variations in grip strength are a marker of brain health. *Neurobiol. Aging* **2018**, *71*, 189–222. [CrossRef]
18. Fritz, N.E.; McCarthy, C.J.; Adamo, D.E. Handgrip strength as a means of monitoring progression of cognitive decline—A scoping review. *Ageing Res. Rev.* **2017**, *35*, 112–123. [CrossRef]
19. Sternäng, O.; Reynolds, C.A.; Finkel, D.; Ernsth-Bravell, M.; Pedersen, N.L.; Dahl Aslan, A.K. Grip strength and cognitive abilities: Associations in old age. *J. Gerontol. Ser. B Psychol. Sci. Soc. Sci.* **2016**, *71*, 841–848. [CrossRef]
20. Choudhary, A.K.; Jiwane, R.; Alam, T.; Kishanrao, S.S. Grip Strength and Impact on Cognitive Function in Healthy Kitchen Workers. *Achiev. Life Sci.* **2016**, *10*, 168–174. [CrossRef]
21. Adamo, D.E.; Anderson, T.; Koochaki, M.; Fritz, N.E. Declines in grip strength may indicate early changes in cognition in healthy middle-aged adults. *PLoS ONE* **2020**, *15*, e0232021. [CrossRef]
22. Yang, L.; Koyanagi, A.; Smith, L.; Hu, L.; Colditz, G.A.; Toriola, A.T.; Felipe López Sánchez, G.; Vancampfort, D.; Hamer, M.; Stubbs, B.; et al. Hand grip strength and cognitive function among elderly cancer survivors. *PLoS ONE* **2018**, *13*, e0197909. [CrossRef] [PubMed]
23. Veronese, N.; Smith, L.; Barbagallo, M.; Yang, L.; Zou, L.; Haro, J.M.; Koyanagi, A. Sarcopenia and fall-related injury among older adults in five low- and middle-income countries. *Exp. Gerontol.* **2021**, *147*, 111262. [CrossRef]
24. Kim, G.R.; Sun, J.; Han, M.; Nam, C.M.; Park, S. Evaluation of the directional relationship between handgrip strength and cognitive function: The Korean Longitudinal Study of Ageing. *Age Ageing* **2019**, *48*, 426–432. [PubMed]
25. Jeong, S.M.; Choi, S.; Kim, K.; Kim, S.M.; Kim, S.; Park, S.M. Association among handgrip strength, body mass index and decline in cognitive function among the elderly women. *BMC Geriatr.* **2018**, *18*, 1–9. [CrossRef]
26. Barha, C.K.; Hsu, C.L.; ten Brinke, L.; Liu-Ambrose, T. Biological Sex: A Potential Moderator of Physical Activity Efficacy on Brain Health. *Front. Aging Neurosci.* **2019**, *11*, 329. [CrossRef]
27. Barha, C.K.; Liu-Ambrose, T. Exercise and the Aging Brain: Considerations for Sex Differences. *Brain Plast.* **2018**, *4*, 53–63. [CrossRef]
28. Liu, X.; Chen, J.; Geng, R.; Wei, R.; Xu, P.; Chen, B.; Liu, K.; Yang, L. Sex- and age-specific mild cognitive impairment is associated with low hand grip strength in an older Chinese cohort. *J. Int. Med. Res.* **2020**, *48*, 1–11. [CrossRef]
29. Barha, C.K.; Best, J.R.; Rosano, C.; Yaffe, K.; Catov, J.M.; Liu-Ambrose, T. Sex-specific relationship between long-term maintenance of physical activity and cognition in the health ABC Study: Potential role of hippocampal and dorsolateral prefrontal cortex volume. *J. Gerontol. Ser. A Biol. Sci. Med. Sci.* **2020**, *75*, 764–770. [CrossRef]
30. Kowal, P.; Chatterji, S.; Naidoo, N.; Biritwum, R.; Fan, W.; Ridaura, R.L.; Maximova, T.; Arokiasamy, P.; Phaswana-Mafuya, N.; Williams, S.; et al. Data resource profile: The world health organization study on global ageing and adult health (SAGE). *Int. J. Epidemiol.* **2012**, *41*, 1639–1649. [CrossRef] [PubMed]
31. Unger, T.; Borghi, C.; Charchar, F.; Khan, N.A.; Poulter, N.R.; Prabhakaran, D.; Ramirez, A.; Schlaich, M.; Stergiou, G.S.; Tomaszewski, M.; et al. 2020 International Society of Hypertension Global Hypertension Practice Guidelines. *Hypertension* **2020**, *75*, 1334–1357. [CrossRef] [PubMed]
32. Ramlagan, S.; Peltzer, K.; Phaswana-Mafuya, N. Hand grip strength and associated factors in non-institutionalised men and women 50 years and older in South Africa. *BMC Res. Notes* **2014**, *7*, 8. [CrossRef]

33. Kim, Y.M.; Kim, S.; Bae, J.; Kim, S.H.; Won, Y.J. Association between relative hand-grip strength and chronic cardiometabolic and musculoskeletal diseases in Koreans: A cross-sectional study. *Arch. Gerontol. Geriatr.* **2021**, *92*, 104181. [CrossRef]
34. Wilke, J.; Stricker, V.; Usedly, S. Free-weight resistance exercise is more effective in enhancing inhibitory control than machine-based training: A randomized, controlled trial. *Brain Sci.* **2020**, *10*, 702. [CrossRef] [PubMed]
35. Zhao, Q.; Guo, Q.; Hong, Z. Clustering and switching during a semantic verbal fluency test contribute to differential diagnosis of cognitive impairment. *Neurosci. Bull.* **2013**, *29*, 75–82. [CrossRef]
36. Gildner, T.E.; Liebert, M.A.; Kowal, P.; Chatterji, S.; Snodgrass, J.J. Associations between sleep duration, sleep quality, and cognitive test performance among older adults from six middle income countries: Results from the study on global ageing and adult health (SAGE). *J. Clin. Sleep Med.* **2014**, *10*, 613–621. [CrossRef]
37. Vancampfort, D.; Stubbs, B.; Lara, E.; Vandenbulcke, M.; Swinnen, N.; Koyanagi, A. Mild cognitive impairment and physical activity in the general population: Findings from six low- and middle-income countries. *Exp. Gerontol.* **2017**, *100*, 100–105. [CrossRef] [PubMed]
38. Armstrong, T.; Bull, F. Development of the World Health Organization Global Physical Activity Questionnaire (GPAQ). *J. Public Health (Bangkok)* **2006**, *14*, 66–70. [CrossRef]
39. Bull, F.C.; Maslin, T.S.; Armstrong, T. Global physical activity questionnaire (GPAQ): Nine country reliability and validity study. *J. Phys. Act. Heal.* **2009**, *6*, 790–804. [CrossRef]
40. Keating, X.D.; Zhou, K.; Liu, X.; Hodges, M.; Liu, J.; Guan, J.; Phelps, A.; Castro-Piñero, J. Reliability and concurrent validity of global physical activity questionnaire (GPAQ): A systematic review. *Int. J. Environ. Res. Public Health* **2019**, *16*, 4128. [CrossRef]
41. Alfaro-Acha, A.; Al Snih, S.; Raji, M.A.; Kuo, Y.F.; Markides, K.S.; Ottenbacher, K.J. Handgrip strength and cognitive decline in older Mexican Americans. *J. Gerontol. Ser. A Biol. Sci. Med. Sci.* **2006**, *61*, 859–865. [CrossRef]
42. Praetorius Bjork, M.; Johansson, B.; Hassing, L.B. I forgot when I lost my grip-strong associations between cognition and grip strength in level of performance and change across time in relation to impending death. *Neurobiol. Aging* **2016**, *38*, 68–72. [CrossRef]
43. Zuo, M.; Gan, C.; Liu, T.; Tang, J.; Dai, J.; Hu, X. Physical Predictors of Cognitive Function in Individuals With Hypertension: Evidence from the CHARLS Basline Survey. *West. J. Nurs. Res.* **2019**, *41*, 592–614. [CrossRef] [PubMed]
44. Barha, C.K.; Davis, J.C.; Falck, R.S.; Nagamatsu, L.S.; Liu-Ambrose, T. Sex differences in exercise efficacy to improve cognition: A systematic review and meta-analysis of randomized controlled trials in older humans. *Front. Neuroendocrinol.* **2017**, *46*, 71–85. [CrossRef] [PubMed]
45. Falck, R.S.; Davis, J.C.; Best, J.R.; Crockett, R.A.; Liu-Ambrose, T. Impact of exercise training on physical and cognitive function among older adults: A systematic review and meta-analysis. *Neurobiol. Aging* **2019**, *79*, 119–130. [CrossRef] [PubMed]
46. Herold, F.; Törpel, A.; Schega, L.; Müller, N.G. Functional and/or structural brain changes in response to resistance exercises and resistance training lead to cognitive improvements—A systematic review. *Eur. Rev. Aging Phys. Act.* **2019**, *16*, 10. [CrossRef] [PubMed]
47. Herold, F.; Behrendt, T.; Törpel, A.; Hamacher, D.; Müller, N.G.; Schega, L. Cortical hemodynamics as a function of handgrip strength and cognitive performance: A cross-sectional fNIRS study in younger adults. *BMC Neurosci.* **2021**, *22*, 1–16. [CrossRef]
48. Herold, F.; Müller, P.; Gronwald, T.; Müller, N.G. Dose–Response Matters!—A Perspective on the Exercise Prescription in Exercise–Cognition Research. *Front. Psychol.* **2019**, *10*, 2338. [CrossRef]
49. Stillman, C.M.; Cohen, J.; Lehman, M.E.; Erickson, K.I. Mediators of physical activity on neurocognitive function: A review at multiple levels of analysis. *Front. Hum. Neurosci.* **2016**, *10*, 626. [CrossRef]
50. Stimpson, N.J.; Davison, G.; Javadi, A.H. Joggin' the Noggin: Towards a Physiological Understanding of Exercise-Induced Cognitive Benefits. *Neurosci. Biobehav. Rev.* **2018**, *88*, 177–186. [CrossRef]
51. Walsh, E.I.; Smith, L.; Northey, J.; Rattray, B.; Cherbuin, N. Towards an understanding of the physical activity-BDNF-cognition triumvirate: A review of associations and dosage. *Ageing Res. Rev.* **2020**, *60*, 101044. [CrossRef] [PubMed]
52. Schmalhofer, M.L.; Markus, M.R.P.; Gras, J.C.; Kopp, J.; Janowitz, D.; Grabe, H.J.; Groß, S.; Ewert, R.; Gläser, S.; Albrecht, D.; et al. Sex-Specific associations of brain-derived neurotrophic factor and cardiorespiratory fitness in the general population. *Biomolecules* **2019**, *9*, 630. [CrossRef] [PubMed]
53. Currie, J.; Ramsbottom, R.; Ludlow, H.; Nevill, A.; Gilder, M. Cardio-respiratory fitness, habitual physical activity and serum brain derived neurotrophic factor (BDNF) in men and women. *Neurosci. Lett.* **2009**, *451*, 152–155. [CrossRef] [PubMed]
54. Erickson, K.I.; Prakash, R.S.; Voss, M.W.; Chaddock, L.; Heo, S.; McLaren, M.; Pence, B.D.; Martin, S.A.; Vieira, V.J.; Woods, J.A.; et al. Brain-derived neurotrophic factor is associated with age-related decline in hippocampal volume. *J. Neurosci.* **2010**, *30*, 5368–5375. [CrossRef] [PubMed]
55. Leckie, R.L.; Oberlin, L.E.; Voss, M.W.; Prakash, R.S.; Szabo-Reed, A.; Chaddock-Heyman, L.; Phillips, S.M.; Gothe, N.P.; Mailey, E.; Vieira-Potter, V.J.; et al. BDNF mediates improvements in executive function following a 1-year exercise intervention. *Front. Hum. Neurosci.* **2014**, *8*, 101044. [CrossRef]
56. Erickson, K.I.; Voss, M.W.; Prakash, R.S.; Basak, C.; Szabo, A.; Chaddock, L.; Kim, J.S.; Heo, S.; Alves, H.; White, S.M.; et al. Exercise training increases size of hippocampus and improves memory. *Proc. Natl. Acad. Sci. USA* **2011**, *108*, 3017–3022. [CrossRef]

57. Firth, J.A.; Smith, L.; Sarris, J.; Vancampfort, D.; Schuch, F.; Carvalho, A.F.; Solmi, M.; Yung, A.R.; Stubbs, B.; Firth, J. Handgrip strength is associated with hippocampal volume and white matter hyperintensities in major depression and healthy controls: A UK biobank study. *Psychosom. Med.* **2020**, *82*, 39–46. [CrossRef]
58. Erickson, K.I.; Prakash, R.S.; Voss, M.W.; Chaddock, L.; Hu, L.; Morris, K.S.; White, S.M.; Wójcicki, T.R.; McAuley, E.; Kramer, A.F. Aerobic fitness is associated with hippocampal volume in elderly humans. *Hippocampus* **2009**, *19*, 1030–1039. [CrossRef]
59. Suo, C.; Singh, M.F.; Gates, N.; Wen, W.; Sachdev, P.; Brodaty, H.; Saigal, N.; Wilson, G.C.; Meiklejohn, J.; Singh, N.; et al. Therapeutically relevant structural and functional mechanisms triggered by physical and cognitive exercise. *Mol. Psychiatry* **2016**, *21*, 1633–1642. [CrossRef]
60. Broadhouse, K.M.; Singh, M.F.; Suo, C.; Gates, N.; Wen, W.; Brodaty, H.; Jain, N.; Wilson, G.C.; Meiklejohn, J.; Singh, N.; et al. Hippocampal plasticity underpins long-term cognitive gains from resistance exercise in MCI. *NeuroImage Clin.* **2020**, *25*, 102182. [CrossRef]
61. Debette, S.; Seshadri, S.; Beiser, A.; Au, R.; Himali, J.J.; Palumbo, C.; Wolf, P.A.; DeCarli, C. Midlife vascular risk factor exposure accelerates structural brain aging and cognitive decline. *Neurology* **2011**, *77*, 461–468. [CrossRef] [PubMed]
62. Shah, C.; Srinivasan, D.; Erus, G.; Schmitt, J.E.; Agarwal, A.; Cho, M.E.; Lerner, A.J.; Haley, W.E.; Kurella Tamura, M.; Davatzikos, C.; et al. Changes in brain functional connectivity and cognition related to white matter lesion burden in hypertensive patients from SPRINT. *Neuroradiology* **2021**, *63*, 913–924. [CrossRef] [PubMed]
63. Muller, M.; Sigurdsson, S.; Kjartansson, O.; Aspelund, T.; Lopez, O.L.; Jonnson, P.V.; Harris, T.B.; Van Buchem, M.; Gudnason, V.; Launer, L.J. Joint effect of mid- and late-life blood pressure on the brain: The AGES-Reykjavik Study. *Neurology* **2014**, *82*, 2187–2195. [CrossRef] [PubMed]
64. Firbank, M.J.; Wiseman, R.M.; Burton, E.J.; Saxby, B.K.; O'Brien, J.T.; Ford, G.A. Brain atrophy and white matter hyperintensity change in older adults and relationship to blood pressure: Brain atrophy, WMH change and blood pressure. *J. Neurol.* **2007**, *254*, 713–721. [CrossRef] [PubMed]
65. Stillman, C.M.; Esteban-Cornejo, I.; Brown, B.; Bender, C.M.; Erickson, K.I. Effects of Exercise on Brain and Cognition Across Age Groups and Health States. *Trends Neurosci.* **2020**, *43*, 533–543. [CrossRef] [PubMed]
66. Selvamani, Y.; Arokiasamy, P.; Chaudhary, M. Himanshu Association of sleep problems and sleep duration with self-rated health and grip strength among older adults in India and China: Results from the study on global aging and adult health (SAGE). *J. Public Health* **2018**, *26*, 697–707. [CrossRef]
67. Wang, T.Y.; Wu, Y.; Wang, T.; Li, Y.; Zhang, D. A prospective study on the association of sleep duration with grip strength among middle-aged and older Chinese. *Exp. Gerontol.* **2018**, *103*, 88–93. [CrossRef] [PubMed]
68. Auyeung, T.W.; Kwok, T.; Leung, J.; Lee, J.S.W.; Ohlsson, C.; Vandenput, L.; Wing, Y.K.; Woo, J. Sleep Duration and Disturbances Were Associated With Testosterone Level, Muscle Mass, and Muscle Strength-A Cross-Sectional Study in 1274 Older Men. *J. Am. Med. Dir. Assoc.* **2015**, *16*, 630.e1–630.e6. [CrossRef]
69. Wilckens, K.A.; Erickson, K.I.; Wheeler, M.E. Physical Activity and Cognition: A Mediating Role of Efficient Sleep. *Behav. Sleep Med.* **2018**, *16*, 569–586. [CrossRef]
70. Li, L.; Yu, Q.; Zhao, W.; Herold, F.; Cheval, B.; Kong, Z.; Li, J.; Mueller, N.; Kramer, A.F.; Cui, J.; et al. Physical Activity and Inhibitory Control: The Mediating Role of Sleep Quality and Sleep Efficiency. *Brain Sci.* **2021**, *11*, 664. [CrossRef]
71. Basile, G.; Catalano, A.; Mandraffino, G.; Crucitti, A.; Ciancio, G.; Morabito, N.; Lasco, A. Cognitive impairment and slow gait speed in elderly outpatients with arterial hypertension: The effect of blood pressure values. *J. Am. Geriatr. Soc.* **2015**, *63*, 1260–1261. [CrossRef] [PubMed]
72. Zhou, B.; Perel, P.; Mensah, G.A.; Ezzati, M. Global epidemiology, health burden and effective interventions for elevated blood pressure and hypertension. *Nat. Rev. Cardiol.* **2021**, 1–18.
73. Ungvari, Z.; Toth, P.; Tarantini, S.; Prodan, C.I.; Sorond, F.; Merkely, B.; Csiszar, A. Hypertension-induced cognitive impairment: From pathophysiology to public health. *Nat. Rev. Nephrol.* **2021**, 1–16.

Article

Using Brain-Breaks® as a Technology Tool to Increase Attitude towards Physical Activity among Students in Singapore

Govindasamy Balasekaran [1,*], Ahmad Arif Bin Ibrahim [1], Ng Yew Cheo [2], Phua Kia Wang [3], Garry Kuan [4], Biljana Popeska [5], Ming-Kai Chin [6], Magdalena Mo Ching Mok [7,8], Christopher R. Edginton [9], Ian Culpan [10] and J. Larry Durstine [11]

1 Physical Education & Sports Science, National Institute of Education, Nanyang Technological University, Singapore 637616, Singapore; arif20091993@gmail.com
2 Sports & Physical Education, Singapore University of Social Sciences, Singapore 599494, Singapore; yewcheo@gmail.com
3 Ministry of Education, Singapore 138675, Singapore; phua_kia_wang@moe.gov.sg
4 Exercise and Sports Science Programme, School of Health Sciences, Universiti Sains Malaysia, Kubang Kerian 16150, Malaysia; garry@usm.my
5 Faculty of Educational Sciences, Goce Delcev University, 2000 Stip, North Macedonia; biljana.popeska@ugd.edu.mk
6 The Foundation for Global Community Health, 1550 W Horizon Ridge Pkwy Ste R #206, Henderson, NV 89012, USA; chinmingkai@yahoo.com
7 Graduate Institute of Educational Information and Measurement, National Taichung University of Education, 140 Minsheng Road, West District, Taichung City 40306, Taiwan; mmcmok@friends.eduhk.hk
8 Assessment Research Centre, Department of Psychology, The Education University of Hong Kong, 10 Lo Ping Road, Taipo, N.T., Hong Kong
9 Department of Health, Recreation and Community Services, University of Northern Iowa, Cedar Falls, IA 50614, USA; christopher.edginton@uni.edu
10 School of Health Sciences, University of Canterbury, Christchurch 8140, New Zealand; ian.culpan@canterbury.ac.nz
11 Department of Exercise Science, University of South Carolina, Columbia, SC 29208, USA; ldurstin@mailbox.sc.edu
* Correspondence: govindasamy.b@nie.edu.sg

Citation: Balasekaran, G.; Ibrahim, A.A.B.; Cheo, N.Y.; Wang, P.K.; Kuan, G.; Popeska, B.; Chin, M.-K.; Mok, M.M.C.; Edginton, C.R.; Culpan, I.; et al. Using Brain-Breaks® as a Technology Tool to Increase Attitude towards Physical Activity among Students in Singapore. *Brain Sci.* **2021**, *11*, 784. https://doi.org/10.3390/brainsci11060784

Academic Editors: Filipe Manuel Clemente and Ana Filipa Silva

Received: 22 May 2021
Accepted: 11 June 2021
Published: 14 June 2021

Publisher's Note: MDPI stays neutral with regard to jurisdictional claims in published maps and institutional affiliations.

Abstract: The purpose of this study was to investigate the effects of classroom-based Brain Breaks® Physical Activity Solution in Southeast Asia Singaporean primary school students and their attitude towards physical activity (PA) over a ten-week intervention. A total of 113 participants (8–11 years old) were randomly assigned to either an experimental (EG) or a control group (CG), with six classes to each group; the Brain Breaks® group (EG: six classes) and the Control group (CG: six classes). All EG members participated in a Brain Breaks® video intervention (three–five min) during academic classes and the CG continued their lessons as per normal. The student's attitudes towards PA in both research conditions were evaluated using the self–reported Attitudes toward Physical Activity Scale (APAS), applied before and after intervention. The effects of the intervention on APAS scores were analysed using a mixed model analysis of variance with Time as within-subject and Group as between-subject factors. The analysis revealed evidence in support of the positive effect of classroom video interventions such as Brain Breaks® on student's attitudes toward benefits, importance, learning, self-efficacy, fun, fitness, and trying to do their personal best in PA. The Brain Breaks® intervention provided a positive significant impact on students in Singapore. This study also revealed that interactive technology tools implemented into the school curriculum benefit students in terms of health and education.

Keywords: video exercises; physical activity; attitudes; online platform; Brain Breaks®

1. Introduction

The World Health Organization [1] has defined a person who has a body mass index (BMI) of over 30 $kg·m^{-2}$ as obese, and ≥ 25 $kg·m^{-2}$ as overweight. Research has concluded that Asians tend to carry a higher percentage of body fat as compared to other racial and ethnic groups of the same BMI [2]. Therefore, the BMI scale for Asians has been lowered (obese: 27.5 kg·m^{-2}, overweight: 23 kg·m^{-2}). Being overweight and obese during childhood years is linked to chronic diseases risk factors such as diabetes and cardiovascular diseases [3,4]. Furthermore, childhood obesity can persist into adulthood [5–7]. Globally, childhood obesity has been on the rise [8].

A possible reason for increased childhood obesity is the availability of current technology. When students use technology, participation in physical activity (PA) is reduced. By the age of 10 years, students have access to at least five different types of screens for viewing at home [9], the use of which is referred to as "screen time". Significant correlations between the rise of screen time and the lack of PA in students are associated with the rise in obesity [10–13]. Maher et al. [14] evaluated 2200 Australian students aged 9 to 16 years old and found a high correlation between screen time and the likelihood for a student to be overweight or obese.

The WHO recommends 60 min of daily PA for students 5 to 17 years old [15]. A recent study reported that students are becoming less physically active and more sedentary [16,17]. These trends were established by tracking students PA. Accelerometers were used to measure students' activity levels during the ages of 6, 9, and 11 years. Students' PA duration decreased from an average of 66 min a day at age 6 to an average of 53 min a day at age 11 [18]. Jago et al.'s [18] study demonstrated that students spend less time doing PA as they age and are well below WHO's recommended duration of daily PA. On average, students lost about 63 min of PA weekly at age 11 compared to age 6. Yearly, this decrease equates to 3276 min of lost PA time for the students. The investigation provided additional evidence, suggesting students are becoming less physically active as they grow older. The loss of PA time has likely led to the rise in students' obesity levels. Jago et al. [18] reported that, at the start of their study, 11% of the students were overweight and 8% were obese. However, by the end of the study, 14% of the students became overweight and 15% were obese. Most students who were overweight or obese at the start of the study remained overweight or obese at the end of the study.

The current generation of students are referred to as "tech-savvy" and show a growing interest in technology. Boone et al. [10] and Lewallen et al. [19] suggest that technology can encourage students to increase their PA levels. Presently, Singaporean students are moving towards digital platforms for learning, gaming, and PA. Singapore educators are actively encouraging the incorporation of technology into lessons or co-curricular activities (CCA) to assist students to cope with the necessary competencies for living in a globalized world. Technology such as HOPSports Brain Break® videos, online streaming, and virtual reality games such as Pokemon GO were developed to increase students' and adults' PA time [20]. Althoff et al. [21], focusing on the influence of Pokemon Go on PA levels, found a significant increase in PA by sedentary users when starting to play using this particular form of technology. Results were calculated by tracking the number of steps the user took before, during, and after playing a game. However, no significant difference was found in users who were already physically active.

Exercise videos such as HOPSports and JumpJam are becoming more popular with students. The current study specifically selected HOPSports Brain Breaks® videos, as these videos utilize a dynamic online platform that is closely aligned to the Whole School, Whole Community, and Whole Child (WSCC) Guidelines [22,23], and the United Nations Sustainable Development Goals (UNSDG) [24,25]. Singapore, a small country with limited land and lack of natural resources, recognizes the challenges of sustainable development. Prime Minister Lee Hsien Loong stated that Singapore is committed to the 2030 Agenda for Sustainable Development [26]. Further, the UNSDG statement encompasses social–emotional learning, nutritional education, PA and education, career education, and environmental education all into one online platform. These values are also in-line with Singapore's Ministry of Education's (MOE) Desired Outcomes of Education and 21st Century Competencies [27]. All classrooms in local schools are equipped with internet connections, a desktop computer, a sound system, and a projector or interactive whiteboard to ensure that students and teachers are able to keep abreast with advancements in technology.

Brain Breaks[®] videos are video exercises which average three–five min duration. Previous studies concluded that the use of Brain Breaks[®] videos help develop positive changes in students' attitudes towards PA [28–32]. Students who completed the intervention did simple aerobic/movement exercises following the video instructions. Also included in these videos was content pertaining to health and nutrition, social learning, character building, and arts and culture [24,33]. The results of these studies indicated a positive change in the intervention group's attitudes and interest towards PA. Krause & Benavidez [34] found that technology presented a more effective way to promote PA as compared to the traditional games and sports. As Singapore is a highly developed country, technology leverage is a more effective way to promote school students' PA. Brain Breaks[®], as an intervention tool, has already been shown to improve students' knowledge, self-awareness, and positive attitude leading towards motivation for increased PA [35].

Thus, the aim of this study was to investigate the use of Brain Breaks[®] videos and the videos' effects on Singaporean students' attitudes towards PA and possible increasing PA participation. This study is the first to examine the use of Brain Breaks[®] videos in the context of Singapore's students and their school system. We hypothesize that Brain Breaks[®] videos will positively impact students' attitudes, which may increase PA participation.

2. Materials and Methods

2.1. Research Design

This study was a two-group (experimental/control) quasi-experimental design. The experimental group (EG) participated in the Brain Breaks[®] intervention program of performing the Brain Breaks[®] video for 10 weeks, averaging three–five min daily during their class time, five days per week. The Brain Breaks[®] video was projected on a screen using a projector in the classroom. The videos featured physical movement activities, accompanied with songs and dance, and movements that can be done safely by maintaining adequate social distance between students (the full content of the program can be retrieved at Kuan et al. [24]. Students were invited to follow the movements shown on the screen. To maintain students' enjoyment and motivation, a variety of videos were played for each of the five days. Online access to the official project website is found at https://brain-breaks.com, (accessed on 1 March 2019) [17,24]. The control group (CG) continued their academic lessons as per normal for 10 weeks consecutively without video intervention. Participants' attitudes toward PA in both groups were measured before and after the intervention using the self-reported Attitudes toward Physical Activity Scale (APAS) questionnaire. Data collection took place before the 10-week intervention in the first week of the school term, and again at the end of the intervention. Participants were obtained from 12 student classes from a local Singapore school system. These 12 intact classes were randomly assigned into either the EG (six classes) or the CG (six classes) groups using a computer-generated randomization (www.randomization.com, accessed on 1 March 2019).

2.2. Ethical Approval

This study obtained approval from the Institutional Review Board from Nanyang Technological University (NTU-IRB Reference Number-2019-01-025), and was a school collaborative research initiative. Parents and students voluntarily signed informed consent forms agreeing to participate in this study.

2.3. Participants

Participants comprised 113 (47 boys, 66 girls) clinically healthy students ranging from 8 to 11 years old (Table 1). According to the intact class to which the student belonged, the classes were separated into either the experimental Brain Breaks[®] group (EG: six classes of 48 total students) or the Control group (CG: six classes of 65 total students). All students in the recruited classes were invited to participate. Students with prior injuries or conditions such as heart problems were excluded. Students who were excused from physical activity as advised by their doctors or had not acquired parental consent were also excluded. The required sample size was estimated using G-Power Version 3.1. Based on the repeated measures ANOVA with two research conditions (experimental and control group) $\times 2$ time points (baseline, and post), statistical power set at 80% with a 95% confidence interval, and an effect size of 0.25 [17], a sample size of 98 was calculated. With a 15% dropout, a total of 113 was judged to be sufficient to detect the hypothesized between-condition differences.

Table 1. General Characteristics of the Participants (n = 113, boys = 47, girls = 66).

Variables	Total (n = 113) Mean $\pm$ SD	EG (n = 48) Mean $\pm$ SD	CG (n = 65) Mean $\pm$ SD
Gender			
Male (n, %)	47 (41.6%)	22 (45.8%)	25 (38.5%)
Female (n, %)	66 (58.4%)	26 (54.2%)	40 (61.5%)
Age (years)	9.68 $\pm$ 0.95	9.71 $\pm$ 0.99	9.66 $\pm$ 0.94
Height (m)	1.38 $\pm$ 8.27	1.37 $\pm$ 0.09	1.39 $\pm$ 0.09
Weight (kg)	35.21 $\pm$ 10.21	34.91 $\pm$ 10.97	35.43 $\pm$ 10.55
Body Mass Index (kg·m^{-2})	18.19 $\pm$ 2.86	18.24 $\pm$ 4.18	18.16 $\pm$ 3.67

Note. EG = Experimental Group, CG = Control Group; Age (years); Height (meters, m); Weight (kilograms, kg); Body Mass Index (kilograms per meter square, kg·m^{-2}). No significant difference between EG and CG was found for all variables above ($p > 0.05$).

2.4. Measures

Students' Attitudes toward Physical Activity Scale (APAS)

The APAS is a self-reported questionnaire used to measure beliefs, attitudes, and self-efficacy towards PA from students. The questionnaire is composed of seven sections using Likert-type scales. An additional section gathers demographic information regarding gender, age, school grade level, body height, and weight. The remaining seven sections referred to: (F1) 'promoting holistic health', 10 items constructed to measure students' attitudes toward the effectiveness of physical activities to promote holistic health. An example item is, "Being physically active helps to give me good health"; (F2) 'importance of exercise habit': five items designed to measure attitudes toward the importance of doing exercise as a lifestyle. An example item is, "It is important to be physically active for my health"; (F3) 'self-efficacy in learning with video exercises': 11 items to measure self-efficacy in learning curriculum content by using video exercises. Example items are, "I learn about art through exercise videos," and "I know how to do physical activity if there is an exercise video to follow"; (F4) 'self-efficacy in selecting video exercises for themselves': four items to measure a student's level of independence when performing their self-selected exercise video. An example item is, "I know how to choose physical activity in the exercise video that suits me." (F5) 'exercise motivation and enjoyment': 14 item scale designed to measure motivation and enjoyment when doing physical exercise. An example item is, "I think physical activity is fun"; (F6) 'self-confidence on physical fitness': eight items constructed to measure self-perception of physical fitness. An example item is, "I am confident with my balance"; (F7) 'trying to do my personal best': five items constructed to measure personal best goal orientation to engage in PA. An example item is, "My target is to go beyond what I have achieved in physical activity".

Questionnaire response options for each item were a four point Likert-type response category including "strongly disagree", "disagree", "agree", and "strongly agree". The seven scales in the original version of this questionnaire were validated for their reliabilities, uni-dimensionality, effectiveness of the response categories, and absence of gender differential item functioning (DIF) by Mok et al. [30] using the Rasch analysis. A subsequent study by Dinc et al. [31] updated the questionnaire and further enhanced internal consistency. The current study reported here made use of the updated version of APAS. The Cronbach's Alpha reliability coefficients for Singapore students ranged from 0.81 to 0.92 (Table 2).

Table 2. Cronbach's Alpha Reliability Analysis of APAS for Singapore Students ($n = 113$).

Scale	Number of Items	Cronbach's Alpha
Promoting Holistic Health (F1)	10	0.84
Importance of Exercise Habit (F2)	5	0.81
Self-efficacy in Learning with Video Exercises (F3)	11	0.93
Self-efficacy in Selecting Video Exercises (F4)	4	0.90
Exercise Motivation and Enjoyment (F5)	14	0.87
Self-confidence on Physical Fitness (F6)	8	0.92
Trying to do Personal Best (F7)	5	0.82

2.5. Data Analysis

Data analysed were completed using the IBM SPSS version 26.0 software (IBM Corp., Armonk, NY, USA). Distribution of the data including mean (M) + standard deviation (SD) of the variables was assessed for normality (Skewness and Kurtosis values were close to 0 and z-values ranged between -1.96 and 1.96). No non-normal distributions were identified. Effects of applied Brain Breaks® intervention on APAS scores were analysed using a two-way 2×2 mixed analysis of variance (ANOVA) with Time (pre-test/ post-test) as the within-subject factor (repeated measures) and Group (EG/CG) as the between-subject factor. The partial eta-squared (η^2) effect sizes for the tests were calculated to indicate the magnitude of the effects. The level of statistical significance was set as $p < 0.05$.

3. Results

A one-way ANOVA Brown–Forsythe test was used to compare pre-test between groups. As presented in Table 3, where CG indicated higher scores; pre-F4 (EG: 3.02 ± 0.62 vs. CG: 3.10 ± 0.60, $p < 0.001$), pre-F6 (EG: 2.97 ± 0.73 vs. CG: 3.05 ± 0.73, $p < 0.001$), pre-F7 (EG: 3.20 ± 0.61 vs. CG: 3.29 ± 0.65, $p < 0.001$). Table 3 presents the mean scores of the APAS obtained by EG and CG groups before and after intervention, the results of the 2×2 mixed ANOVA with one within-subject factor (Time: before or after intervention) and one between-subject factor (Group: experimental or control), as well as the effect sizes (Time η^2, Group η^2 and Time*Group η^2). A significant increase in the mean scores of all APAS scales, for EG and CG, between pre-and post-10-weeks intervention was found (Table 3). The main Time effect was significant ($p < 0.05$) for all APAS scales. Effect sizes of the Time factor ranged from 0.04 to 0.19.

The results found in Table 3 revealed significant main effects on Group for importance ($p = 0.012$), learning ($p < 0.001$), self-efficacy ($p < 0.001$), fun ($p = 0.034$), and fitness ($p = 0.022$) scales. Effect sizes for the Group main effect were relatively small ($\eta^2 < 0.12$) for all scales, except for Learning ($\eta^2 = 0.23$).

Time*Group interaction effects were significant for all APAS scales ($p < 0.05$) (Table 3). Effect sizes of the Time*Group interaction effect ranged from 0.08 (Personal Best scale) to 0.37 (Learning scale). EG showed an increase for all pre-test to post-test APAS scale scores when compared to CG (Figure 1). The one-way ANOVA Brown–Forsythe test was also used to compare between genders for EG. A significant difference was found for post-F3, where boys, when compared to girls, had higher self-efficacy scores in learning curriculum subjects through video exercises (EG boys: 3.76 ± 0.28 vs. EG girls: 3.54 ± 0.39, $p = 0.04$) (Figure 2).

Table 3. Descriptive Statistics and ANOVA at pre-test/post-test for students in the Experimental Group (EG; *n* = 48) and Control Group (CG; *n* = 65).

Variables on Physical Activity	Group	Pretest Mean (SD)	Posttest Mean (SD)	Time F	p	η^2	Group F	p	η^2	Time*Group F	p	η^2
Benefits (F1)	CG	3.19 (0.55)	3.13 (0.64)	18.87	<0.001	0.15	3.86	0.052	0.03	29.29	<0.001	0.21
	EG	3.06 (0.51)	3.61 (0.37)									
Importance (F2)	CG	3.28 (0.61)	3.32 (0.60)	17.89	<0.001	0.14	6.49	0.012	0.06	12.57	0.001	0.10
	EG	3.29 (0.52)	3.74 (0.31)									
Learning (F3)	CG	2.77 (0.78)	2.32 (0.88)	9.35	0.003	0.08	32.55	<0.001	0.23	66.08	<0.001	0.37
	EG	2.65 (0.67)	3.64 (0.36)									
Self-efficacy (F4)	CG	3.10 (0.60)	2.82 (0.74)	4.38	0.039	0.04	14.89	<0.001	0.12	33.45	<0.001	0.23
	EG	3.02 (0.62)	3.62 (0.51)									
Fun (F5)	CG	3.18 (0.58)	3.19 (0.66)	13.70	<0.001	0.11	4.59	0.034	0.04	13.27	<0.001	0.11
	EG	3.17 (0.48)	3.58 (0.38)									
Fitness (F6)	CG	3.05 (0.73)	3.07 (0.69)	25.92	<0.001	0.19	5.42	0.022	0.05	22.89	<0.001	0.17
	EG	2.97 (0.73)	3.65 (0.39)									
Personal Best (F7)	CG	3.29 (0.65)	3.32 (0.63)	11.59	<0.001	0.10	2.08	0.153	0.02	9.36	0.003	0.08
	EG	3.20 (0.61)	3.66 (0.43)									

Note. EG = Experimental Group, CG = Control Group. * Time and group interactions.

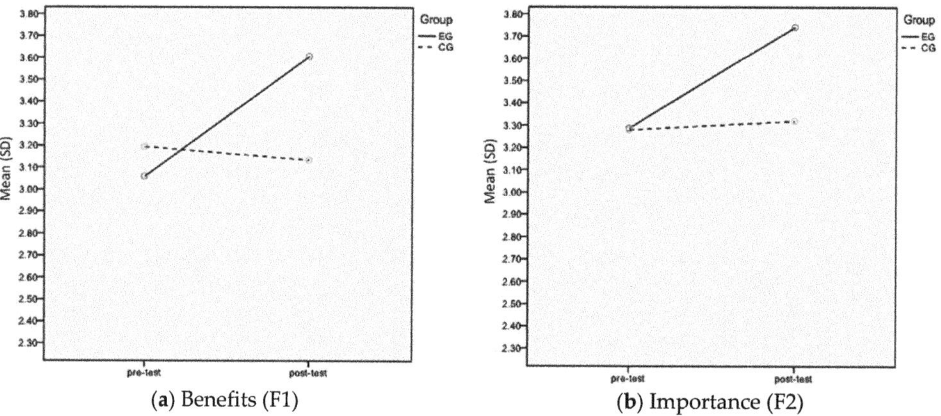

(**a**) Benefits (F1)　　(**b**) Importance (F2)

Figure 1. *Cont.*

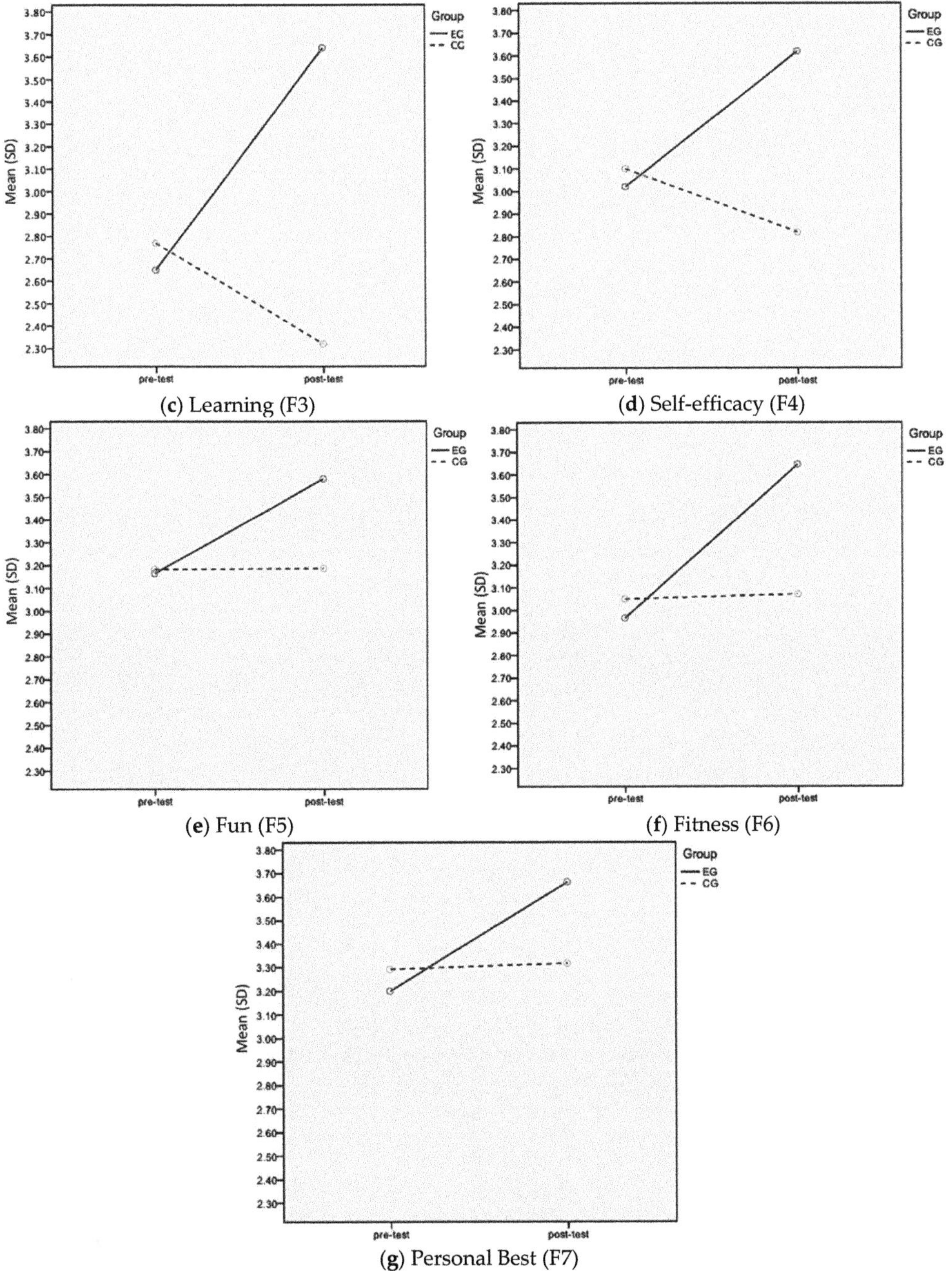

(**c**) Learning (F3)

(**d**) Self-efficacy (F4)

(**e**) Fun (F5)

(**f**) Fitness (F6)

(**g**) Personal Best (F7)

Figure 1. Scale Mean Values of the Experimental and the Control Groups at Pre-test and Post-test.

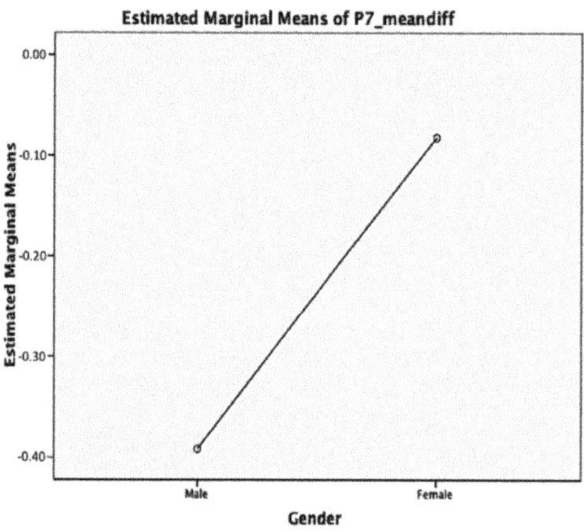

Figure 2. Post-Self–efficacy Scores in Learning Curriculum Subjects through Video Exercises of the Experimental Group.

4. Discussion

The purpose of this study was to determine whether using Brain Breaks® videos would positively change Singaporean students' attitudes towards PA and increase their PA participation. Empirical evidence in support of the positive effect of Brain Breaks® videos of three–five min a day, five times a week for 10 weeks, to enhance students' attitude towards PA was found. This study is the first conducted in Southeast Asia among Singaporean students. The findings from this study are similar to the findings from previous studies investigating the impact of Brain Breaks® videos on students aged 9 to 11 years old in China [25], Malaysia [16], Turkey [31], Lithuania [33], Poland [28] and Macedonia [32], students 12 years old [36], and students in higher education [31]. These previous studies found that students utilizing Brain Breaks® videos had positive changes in their attitudes towards PA. This difference could be due to the fun element, with movement and music, found in the exercise videos that the students performed during their short breaks. In addition, no significant differences were found between the scores of the experimental and control groups before the intervention. These earlier studies report that short amounts of exercise completed in the classroom are linked to increases in students' on-task behavior and PA [28,33], improved self-awareness, self-efficiency in using video exercises [32], fun, and effort to do their personal best [25] while participating in Brain Break® activities. Other studies have shown that students who are more physically active perform better in their academic studies [37,38]. One potential explanation for better academic achievement is that PA increases the blood flow to the brain and may be the link to better academic performance. With more blood flowing to the brain, oxygen delivery is increased, which potentially has a positive impact on brain function [38].

As the results from the current study show a positive increase in Singapore students' attitudes towards PA, Singapore school systems should consider the integration of daily Brain Breaks® into the school curriculum. This internet format is stable and requires state-of-the-art infrastructure (e.g., projectors, sound system, etc.) and technology (e.g., internet, fast wireless network, etc.). Utilizing such a system would assist students during long periods of continuous lessons where teachers and students can experience intellectual or mental fatigue that can reduce their ability to focus and concentrate. Short PA breaks between lessons would help teachers and students to refocus for the next lesson. Alternatively, Brain Breaks® videos can be integrated prior to the end of the students' recess or snack time.

In 2014, Singapore launched a project named the Smart Nation Initiative [39] to address topics such as health/well-being. Experiences gained from this collaboration enabled Singapore to grow and move the city's economy towards greater technology for services-based applications. These cooperative relationships allowed Singapore to make significant gains towards their smart and sustainable city goals by creating a system that provided the backbone to achieve other sustainable goals. These goals are, in part, attained by using either free or affordable high-speed internet. Globally, cities need to follow the example of Singapore in using advanced systems to achieve educational sustainability goals. Presently, almost every Singaporean household has access to the internet, and almost every student owns at least one technological device (e.g., smart phone, desktop, tablets, etc.). Hence, students are able to access the Brain Breaks® videos at home and carry out PA independently.

With the recent COVID-19 pandemic, the Ministry of Education (MOE) developed home-based learning (HBL) for all schools. Students can login into the school's system and attend online lessons. As students become accustomed and proficient with HBL, they can access Brain Breaks® videos. Hence, future studies can investigate whether greater access to exercise videos leads to increased students' motivation for learning while simultaneously keeping them physically active. Online PA programs do increase PA accessibility, are available, and should be used during and after the COVID-19 pandemic [40–45]. In this regard, development of online programs is needed, and these programs must be consistent with the United Nations Sustainable Development Goal 3 (Health and Well-Being) and Goal 11 (Sustainable Cities and Communities) [42,45–47].

Effective implementation of classroom-based PA intervention is highly dependent on the cooperation and interaction of children and teachers throughout the intervention [17]. The education system in Singapore is based on a systematic design and structured timetable. Furthermore, the teachers' creativity, behavior, cooperation, and personal motivation are essential in implementing the program. The teachers' knowledge about the benefits and advantages of the program do play an important role in the students' motivation. In addition, the school principal's support for the program is also an important factor in monitoring and ensuring that the students carry out the Brain Breaks® PA program in a safe environment.

The self-reported nature of APAS is one study limitation. For future studies, APAS could be supplemented by using accelerometers or pedometers to track the students' PA level. The added information from these measures would provide quantifiable data pertaining to students' PA, and provide insights into the physical responses of students other than their perceived attitudes toward PA. Another possibility is that the positive increase in student attitude to PA found in this study was from their Physical Education portion of the schools' curriculum. Therefore, future studies could evaluate the interaction of Brain Breaks® and current physical education curriculums.

Besides measuring attitudes toward PA that affect behavior and academic achievement, other variables such as changes in fundamental movement and motor abilities, students' effects on-time reaction, and sense of rhythm are also important elements that can be incorporated as part of future investigations to improve Brain Breaks® videos. This study is the first to assess the effects of Brain Breaks® on Singapore's primary school children. This acknowledgement, along with the study's experimental design and implementation in a real-world classroom setting, are distinct study strengths. These findings provide compelling evidence for the potential use of classroom-based PA and increased awareness of technological solutions that can increase Singapore's children PA engagement.

5. Conclusions

The results of this Southeast Asian study provide further international evidence on the positive benefits of Brain Breaks® Physical Activity Solutions on students' learning in support of evidence found by other international studies completed in China, Malaysia, Poland, North Macedonia, South Africa, and Turkey. This study adds to the understanding that investing as little as three–five min a day is enough time to improve students' perception of PA and subsequently lead students to engage in more PA during their free time. What remains unclear is whether these behaviors are carried into adulthood. From this study comes the recommendation to utilize and implement exercise videos as an interactive technology tool in the Singapore school system curriculum. The addition of exercise videos will benefit students in terms of enhancing their health and facilitating the education process.

Author Contributions: Conceptualization, G.B. and M.-K.C.; methodology, G.B., A.A.B.I., N.Y.C., P.K.W., G.K., B.P., and M.M.C.M.; formal analysis, G.B., G.K., M.M.C.M.; writing—original draft preparation, G.B., A.A.B.I., N.Y.C., P.K.W., G.K., B.P., M.-K.C., M.M.C.M., C.R.E., I.C., and J.L.D.; writing—review and editing, G.K., B.P., M.-K.C., M.M.C.M., C.R.E., I.C., and J.L.D.; supervision, G.B.; project administration, G.B.; funding acquisition, G.B. All authors have read and agreed to the published version of the manuscript.

Funding: This research received no external funding.

Institutional Review Board Statement: The study was conducted according to the guidelines of the Declaration of Helsinki and obtained approval from the Institutional Review Board from Nanyang Technological University (NTU-IRB Reference Number-2019-01-025).

Informed Consent Statement: Informed assent consent was obtained from all subjects and their parents before participating in the study.

Data Availability Statement: Data is available upon request from the corresponding author.

Acknowledgments: The cooperation of the Singaporean primary school and the school authorities, teachers, students and parents are greatly appreciated. The authors gratefully acknowledge especially classroom teachers for their assistance in implementing Break Breaks® intervention and their support with data collection for the study. The authors would like to also acknowledge HOPSports® Inc. for technical support.

Conflicts of Interest: Ming-Kai Chin is employed by HOPSports. But the structure of research data collection, analysis, and write-up responsibilities precluded this relationship from interfering with the project overall research implications and data interpretation. No other authors declare a conflict of interest.

References

1. World Health Organization. Commission on Ending Childhood Obesity. 2017. Available online: https://www.who.int/end-childhood-obesity/facts/en/ (accessed on 21 April 2021).
2. Poh, B.K.; Safiah, M.Y.; Tahir, A.; Haslinda, M.D.S.; Norazlin, N.S.; Norimah, A.K.; Manan, W.W.; Mirnalini, K.; Zalilah, M.S.; Azmi, M.Y.; et al. Physical Activity Pattern and Energy Expenditure of Malaysian Adults: Findings from the Malaysian Adult Nutrition Survey (MANS). *Malays. J. Nutr.* **2010**, *16*, 13–37. [PubMed]
3. Bacha, F.; Gidding, S. Cardiac abnormalities in youth with obesity and type 2 diabetes. *Curr. Diabetes Rep.* **2016**, *16*, 62. [CrossRef] [PubMed]
4. Hidrus, A.; Kueh, Y.C.; Norsaádah, B.; Chang, Y.-K.; Hung, T.-M.; Naing, N.N.; Kuan, G. Effects of brain-breaks videos on the motives for the physical activity of Malaysians with type-2 diabetes mellitus. *Int. J. Environ. Res. Public Health* **2020**, *17*, 2507. [CrossRef]
5. Kumar, S.; Kelly, A.S. Review of Childhood Obesity: From Epidemiology, Etiology, and Comorbidities to Clinical Assessment and Treatment. In *Mayo Clinic Proceedings*; Elsevier: Amsterdam, The Netherlands, 2017; Volume 92, pp. 251–265.
6. Lindberg, L.; Danielsson, P.; Persson, M.; Marcus, C.; Hagman, E. Association of childhood obesity with risk of early all-cause and cause-specific mortality: A Swedish prospective cohort study. *PLoS Med.* **2020**, *17*, e1003078. [CrossRef]
7. Ward, Z.J.; Long, M.W.; Resch, S.C.; Giles, C.M.; Cradock, A.L.; Gortmaker, S.L. Simulation of growth trajectories of childhood obesity into adulthood. *N. Engl. J. Med.* **2017**, *377*, 2145–2153. [CrossRef]
8. World Health Organization. Taking Action on Childhood Obesity. 2018. Available online: https://apps.who.int/iris/bitstream/handle/10665/274792/WHO-NMH-PND-ECHO-18.1-eng.pdf (accessed on 21 April 2021).
9. Sigman, A. Time for a view on screen time. *Arch. Dis. Child.* **2012**, *97*, 935–942. [CrossRef]
10. Boone, J.E.; Gordon-Larsen, P.; Adair, L.S.; Popkin, B.M. Screen time and physical activity during adolescence: Longitudinal effects on obesity in young adulthood. *Int. J. Behav. Nutr. Phys. Act.* **2007**, *4*, 26. [CrossRef] [PubMed]
11. Brindova, D.; Veselska, Z.D.; Klein, D.; Hamrik, Z.; Sigmundova, D.; Van Dijk, J.P.; Reijneveld, S.A.; Geckova, A.M. Is the association between screen-based behaviour and health complaints among adolescents moderated by physical activity? *Int. J. Public Health* **2015**, *60*, 139–145. [CrossRef] [PubMed]

12. Cai, Y.; Zhu, X.; Wua, X. Overweight, obesity, and screen-time viewing among Chinese school-aged children: National prevalence estimates from the 2016 Physical Activity and Fitness in China—The youth study. *J. Sport Health Sci.* **2017**, *6*, 404–409. [CrossRef] [PubMed]

13. Stiglic, N.; Viner, R.M. Effects of screen time on the health and well-being of children and adolescents: A systematic review of reviews. *BMJ Open* **2019**, *9*, e023191. [CrossRef]

14. Maher, C.; Olds, T.; Eisenmann, J.; Dollman, J. Screen time is more strongly associated than physical activity with overweight and obesity in 9- to 16-year-old Australians. *Acta Paediatr.* **2012**, *101*, 1170–1174. [CrossRef]

15. World Health Organization. Physical Activity. 2020. Available online: https://www.who.int/news-room/fact-sheets/detail/physical-activity (accessed on 26 November 2020).

16. Hajar, M.S.; Rizal, H.; Kueh, Y.C.; Muhamad, A.S.; Kuan, G. The effects of brain-breaks on motives of participation in physical activity among primary school children in Malaysia. *Int. J. Environ. Res. Public Health* **2019**, *16*, 2331. [CrossRef]

17. Rizal, H.; Hajar, M.S.; Muhamad, A.S.; Kueh, Y.C.; Kuan, G. The effect of brain breaks® on physical activity behavior among primary school children: A transtheoretical perspective. *Int. J. Environ. Res. Public Health* **2019**, *16*, 4283. [CrossRef]

18. Jago, R.; Salway, R.; Emm-Collison, L.; Sebire, S.J.; Thompson, J.L.; Lawlor, D.A. Association of BMI category with change in children's physical activity between ages 6 and 11 years: A longitudinal study. *Int. J. Obes.* **2019**, *44*, 104–113. [CrossRef]

19. Lewallen, T.C.; Hunt, H.; Potts-Datema, W.; Zaza, S.; Giles, W. The whole school, whole community, whole child model: A new approach for improving educational attainment and healthy development for students. *J. Sch. Health* **2015**, *85*, 729–739. [CrossRef] [PubMed]

20. Ginsburg, R.D.; Durant, S.; Baltzell, A. *Whose Game Is It, Anyway? A Guide to Helping Your Child Get the Most from Sports, Organized by Age and Stage*; Mariner Books: New York, NY, USA, 2006.

21. Althoff, T.; White, R.W.; Horvitz, E. Influence of Pokémon Go on physical activity: Study and implications. *J. Med. Internet Res.* **2016**, *18*, e315. [CrossRef] [PubMed]

22. Centers for Disease Control and Prevention. Whole School, Whole Community, Whole Child (WSCC). 2020. Available online: https://www.cdc.gov/healthyschools/wscc/index.htm (accessed on 12 January 2021).

23. Shields, M.K.; Behrman, R.E. Children and computer technology: Analysis and recommendations. *Future Child.* **2000**, *10*, 4–30. [CrossRef] [PubMed]

24. Kuan, G.; Rizal, H.; Hajar, M.S.; Chin, M.K.; Mok, M.M.C. Bright sports, physical activity investments that work: Implementing brain breaks in Malaysia primary schools. *Br. J. Sports Med.* **2019**, *53*, 905–906. [CrossRef]

25. Zhou, K.; He, S.; Zhou, Y.; Popeska, B.; Kuan, G.; Chen, L.; Chin, M.-K.; Mok, M.M.C.; Edginton, C.R.; Culpan, I.; et al. Implementation of brain breaks in the classroom and its effects on attitude towards physical activity in Chinese school setting. *Int. J. Environ. Res. Public Health* **2021**, *18*, 272. [CrossRef]

26. Ministry of Education (MOE), Singapore. 2020. Available online: https://beta.moe.gov.sg/education-in-SG/ (accessed on 30 October 2020).

27. Ministry of Foreign Affairs (MFA). Towards a Sustainable and Resilient Singapore. 2018. Available online: https://sustainabledevelopment.un.org/content/documents/19439Singapores_Voluntary_National_Review_Report_v2.pdf (accessed on 2 November 2020).

28. Glapa, A.; Grzesiak, J.; Laudanska-Krzeminska, I.; Chin, M.K.; Edginton, C.R.; Mok, M.M.C.; Bronikowski, M. The impact of brain breaks classroom-based physical activities on attitudes toward physical activity in Polish school children in third to fifth grade. *Int. J. Environ. Res. Public Health* **2018**, *15*, 368. [CrossRef]

29. Hajar, M.S.; Rizal, H.; Muhamad, A.S.; Kuan, G. The effects of Brain-Breaks on Short-Term Memory among Primary School Children in Malaysia. In *Enhancing Health and Sports Performance by Design*, 1st ed.; Hassan, M.H.A., Muhamed, A.M.C., Ali, N.F.M., Lian, D.K.C., Yee, K.L., Safii, N.S., Yusof, S.M., Fauzi, N.G.M., Eds.; Springer: Singapore, 2020; pp. 1–12.

30. Mok, M.M.C.; Chin, M.K.; Emeljanovas, A.; Mieziene, B.; Bronikowski, M.; Laudanska-Krzeminska, I.; Milanovic, I.; Pasic, M.; Balasekaran, G.; Phua, K.W.; et al. Psychometric properties of the attitudes towards physical activity scale: A Rasch analysis based on data from five location. *J. Appl. Meas.* **2015**, *16*, 379–400.

31. Dinc, S.C.; Uzunoz, F.S.; Mok, M.M.C.; Chin, M.-K. Adaptation of the attitudes toward physical activity scale for higher education students in Turkey. *J. Educ. Learn.* **2019**, *8*, 95–101. [CrossRef]

32. Popeska, B.; Jovanova-Mitkovska, S.; Chin, M.K.; Edginton, C.R.; Mok, M.M.C.; Gontarev, S. Implementation of brain breaks® in the classroom and effects on attitudes toward physical activity in a Macedonian school setting. *Int. J. Environ. Res. Public Health* **2018**, *15*, 1127. [CrossRef]

33. Uzunoz, F.S.; Chin, M.K.; Mok, M.M.C.; Edginton, C.R.; Podnar, H. The Effects of Technology Supported Brain Breaks on Physical Activity in School Children. In *Passionately Inclusive: Towards Participation and Friendship in Sport*; Dumon, D., Hofmann, A.R., Diketmuller, R., Koenen, K., Bailey, R., Zinkler, C., Eds.; Festschrift für Gudrun Doll-Tepper: Münster, NY, USA, 2018; pp. 87–104.

34. Krause, J.M.; Benavidez, E.A. Potential influences of exergaming on self-efficacy for physical activity and sport. *J. Phys. Educ. Recreat. Danc.* **2014**, *85*, 15–20. [CrossRef]

35. Mok, M.M.C.; Chin, M.-K.; Korcz, A.; Popeska, B.; Edginton, C.R.; Uzunoz, F.S.; Podnar, H.; Coetzee, D.; Georgescu, L.; Emeljanovas, A.; et al. Brain Breaks® Physical Activity Solutions in the Classroom and on Attitudes toward Physical Activity: A Randomized Controlled Trial among Primary Students from Eight Countries. *Int. J. Environ. Res. Public Health* **2020**, *17*, 1666. [CrossRef]

36. Bonnema, J.; Coetzee, D.; Lennox, A. Effect of a three-month HOPSports Brain Breaks® intervention programme on the attitudes of Grade 6 learners towards physical activities and fitness in South Africa. *J. Physic. Educ. Sport* **2020**, *20*, 196–205. [CrossRef]
37. Hajar, M.S.; Rizal, H.; Kuan, G. Effects of physical activity on sustained attention: A systematic review. *Sci. Med.* **2019**, *29*, e32864. [CrossRef]
38. Perrey, S. Promoting motor function by exercising the brain. *Brain Sci.* **2013**, *3*, 101–122. [CrossRef] [PubMed]
39. Singapore Unveils Plan in Push to Become Smart Nation. 2014. Available online: https://www.zdnet.com/article/singapore-unveils-plan-in-push-to-become-smart-nation/ (accessed on 21 April 2021).
40. Ammar, A.; Brach, M.; Trabelsi, K.; Chtourou, H.; Boukhris, O.; Masmoudi, L.; Bouaziz, B.; Bentlage, E.; How, D.; Ahmed, M.; et al. Effects of COVID-19 home confinement on eating behaviour and physical activity: Results of the ECLB-COVID19 international online survey. *Nutrients* **2020**, *12*, 1583. [CrossRef] [PubMed]
41. Constandt, B.; Thibaut, E.; De Bosscher, V.; Scheerder, J.; Ricour, M.; Willem, A. Exercising in times of lockdown: An analysis of the impact of COVID-19 on levels and patterns of exercise among adults in Belgium. *Int. J. Environ. Res. Public Health* **2020**, *17*, 4144. [CrossRef] [PubMed]
42. Ding, K.; Yang, J.; Chin, M.K.; Sullivan, L.; Durstine, J.L.; Violant-Holz, V.; Demirhan, G.; Oliveira, N.R.C.; Popeska, B.; Kuan, G.; et al. Physical Activity among Adults residing in 11 countries during the COVID-19 Pandemic Lockdown. *Int. J. Environ. Res. Public Health* **2021**. Under review. [CrossRef] [PubMed]
43. Dwyer, M.J.; Pasini, M.; De Dominicis, S.; Righi, E. Physical activity: Benefits and challenges during the COVID-19 pandemic. *Scand. J. Med. Sci. Sports* **2020**, *30*, 1291–1294. [CrossRef] [PubMed]
44. Khoramipour, K.; Basereh, A.; Hekmatikar, A.A.; Castell, L.; Ruhee, R.T.; Suzuki, K. Physical activity and nutrition guidelines to help with the fight against COVID-19. *J. Sports Sci.* **2021**, *39*, 101–107. [CrossRef] [PubMed]
45. Schnitzer, M.; Schöttl, S.E.; Kopp, M.; Barth, M. COVID-19 stay-at-home order in Tyrol, Austria: Sports and exercise behaviour in change? *Public Health* **2020**, *185*, 218–220. [CrossRef]
46. Dhingra, M.; Chattopadhyay, S. Advancing smartness of traditional settlements-case analysis of Indian and Arab old cities. *Int. J. Sustain. Built Environ.* **2016**, *5*, 549–563. [CrossRef]
47. World Health Organization. Sustainable Development Goals (SDGs). 2020. Available online: https://www.who.int/health-topics/sustainable-development-goals#tab=tab_1 (accessed on 21 April 2021).

Systematic Review

Active School Breaks and Students' Attention: A Systematic Review with Meta-Analysis

Álvaro Infantes-Paniagua [1,*], Ana Filipa Silva [2,3], Rodrigo Ramirez-Campillo [4,5], Hugo Sarmento [6,7], Francisco Tomás González-Fernández [8], Sixto González-Víllora [9] and Filipe Manuel Clemente [10,11]

1 Department of Physical Education, Arts Education, and Music, Faculty of Education of Albacete, University of Castilla-La Mancha, 02071 Albacete, Spain
2 N2i, Polytechnic Institute of Maia, 4475-690 Maia, Portugal; anafilsilva@gmail.com
3 The Research Centre in Sports Sciences, Health Sciences and Human Development (CIDESD), 5001-801 Vila Real, Portugal
4 Department of Physical Activity Sciences, Universidad de Los Lagos, Santiago 8320000, Chile; r.ramirez@ulagos.cl
5 Centro de Investigación en Fisiología del Ejercicio, Facultad de Ciencias, Universidad Mayor, Santiago 7500000, Chile
6 University of Coimbra, 3004-531 Coimbra, Portugal; hg.sarmento@gmail.com
7 Research Unit for Sport and Physical Activity, Faculty of Sport Sciences and Physical Education, 3004-531 Coimbra, Portugal
8 Centro de Estudios Superiores Alberta Giménez, Department of Physical Activity and Sport Sciences, Pontifical University of Comillas, 07013 Palma, Spain; francis.gonzalez.fernandez@gmail.com
9 Department of Physical Education, Arts Education, and Music, Faculty of Education of Cuenca, University of Castilla-La Mancha, 17071 Cuenca, Spain; sixto.gonzalez@uclm.es
10 Escola Superior Desporto e Lazer, Instituto Politécnico de Viana do Castelo, Rua Escola Industrial e Comercial de Nun'Álvares, 4900-347 Viana do Castelo, Portugal; Filipe.clemente5@gmail.com
11 Instituto de Telecomunicações, Delegação da Covilhã, 1049-001 Lisboa, Portugal
* Correspondence: Alvaro.Infantes@uclm.es; Tel.: +34-967-599-200 (ext. 2564)

Citation: Infantes-Paniagua, Á.; Silva, A.F.; Ramirez-Campillo, R.; Sarmento, H.; González-Fernández, F.T.; González-Víllora, S.; Clemente, F.M. Active School Breaks and Students' Attention: A Systematic Review with Meta-Analysis. *Brain Sci.* **2021**, *11*, 675. https://doi.org/10.3390/brainsci11060675

Received: 19 April 2021
Accepted: 18 May 2021
Published: 21 May 2021

Publisher's Note: MDPI stays neutral with regard to jurisdictional claims in published maps and institutional affiliations.

Abstract: School physical activity breaks are currently being proposed as a way to improve students' learning. However, there is no clear evidence of the effects of active school breaks on academic-related cognitive outcomes. The present systematic review with meta-analysis scrutinized and synthesized the literature related to the effects of active breaks on students' attention. On January 12th, 2021, PubMed, PsycINFO, Scopus, SPORTDiscus, and Web of Science were searched for published interventions with counterbalanced cross-over or parallel-groups designs with a control group, including school-based active breaks, objective attentional outcomes, and healthy students of any age. Studies' results were qualitatively synthesized, and meta-analyses were performed if at least three study groups provided pre-post data for the same measure. Results showed some positive acute and chronic effects of active breaks on attentional outcomes (i.e., accuracy, concentration, inhibition, and sustained attention), especially on selective attention. However, most of the results were not significant. The small number of included studies and their heterogeneous design are the primary limitations of the present study. Although the results do not clearly point out the positive effects of active breaks, they do not compromise students' attention. The key roles of intensity and the leader of the active break are discussed. INPLASY registration number: 202110054.

Keywords: physical activity; exercise; attention; attentional bias; arousal; randomized controlled trials; non-randomized controlled trials; cross-over studies; systematic review; meta-analysis

1. Introduction

Active breaks (ABs) are currently gaining attention within the educational context [1]. ABs consist of short periods (usually between five and 15 min) of classroom-based physical activity (PA) [2], which are integrated into the routine of the class [1,3]. These can be implemented by the teacher [2] during or between academic instructions [4]. Compared to other kinds of school-based PA interventions, ABs show some advantages. For example, (i) they do not require special spaces or equipment, (ii) teachers can choose when to utilize ABs according to their lessons' necessities [2], and (iii) they are not too time-consuming for practical use [5].

Some authors consider ABs to be an effective approach to promote PA with the final aim of improving students' health since school time represents an ideal setting for such purposes [6]. There is evidence confirming that school-based PA interventions increase students' PA levels [7,8]. In fact, the scientific literature suggests that the brain learns better when active methodologies (active role of students) are implemented instead of passive methodologies or traditional lessons [9]. This is of special interest nowadays because most young students do not meet the PA guidelines recommended by the World Health Organization [10]. Additionally, there is extensive research showing that PA interventions can improve students' cognitive, metacognitive, and academic outcomes, such as working memory, attention, processing speed, and academic performance [11,12]. Both the acute and chronic effects of school-based PA interventions on cognitive and academic performance have been extensively reviewed [1,3,12,13]. Although their positive effects are not completely clear for all those variables (e.g., attention, processing speed, or academic performance), it seems that increasing the amount of school time spent on PA does not compromise students' cognitive or academic performance. Therefore, school-based PA interventions are promising practices when appropriately implemented [14].

Among the cognitive outcomes addressed, attention is of great relevance for students since it plays a key role in learning [15] and academic achievement [16]. Conceptualizing attention is not easy due to the myriad of concepts that it involves. Therefore, in the present study, following Janssen et al. (2014) [17], we did not focus on a single measure of attention but instead considered attentional outcomes objectively measured within AB research.

Some authors perceive attention as a process of exerting mental effort on specific stimuli [18], while for others, it is like a "gate" that manages the input of information into conscious awareness [6]. Notwithstanding, most researchers agree with the multicomponent nature of attention [19]. This is reflected in the numerous different tasks used to measure attention in previous research [17], such as concentration tasks, time-on-task behaviors, or even electroencephalography.

It has been hypothesized that the effects of PA on attention, both from acute and chronic points of view, have a physiological basis (e.g., cardiovascular hypothesis, intensity of PA, or increases in cerebral blood flow and the number of neurotransmitters) [20]. There is sufficient evidence to suggest that ABs improve cognition, especially attention [20–22]. Furthermore, previous evidence has shown that ABs can improve students' attention [7,8,23]. However, results are still heterogeneous [2] and require further confirmation [3].

The inconsistent results presented in the literature could perhaps be explained by differences in the factors considered from a study design, how attention is measured, or the inclusion of samples representing different age groups [6] or cardiovascular fitness levels [6], just as happens in research on overall cognitive performance [2,12]. Moreover, in one study [24], it was suggested that different results on cognitive effects arise from differences in AB characteristics (e.g., cognitive engagement or complexity, intensity, duration).

Regarding the acute effects of AB duration, a recent meta-regression analysis suggested that shorter PA bouts may be more effective than longer ones for improving attention [12]. However, other research has indicated that longer bouts (i.e., >20 min) showed greater effects [20]. Since other studies did not find differences regarding the duration of PA bout [25], this topic requires further research to investigate the optimal duration of ABs.

Researchers have already highlighted the importance of investigating the duration of the cognitive benefits that remain after a PA bout, which is difficult to establish since post-test timings vary widely across studies, and most studies do not correctly report this information [1].

Finally, the person responsible for delivering the AB might also influence the characteristics of PA, especially regarding the intensity and student engagement [26,27]. In their review, Daly-Smith et al. (2019) [1] reported that the highest proportion of time spent on moderate-vigorous physical activity (MVPA) in active lessons was associated with the researcher-led intervention. Similarly, Watson et al. (2017) [3] pointed out that programs presented a higher fidelity to the required intensity when research staff was responsible for the intervention. These studies highlight the importance of the intervention deliverers' qualifications, which has not yet been clarified in the AB literature.

For more than a decade, students' attention deficits have been a significant concern of teachers [28]. ABs seem to be a promising way to enhance students' attentional levels in the class. However, several questions remain to be answered. To the best of our knowledge, there is only one previous meta-analysis that examined this issue [2], and it focused only on overall cognitive- or academic-related outcomes in 6- to 9-year-old students.

Therefore, the aim of this systematic review with meta-analysis was to scrutinize and synthesize the literature related to the effects of ABs (compared to control conditions) on the attention of students (of any age). We also addressed some possible moderators that previous research pointed out as relevant to the effects of ABs on cognition.

2. Materials and Methods

This systematic review (with meta-analysis) followed established international guidelines [29,30]. The protocol was published in INPLASY (International Platform of Registered Systematic Review and Meta-analysis Protocols) with the identification number of 202,110,054 and DOI 10.37766/inplasy2021.1.0054.

2.1. Eligibility Criteria

According to the Participants, Intervention, Comparators, Outcomes, and Study design (P.I.C.O.S.) approach, the inclusion and exclusion criteria for this systematic review and meta-analysis can be found in Table 1.

Table 1. Inclusion and exclusion criteria following the P.I.C.O.S. approach.

PICOS	Inclusion Criteria	Exclusion Criteria
Population	Healthy students of any age and of any sex from elementary to college educational levels.	Populations other than students (e.g., workers, athletes). Students with a diagnosed mental disease.
Intervention	ABs consisting of short bouts of exercise in class during or between academic lessons (e.g., structured exercises, free exercise).	No ABs (e.g., physical education classes; playing with instruments without allowing PA).
Comparator	Control conditions (passive or non-active breaks with limited PA).	Other forms of physical activity interventions (e.g., physical education lessons).
Outcome	Attentional outcomes (e.g., focused or selective attention, vigilance, inhibitory control) measured before (pre-) and after (post-) ABs or a chronic intervention of ABs.	Outcomes other than attention. No pre-post comparison. Inaccessible pre- or post-intervention data.
Study design	Counterbalanced cross-over design and parallel-groups design.	Study designs that do not allow within-subjects comparisons for both control and AB conditions.
Additional criteria	Original and full-text studies written in English.	Non-original articles (e.g., reviews, letters to editors, trial registrations, proposals for protocols, editorials, book chapters, conference abstracts).

2.2. Information Sources

Five electronic databases (PubMed, PsycINFO, Scopus, SPORTDiscus, and Web of Science) were searched for relevant publications prior to 12 January 2021. Keywords and synonyms were entered in various combinations: ("activ* break*" OR "physical break*" OR "physical activity break*" OR "exercise break*" OR "brain break" OR "brain hacking" OR "movement learning" OR "active learning") AND (student* OR class* OR school*) AND attent*. Additionally, the reference lists of included studies were manually searched to identify potentially eligible studies not captured by the electronic searches. All records were screened by two researchers (AIP and FTGF).

2.3. Data Extraction

A data extraction was prepared in Microsoft Excel sheet (Microsoft Corporation, Readmon, WA, USA), similar to the Cochrane Consumers and Communication Review Group's data extraction template (Group, 2016). The Excel sheet was used to assess inclusion and exclusion requirements and subsequently tested for all selected studies. The process was independently conducted by two authors (AIP and HS). Any disagreement regarding study eligibility was resolved in a discussion with a third author (FTGF) when necessary. Full text articles excluded and the reasons for doing so were recorded (see Table A1 in Appendix A). All the records were stored in the sheet.

2.4. Data Items

The following categories of information were extracted from included articles: (i) randomization unit, design, number of participants (n), age group (schoolchildren, young adults or both), sex (men, women or both); (ii) fitness of participants; (iii) identification of ABs (time, duration, weekly and/or daily frequency, intensity and type of PA, academic content, the person who is responsible for the AB, and protocol), (iv) treatment fidelity, (v) measurement of attention (i.e., task), (vi) time of measurements (pre and post) and (vii) effect measured (i.e., acute effects vs. chronic effects).

2.5. Assessment of Methodological Quality

The methodological quality of studies was assessed using the *Revised Cochrane risk-of-bias tool for randomized trials* (RoB 2) for randomized controlled trials (RCTs) [31], as well as the supplements for cluster randomized trials (CRTs) [32] and for cross-over trials [33]. For non-RCT, the *Cochrane* risk of bias tool for non-randomized studies of interventions (ROBINS-I) scale was used [34]. These tools include a minimum of 21 items that enable the assessment of the risk of bias (i.e., "low risk", "some concerns", or "high risk") of several dimensions that vary according to the study design (namely, bias arising from the randomization process, bias due to deviations from intended interventions, bias due to missing outcome data, bias in measurement of the outcome, and bias in selection of the reported result). An 'intention-to-treat' effect approach was followed for all the assessments, which implies that the interest focused on the effect of assignment to the interventions. This approach was followed because there was a wide variety of study designs and protocols could not be reviewed in most of the cases. Altogether an overall level of risk of bias per study was computed. Risk of bias assessments were based on the published articles, which were accompanied with the trial protocols in two studies [6,35]. Two of the authors (AIP and HS) independently screened and assessed the included articles. Discrepancies were solved by consensus between the two authors without the need for assistance from a third author.

2.6. Statistical Analyses

Meta-analyses were performed if at least three study groups provided pre-post AB-related data for the same measure. Using a random-effects model, the means and standard deviations (SD) for dependent variables were used to calculate effect sizes (ES; Hedges' g) for each outcome in AB treatments and control conditions. When means and SDs were not available, they were obtained from 95% confidence intervals (CIs) or standard error of mean (SEM), using Cochrane recommended formulas. Data were standardized using post-intervention SD values. The ES values are presented with 95% confidence intervals (CI). Calculated ES were interpreted using the following scale: <0.2, trivial; 0.2–0.6, small; >0.6–1.2, moderate; >1.2–2.0, large; >2.0–4.0, very large; >4.0, extremely large [36]. Heterogeneity was assessed using the I^2 statistic, with values of <25%, 25–75%, and >75% considered to represent low, moderate, and high levels of heterogeneity, respectively [37]. The risk of bias was explored using the extended Egger's test [38]. To adjust for publication bias, a sensitivity analysis was conducted using the trim and fill method [39], with L0 as the default estimator for the number of missing studies [40]. All analyses were carried out using the Comprehensive Meta-Analysis software (version 2; Biostat, Englewood, NJ, USA). Statistical significance was set at $p \leq 0.05$.

3. Results

3.1. Study Identification and Selection

The database search retrieved 1809 titles, which were exported to reference manager software (EndNoteTM X9, Clarivate Analytics, Philadelphia, PA, USA). Duplicates (520 references) were subsequently removed either automatically or manually. The remaining 1289 articles were screened for their relevance based on titles and abstracts, resulting in the removal of a further 1244 studies. Following the screening procedure, 45 articles were selected for in-depth reading and analysis. After reading full texts, a further 36 studies were excluded due to not meeting the eligibility criteria (Table A1). Finally, nine studies were selected for the further analysis together with another seven studies that were identified from other sources, reaching a total of 16 included studies (Figure 1), involving 3383 participants between 6 and 13 years old. Due to the limited number of studies included into the review for each attentional outcome (e.g., global attention, selective attention, inhibition, etc.), results from participants of all ages included were grouped together despite this age range involves different stages of development.

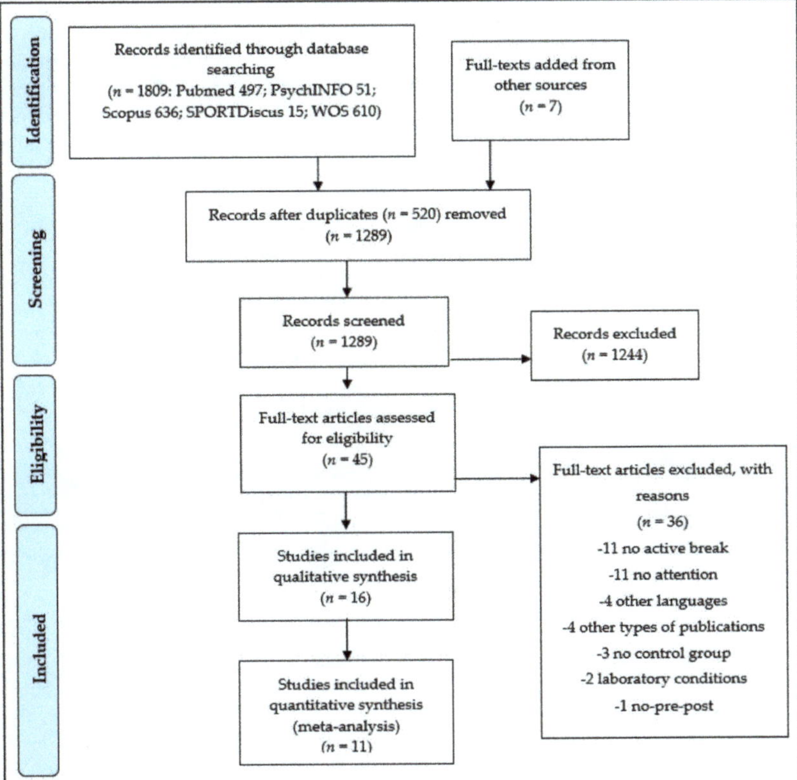

Figure 1. Preferred Reporting Items for Systematic Reviews and Meta-Analyses (PRISMA) flow diagram highlighting the selection process for the studies included in the systematic review.

3.2. Study Characteristics

Eleven studies followed a parallel-groups design with seven RCTs [19,35,41–45], three CRTs [24,46,47], and one non-randomized [48] (Table 2). Additionally, the remaining five studies followed a cross-over design (three single-group design studies [6,23,49] and two CRTs [50,51]) (Table 3).

The protocols' characteristics for all studies can be found in Table 4. Most of the studies addressed the acute effects of ABs [6,19,23,35,41–45,49–51], while only four addressed chronic effects [24,46–48]. Considering both types of effects, AB duration varied from 4 [23] to $\approx$ 25 min [41]. Interventions addressing chronic effects were applied to two ABs per week over two weeks [24] to five ABs per week over ten weeks [46]. Overall, ABs consisted of aerobic or coordinative moderate PA (MPA) [6,23,35,41,46,50], vigorous PA (VPA) [6,44,45], or MVPA [19,41–43,47,49]. Two studies reported that the registered intensity was lower (i.e., light or light-to-moderate) than expected [24,51], and one did not report intensity-related data [48]. Regarding the type of PA, nine studies included cognitively engaging PA conditions (i.e., combined activities, games, dancing, or coordinative exercises) [19,24,41–43,47–49,51], with two of them relating the PA to academic contents [24,49]; the other interventions included aerobic PA. Most of the interventions were delivered by the researchers [6,19,23,24,35,41–43,45,51]. Five of them were by the classroom teachers [6,47–50], and four of them also relied on videos to guide the ABs [24,35,48,51].

3.3. Methodological Quality

The overall methodological quality of the intervention studies can be found in Table A2. Nine studies were assessed as having some concerns in their overall RoB 2 quality scale, and eight were assessed as high risk of bias. The score for the only study assessed by ROBINS-I was critical [48]. None of the studies achieved low risk of bias. Methodological assessment revealed issues on the quality of the bias in the information reported on the randomization process, the reporting of possible deviations from the intended interventions and the selection of the reported result.

3.4. Active Breaks: Effects on Attention

Due to the multi-component nature of attention, a wide variety of attention-related outcomes were reported among included studies (e.g., global attention; selective attention). Table A3 shows a synthesis of the outcomes according to the task and their scoring. Overall, results from the 11 parallel groups design studies (Table 2) showed that the effects of ABs on attention were mainly positive or non-significant; no negative effect of ABs was found. In addition, results from the five cross-over design studies (Table 3) showed similar results, with positive or no effects on attentional outcomes and without negative results. The results for each outcome are synthesized in the following sections.

3.4.1. Effects on Accuracy

Accuracy was only measured with the d2 test. Acute positive effects were found only in an aerobic MPA AB intervention on 9 to 11-year-old students [23]. No other acute [19,45] nor chronic effects were found [48]. Meta-analyses could not be run since there were less than three studies per analysis.

Table 2. Characteristics of the selected studies with a parallel-groups design.

Study / Risk of Bias	Design and AB Type	Age (y.o.) Mean ± SD (Range) / Academic Level	Sample Size (*n*) and Sex	Attentional Outcomes (Instrument)	Fitness Level	Results
Altenburg et al. (2016) [35] / Some concerns	RCT: two IG and one CG. Acute.	NR (10–13) / NR	All: 52 (29♀33♂) IG1: 17 (5♀12♂) IG2: 20 (9♀11♂) CG: 19 (12♀7♂)	Selective attention (Sky Search in TEA-Ch)	NR	Children in IG2 (two ABs) had better selective attention than children in IG1 (one AB) or CG. There was no difference between IG1 and CG.
Buchele et al. (2018) [48] / Critical risk	Non-randomized quasi-experimental: two IGs and one CG. Chronic.	NR (≈10–11) / 5th grade	All: 116 (59♀57♂) IG1: 31 (14♀17♂) IG2: 29 (10♀19♂) CG: 56 (35♀21♂)	Accuracy (d2) * Concentration (d2) Selective attention (d2) Sustained attention (d2) *	NR	The IG1 increased all attentional outcomes (except accuracy) compared to the CG and concentration and sustained attention compared to IG2 (no AB). There were no differences between IG2 and CG.
Egger et al. (2018) [41] / Some concerns	RCT: three IGs and one CG. Acute.	All: 7.94 ± 0.44 (7–9) IG1: 7.99 ± 0.38 IG2: 7.93 ± 0.45 IG3: 7.96 ± 0.50 CG: 7.90 ± 0.44 2nd grade	All: 216 (~106♀110♂) IG1: 59 IG2: 53 IG3: 50 CG: 54	Inhibition reaction time (ms) (flanker task) Shifting reaction time (ms) (flanker task additional block)	Multistage 20m-SRT: IG1: 304.58 ± 123.18. IG2: 284.27 ± 141.16. IG3: 306.43 ± 144.23. CG: 278.55 ± 129.13	A significant, negative effect was found for the CE factor in shifting. No effects were found for the PA factor or the interaction between PA and CE.
Jäger et al. (2014) [42] / Some concerns	RCT: one IG and one CG. Acute.	7.91 ± 5.05 (months) (6.83–8.92) 2nd grade	All: 104 (57♀53♂) IG: 51 (27♀24♂) CG: 53 (30♀23♂)	Inhibition reaction time (ms) (flanker task) Shifting reaction time (ms) (flanker task additional block)	Motor fitness: 20m-SRT, 20m sprint test and jump side-to-side.	The AB improved only inhibition, and its effects remained for less than 40 min after the AB. The improvements were suggested to be independent of the participants' characteristics and stronger among those with higher increases in cortisol.

Table 2. *Cont.*

Study Risk of Bias	Design and AB Type	Age (y.o.) Mean $\pm$ SD (Range) Academic Level	Sample Size (n) and Sex	Attentional Outcomes (Instrument)	Fitness Level	Results
Jäger et al. (2015) [43] High risk	RCT: three IGs and one CG. Acute.	11.29 $\pm$ 6.53 (months) (10.33–12.33) NR	All: 217 (120♀97♂) IG1: 54 (35♀19♂) IG2: 62 (28♀34♂) IG3: 60 (30♀30♂) CG: 58 (33♀25♂)	Inhibition reaction time (ms) (flanker task) Shifting reaction time (ms) (flanker task additional block)	18-mSRT: VO$_2$max (ml/kg/min): Posttest: IG1: 46.77 $\pm$ 6.73, IG2: 47.98 $\pm$ 6.01, IG3: 46.77 $\pm$ 5.96), CG: 47.58 (6.12)	No effects of AB (with and without considering CE) were found. Fitness did not moderate the effects.
Niemann et al. (2013) [44] High risk	RCT: one IG and one CG. Acute.	9.69 $\pm$ 0.44 (9–10) IG: 9.65 $\pm$ 0.41 CG: 9.74 $\pm$ 0.48 4th grade	All: 42 IG: 27 (13♀14♂) CG: 15 (7♀8♂)	Concentration (d2)	NR	The IG showed better concentration than CG, although both groups improved from pre- to post-test. There was an interaction between group (IG, CG) test (pre, post), and PA level (high, low).
Ordóñez et al. (2019) [46] High risk	CRT: one IG and one CG. Chronic.	11.1 (11-12) 6th grade (Spanish Elementary Education)	All: 89 IG: 45 CG: 44	Concentration (FACES) Selective attention (FACES)	ALPHA. Lower-limb muscle strength (meters): Pretest: IG: 1.36 $\pm$ 0.21; CG: 1.38 $\pm$ 0.20. Posttest: IG: 1.42 $\pm$ 0.21; CG: 1.40 $\pm$ 0.21. Coordination (no. jumps): Pretest: IG: 28.33 $\pm$ 6.89; CG: 26.40 $\pm$ 5.68. Posttest: IG: 30.87 $\pm$ 5.68; CG: 27.33 $\pm$ 5.90. Cardiorespiratory capacity (min): Pretest: IG: 6.42 $\pm$ 0.75; CG:6.46 $\pm$ 0.83. Posttest: IG: 5.61 $\pm$ 0.68; CG: 6.20 $\pm$ 0.75	Significant differences between groups with higher levels of attention in the IG.

Table 2. *Cont.*

Study Risk of Bias	Design and AB Type	Age (y.o.) Mean $\pm$ SD (Range) Academic Level	Sample Size (n) and Sex	Attentional Outcomes (Instrument)	Fitness Level	Results
Schmidt et al. (2016) [19] Some concerns	RCT: three IGs and one CG. Acute.	11.77 ± 0.41 (11.01–12.98) 5th grade	All: 92 (42♀50♂) IG1: 25 (~12♀23♂) IG2: 22 (10♀12♂) IG3: 25 (11♀14♂) CG: 20 (9♀11♂)	Accuracy (d2) * Concentration (d2)	NR	No significant effects of ABs or their interactions with CE were found concerning attention. However, high CE interventions had a positive effect on focused attention, and positive affect had a mediational role between CE factor, accuracy, and focused attention, but not for PA.
Schmidt et al. (2019) [24] Some concerns	CRT: two IGs and one CG. Chronic.	9.04 ± 0.70 3rd grade	All: 104 (50♀54♂) IG1: 34 IG2: 37 CG: 33	Concentration (d2) (measured after 3rd AB)	NR	Focused attention did not differ between the three groups after controlling for age, step counts, and attention at pretest.
Tine et al. (2012) [45] High risk	RCT: one IG and one CG. Acute	NR (10.33–13.5) 6th–7th grade	All: 164 IG:86 (45♀41♂) CG: 78 (40♀38♂) (divided by income)	Accuracy (d2) * Selective attention (d2)	NR	The IG improved only regarding selective attention. Moreover, lower-income children exhibited greater improvements than higher-income children.
Van den Berg et al. (2019) [47] For most outcomes: Some concernsFor d2: High risk	CRT: one IG and one CG. Chronic.	IG: 10.8 ± 0.6 CG: 10.9 ± 0.7 (9–12) 5th–6th grade	All: 510 (448 to 467, depending on the outcome). IG: 100 (46♀54♂) CG: 100 (47♀53♂)	Alerting reaction time (ms) and accuracy (%) (ANT) *Concentration (d2) Inhibition reaction time (ms) and accuracy (%) (ANT [a] and Stroop Color Word Task *) Orienting reaction time (ms) and accuracy (%) (ANT) *	18-mSRT: VO$_2$max (ml/kg/min): Pretest: IG: 48.1 ± 5.0; CG: 48.0 ± 5.0. Posttest: IG: 48.9 ± 0.2; CG: 48.8 ± 0.2	No intervention effects were detected on any outcome after controlling for pretest score, age, arithmetic performance, class, and school. The IG spent more time in MVPA, but their fitness levels were similar to students in the CG.

AB: active break, ANT: attentional network test; BMI: body mass index, CE: cognitive exertion, CG: control group, CRT: cluster randomized trial, EF: executive functions, IG: intervention group, NR: not reported, PA: physical activity, RCT: randomized controlled trial, SES: socioeconomic status, SRT: shuttle run test; TEA-Ch: Test of Selective Attention in Children. [a] Executive control, but it is similar to the other inhibition tasks. Therefore, results were treated as inhibition. * Not included in the meta-analysis.

Table 3. Characteristics of the selected studies with a cross-over design *.

Study Risk of Bias	Design and Type of AB	Age (y.o.) Mean $\pm$ SD (Range) Academic Level	Sample Size (n)/Sex	Outcomes (Instruments/Tasks)	Fitness Level	Results
Hill et al. (2010) [50] Some concerns	CRT counterbalanced with two conditions. Acute.	NR (8-11) 4th–7th grade (Scottish)	All: 1224 (1074 completed three or more of the tests on both weeks)	Global attention: overall performance of different executive functions tests)	-	AB improved attention only among participants who received the intervention in the second period. Improvements were moderated by test and age.
Janssen et al. (2014) [6] High risk	Single group. Three randomized conditions at the group level. Acute.	10.4 $\pm$ 0.59 (10–11) 5th grade	All: 123 (61♀62♂)	Selective attention (Sky Search in TEA-Ch)	20-mSRT (dichotomized into high or low)	Attention was significantly better in all the conditions than in the 'no break' condition. Attention scores were best after the MPA AB. Attention after VPA breaks was better than after no break but was no different than after the passive break. No moderation effect of fitness was detected.
Ma et al. (2015) [23] High risk	Single group (divided). Two randomized conditions at the group level. Acute (mean of several acutes).	NR (9–11) 3–5th grade	All: 88 (44♀44♂)	Accuracy (d2) Concentration (d2) Selective attention (d2)	-	Better processing speed scores were reported after no AB. Accuracy improved after the AB. No effects on selective attention were observed following the AB, although accuracy improved.

Table 3. *Cont.*

Study Risk of Bias	Design and Type of AB	Age (y.o.) Mean ± SD (Range) Academic Level	Sample Size (n)/Sex	Outcomes (Instruments/Tasks)	Fitness Level	Results
van den Berg et al. (2016) [51] Some concerns	CRT counterbalanced with two conditions in three different groups. Acute.	11.7 ± 0.7 (10–13) 5th–6th grade	All: 184 (46♀54♂) IC1: 66 (47♀53♂) IC2: 71 (44♀56♂) IC3: 47 (49♀51♂)	Concentration (d2)	-	No effects of ABs (LMPA) on attention were reported nor were differential effects of exercise type, after controlling for age and session order. Scores for both conditions improved from day 1 to day 2.
Wilson et al. (2016) [49] Some concerns	Single group. Two randomized conditions at the group level. Acute (mean of several acutes).	11.2 ± 0.6 (≈10–12) 5th–6th grade	All: 58 ♂	Vigilance Mean Reaction Time (ms) and lapses (%) (PVT)	-	There were no significant differences between the AB and no-AB conditions.

AB: active break, BMI: body mass index, CC: control condition, CE: cognitive exertion, CRT: cluster randomized trial, EF: executive functions, IC: intervention condition, NR: not reported, PA: physical activity, PVT: psychomotor vigilance task; RCT: randomized controlled trial, SES: socioeconomic status, SRT: shuttle run test; TEA-Ch: Test of Selective Attention in Children. * Not included in the meta-analysis.

Table 4. Protocols of interventions.

Study ID	Type of AB	CG/CC Activity	AB Duration (Min)	Duration and Weekly/Daily Freq.	Time of AB	Intensity and Type of PA	Responsible	Timing of Pre-Test and Post-Test	Fidelity
Altenburg et al. (2016) [35]	**IG1**: One 20-min AB. **IG2**: Two 20-min PA bouts.	**CG**: No PA. Sitting all morning working on simulated school tasks.	20	NA IG1: 1 t-d; IG2: 2 t-d	IG1: after 90 min of sitting. IG2: one AB at the start and another after 90 min of sitting.	MPA. Aerobic. No AC.	Supervising research staff with videos	Pre: At baseline (T0). Post: After 20 min of school, after 130 min; and after 220 min.	HR monitor
Buchele et al. (2018) [48]	**IG1** "Coordinated bilateral PA". **IG2** * "Fitbit Only": Participants wore HR monitors on weekly days with no addition instructions.	**CG**: Usually scheduled school academic instruction periods while wearing plastic wristbands.	6	4 weeks 5 d-w/1 t-d	After 20 min of sedentary behavior.	NA IG1 Coordination.IG2: no PA. No AC.	Teachers with videos	Pre: The previous week the intervention. Post: The week after the intervention.	NR
Egger et al. (2018) [41]	**IG1** "Combo: high CE + high PA": Running while listening to a song with keywords to perform specific actions and inhibit others. **IG2** * "Cognition: high CE + low PA": Sitting while listening and reacting to a song. **IG3** "Aerobic: low CE + high PA": Running while listening to a song, but without changing the actions performed.	**CG** "Low CE + low PA": Participants sat comfortably in a circle and listened to an age-appropriate audio-book for 20 min.	≈25 [a]	NA	Morning (9:25–9:50 a.m.)	MVPA. IG1: Cognitive. IG2: no PA. IG3: Aerobic.No AC.	Researcher	Pre: Before the AB (9:05–9:25 a.m.). Post: Immediately after the AB (9:50–10:10 a.m.).	HR monitors and Borg RPE scale. Perceived CE was also assessed

Table 4. *Cont.*

Study ID	Type of AB	CG/CC Activity	AB Duration (Min)	Duration and Weekly/Daily Freq.	Time of AB	Intensity and Type of PA	Responsible	Timing of Pre-Test and Post-Test	Fidelity
Hill et al. (2010) [50]	**IC**: Stretching and aerobic PA (e.g., running on the spot, hopping sequences to music).	**CC**: Normal curriculum plan.	10–15	2 weeks 5 d-w/1 t-d	≈30 min after lunch.	MPA. Aerobic. No AC.	Trained teachers	Pre: NR. Post: At the end of the school day.	Teachers' control
Jäger et al. (2014) [42]	**IG** "EF-specific cognitive engaging PA": Warm-up with a song, playing tag, and balancing on various objects.	**CG**: 15 min seated on a mat while listening to an age-appropriate story. The last 5 min were spent answering easy questions.	≈20	NA	10:00–10:20 a.m.	MVPA. Cognitive. No AC.	Researcher	Pre: Prior the intervention. Post: Just after the AB. Follow-up 40 min after.	HR monitors
Jäger et al. (2015) [43]	**IG1** "Physical games: PA + CE": three different cooperative and competitive PA games involving EF. **IG2** "Aerobic exercise": Short tasks and games with different forms of running. **IG3** * "Cognitive games: Sedentary + CE": card game.	**CG**: Sedentary without CE: Participants sat comfortably on a mat and listened to an age-appropriate story.	20	NA	NR	MVPA. IG1: Cognitive IG2: Aerobic. IG3: no PA. No AC.	Researcher	Pre: Just before the intervention. Post: Immediately after the intervention.	HR monitors

168

Table 4. *Cont.*

Study ID	Type of AB	CG/CC Activity	AB Duration (Min)	Duration and Weekly/Daily Freq.	Time of AB	Intensity and Type of PA	Responsible	Timing of Pre-Test and Post-Test	Fidelity
Janssen et al. (2014) [6]	IC1 "MPA-AB": Walking to and from the PE classroom, jogging, and passing and dribbling a ball. IC2 "VPA-AB": Running to and from the PE classroom, running, jumping, and rope skipping.	CC1: No break. Participants were not allowed to ask the teacher for help or go to the toilet. CC2: Passive break. The teacher read a story to the participants.	15	NA	After an hour of regular cognitive tasks (9:30–10:00 a.m.)	IC1: MPA: Aerobic. IC2: VPA: Aerobic. No AC.	Two researchers and the classroom teacher	Pre: Before and after each experimental break in the classroom. Post: After each experimental break in the classroom.	Accelerometry
Ma et al. (2015) [23]	IC "FUNtervals": eight 20 s periods of VPA (i.e., squats, jumping jacks, scissor kicks, jumping on the spot) separated by 10-s rest periods.	CC: 10-min lecture separated from recess by at least 20 min of normal classroom instruction.	10 ª (4)	3 weeks. On two separate days in random.	After at least 20 min of normal classroom instruction following the recess.	MPA. Aerobic No AC.	Researcher	Pre: In week 1, familiarization. Post: after 10-min researcher-delivered lecture following AB.	Teachers' control
Niemann et al. (2013) [44]	IG: Running on a 400 m track. Participants were not allowed to talk to each other and remained silent.	CG: Participants performed sedentary behavior while watching non-arousing scenes. Participants were not allowed to talk to each other and remained silent.	12	NA	After 11:30 a.m.	VPA. Aerobic. No AC.	NR	Pre: After four normal school lessons just before AB. Post: 5 min after AB.	Control of prior PA in interventions days

169

Table 4. *Cont.*

Study ID	Type of AB	CG/CC Activity	AB Duration (Min)	Duration and Weekly/Daily Freq.	Time of AB	Intensity and Type of PA	Responsible	Timing of Pre-Test and Post-Test	Fidelity
Ordóñez et al. (2019) [46]	**IG**: The first two weeks: running a 250-m circuit inside the school; the next four weeks: 500 m; and in the last four weeks: 750 m.	**CG**: No AB.	NA	10 weeks 5 d-w/1 t-d	Between the 2nd and 3rd lesson in the morning.	MPA. Aerobic. No AC.	NR	Pre: At the same time with both groups, just before AB. Post: NR.	Prior familiar-ization for maintaining MPA
Schmidt et al. (2016) [19]	**IG1** "Combo: high CE + high PA": PA-based activity of adding numbers. **IG2** "Cognition: high CE + low PA": A paper-and-pencil trail-making test. **IG3** "Aerobic: low CE + high PA": Running at different speeds.	**CG** "sedentary + low CE": Students remained at their desks in the classroom and listened to an age-appropriate story for 10 min to relax and enjoy.	10	NA	After 20 min of German language class (11:15–11:30 a.m.)	MVPA. IG1: Cognitive IG2: no PA. IG3: Aerobic. No AC.	Researchers	Pre: Before AB (10:45–10:55 a.m.). Post: Immediately after AB (11:30–11:40 a.m.).	HR monitors, Borg scale, and self-perceived CE
Schmidt et al. (2019) [24]	**IG1** "Embodied learning condition": PA-based learning French vocabulary. **IG2** "PA condition": Movements at the same intensity without academic content.	**CG**: Sedentary teaching style (words were repeated equally as under other conditions).	10	2 weeks 2 d-w/1 t-d	10:00 am–12:00 p.m.	LPA IG1: Cognitive IG2: Aerobic. AC: IG1: earning animals in French; IG2: No.	Trained research student with a video	Pre: Before the beginning of the first learning session. Post: Immediately after the third learning session.	Accelerometry

Table 4. *Cont.*

Study ID	Type of AB	CG/CC Activity	AB Duration (Min)	Duration and Weekly/Daily Freq.	Time of AB	Intensity and Type of PA	Responsible	Timing of Pre-Test and Post-Test	Fidelity
Tine et al. (2012) [45]	**IG**: Running around an indoor track.	**CG**: Students remained seated and viewed a 12-min film video.	12	NA	2 sessions on separate days during usual gym classes.	VPA. Aerobic. No AC.	Researchers	Pre: Just before AB. Post: One minute after AB.	HR monitors
van den Berg et al. (2016) [51]	**IC1** "Aerobic": Easy and repetitive movements. **IC2** "Coordination": Complex movements stressing coordinative skills. **IC3** "Strength": Dynamic and static body-weight exercises adjusted to the age.	**CC**: 12 min of sitting and listening to an educational lesson about exercise and movement.	12	NA	8:30–10:00 a.m.	LMPA (target: MVPA). IC1: Aerobic. IC2: Coordination. IC3: Strength. No AC.	Researcher, three research assistants, and standardized movie	Pre: Just before AB. Post: Immediately after AB.	HR monitor, familiarization and control of previous bedtime, breakfast, and transport to school
van den Berg et al. (2019) [47]	**IG**: Following three "Just Dance" videos.	**CG**: Nine 10–15 min educational lessons once a week.	10	9 weeks 5 d-w/1 t-d	NR	MVPA. Dancing. No AC.	Teachers	Pre: The week before the intervention started. Post: The following week after the intervention.	Accelerometry Teachers' control.
Wilson et al. (2016) [49]	**IC** "Active Lesson Breaks" outside the regular classroom, including tag/chasing games, or invasion-type games.	**CC**: Passive lesson break: Participants spent 10 min sitting outside their classroom reading.	10	8 weeks each + 2 weeks of washout. 3 d-w/1 t-d	-	MVPA. Cognitive. AC: based on Take10! and Energizers, or Texas I CAN.	Trained teacher	Pre: 5 min before AB. Post: Immediately after AB.	Accelerometry

AB: active break, AC: academic content; CC: control condition, CE: cognitive engagement, CG: control group, HR: heart rate, IC: intervention condition, IG: intervention group, LMPA: light-to-moderate physical activity, LPA: light physical activity, MPA: moderate physical activity, MVPA: moderate-to-vigorous physical activity, NA: not reported, PA: physical activity; PE: physical education. [a] Including time of preparation. * Not AB: Intervention group or condition that did not include ABs. Groups and conditions are reported in bold letters to improve legibility.

3.4.2. Effects on Inhibition

Inhibition was measured in four studies through the flanker task, the Stroop task, and the ANT tests. Only one study found acute favorable effects after a 20-min cognitively engaged AB intervention [42]. The remaining studies on acute [41,43] or chronic [47] effects found no significant results.

Regarding the meta-analysis, four studies provided data for inhibition (i.e., reaction time results from flanker and ANT) involving six experimental and four control groups (pooled n = 900). There was a trivial effect of AB on inhibition (ES = 0.08; 95% CI = −0.07 to 0.23; p = 0.293; I^2 = 12.0%; Egger's test p = 0.576; relative weight of each group: 9.9 to 44.7%; Figure 2). After study-by-study and group-by-group sensitivity analyses, no significant changes in results were noted (i.e., p-value remained at > 0.05, mean ES = 0.02 to 0.14). No significant sub-group differences (p = 0.342) were identified between acute (ES = 0.14; 95% CI = −0.07 to 0.35; within-group I^2 = 13.8%, five study groups) and chronic effects (ES = 0.01; 95% CI = −0.18 to 0.19; within-group I^2 = 0.0%, one study group).

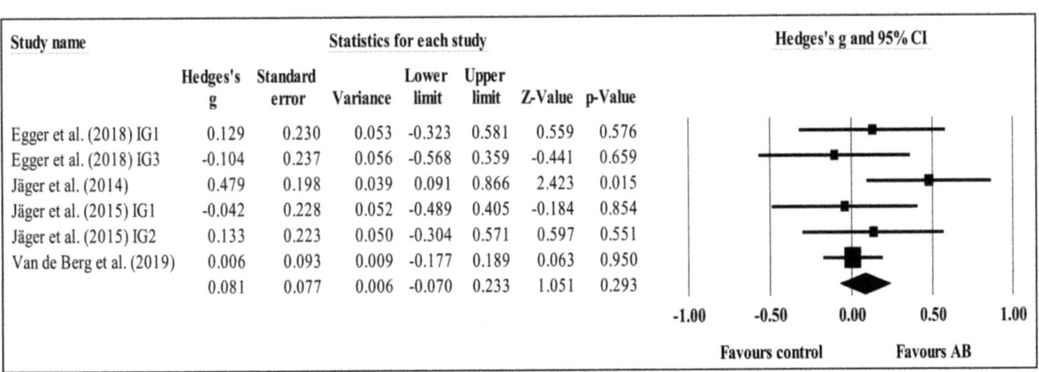

Study name	Statistics for each study							Hedges's g and 95% CI
	Hedges's g	Standard error	Variance	Lower limit	Upper limit	Z-Value	p-Value	
Egger et al. (2018) IG1	0.129	0.230	0.053	-0.323	0.581	0.559	0.576	
Egger et al. (2018) IG3	-0.104	0.237	0.056	-0.568	0.359	-0.441	0.659	
Jäger et al. (2014)	0.479	0.198	0.039	0.091	0.866	2.423	0.015	
Jäger et al. (2015) IG1	-0.042	0.228	0.052	-0.489	0.405	-0.184	0.854	
Jäger et al. (2015) IG2	0.133	0.223	0.050	-0.304	0.571	0.597	0.551	
Van de Berg et al. (2019)	0.006	0.093	0.009	-0.177	0.189	0.063	0.950	
	0.081	0.077	0.006	-0.070	0.233	1.051	0.293	

Figure 2. Forest plot of changes in inhibition in school-age students participating in active breaks (AB) compared to controls. Values shown are effect sizes (Hedges' g) with 95% confidence intervals (CI). The size of the plotted squares reflects the statistical weight of each study. The black diamond reflects the overall result. This trend is not statistically significant. IG: intervention group.

3.4.3. Effects on Concentration

Eight studies reported an index of concentration performance, which has been mainly measured by the d2 and FACES tests. Regarding the acute effects, one of them found positive effects of 12-min VPA on 9 to 10-year-old students' concentration [44]. On the other hand, positive chronic effects after four and 10 weeks of intervention were found in two of four studies [46,48]; however, the four-week study of these was assessed as high risk at ROBINS-I [48]. No other significant effects were found.

Six studies provided data for concentration, involving nine experimental and six control groups (pooled n = 881). There was a trivial effect of AB on concentration (ES = 0.19; 95% CI = −0.08 to 0.46; p = 0.161; I^2 = 63.5%; Egger's test p = 0.581; relative weight of each group: 8.2 to 17.9%; Figure 3). A sensitivity analysis according to ROBINS-I was conducted, removing the study of Buchele et al. (2018) [48], with no significant changes in results. However, a study-by-study sensitivity analysis, removing the study of Schmidt et al., (2019) [24], revealed a small effect of AB on concentration performance (ES = 0.34; 95% CI = 0.20 to 0.48; p < 0.001). No significant sub-group differences (p = 0.627) were identified between acute (ES = 0.27; 95% CI = −0.07 to 0.60; within-group I^2 = 0.0%, four study groups) and chronic effects (ES = 0.14; 95% CI = −0.29 to 0.56; within-group I^2 = 81.2%, five study groups).

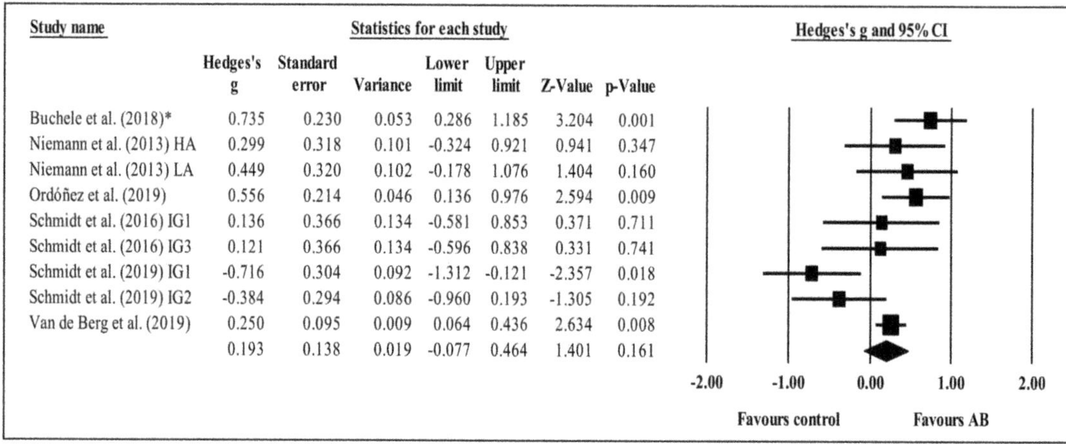

Figure 3. Forest plot of changes in concentration in school-age students participating in active breaks (AB) compared to controls. Values shown are effect sizes (Hedges' g) with 95% confidence intervals (CI). The size of the plotted squares reflects the statistical weight of each study. The black diamond reflects the overall result. This trend is not statistically significant. IG: intervention group; HA: high-active subgroup; LA: low-active subgroup. * Critical risk in ROBINS-I.

3.4.4. Effects on Selective Attention

Six studies measured selective attention with the d2, FACES and the Sky-Search task in TEA-Ch tests. With the exception of one study [23], the other three studies that measured acute effects reported positive results, with ABs varying between 12 and 20 min of MPA or VPA. Of note, one of the interventions only reported favorable results on the group that participated in two ABs during the same morning and no differences between one AB and no AB [35], while other reported only benefits for the AB of MPA and not for the VPA AB [6]. In addition, greater benefits were reported among low-income students [45]. Regarding the chronic effects, results were similar to the concentration's results.

Four studies provided data for selective attention, involving five experimental and four control groups (pooled n = 395). There was a moderate effect of AB on selective attention (ES = 0.61; 95% CI = 0.41 to 0.82; $p < 0.001$; I^2 = 0.0%; Egger's test p = 0.036 (corrected values: ES = 0.67, 95%CI 0.45 to 0.88); relative weight of each group: 7.2 to 41.1%; Figure 4). A sensitivity analysis according to ROBINS-I was conducted, removing the study of Buchele et al. (2018) [48], with no significant changes in results (i.e., p-value remained at <0.001, mean ES = 0.59). Similarly, after study-by-study and group-by-group sensitivity analyses, no significant changes in results were noted (i.e., p-value remained at <0.001, mean ES = 0.48 to 0.65). No significant sub-group differences (p = 0.963) were identified between acute (ES = 0.55; 95% CI = 0.10 to 1.00; within-group I^2 = 43.5%, three study groups) and chronic effects (ES = 0.54; 95% CI = 0.23 to 0.84; within-group I^2 = 0.0%, two study groups).

Study name	Hedges's g	Standard error	Variance	Lower limit	Upper limit	Z-Value	p-Value
Altenburg et al. (2016) IG2	0.401	0.387	0.150	-0.358	1.159	1.035	0.301
Altenburg et al. (2016) IG1	0.089	0.392	0.154	-0.680	0.858	0.227	0.821
Buchele et al. (2018)*	0.547	0.226	0.051	0.104	0.991	2.417	0.016
Ordóñez et al. (2019)	0.531	0.214	0.046	0.112	0.951	2.484	0.013
Tine et al. (2012).	0.821	0.162	0.026	0.503	1.138	5.062	0.000
	0.613	0.104	0.011	0.409	0.817	5.900	0.000

Figure 4. Forest plot of changes in selective attention in school-age students participating in active breaks (AB) compared to controls. Values shown are effect sizes (Hedges' g) with 95% confidence intervals (CI). The size of the plotted squares reflects the statistical weight of each study. The black diamond reflects the overall result. This is a statistically significant result. IG: intervention group. * Critical risk in ROBINS-I.

3.4.5. Effects on Shifting

Shifting was only assessed in three studies [41–43] that used the flanker task to test acute effects. None of them found any acute effect after ABs. In the meta-analyses, the three studies provided data, involving five experimental and three control groups (pooled n = 441). There was a trivial effect of AB on shifting (ES = −0.18; 95% CI = −0.52 to 0.15; p = 0.286; I^2 = 65.7%; Egger's test p = 0.229; relative weight of each group: 19.1 to 21.8%; Figure 5). After study-by-study and group-by-group sensitivity analyses, no significant changes in results were noted (i.e., p-value remained at >0.05, mean ES = −0.29 to 0.27).

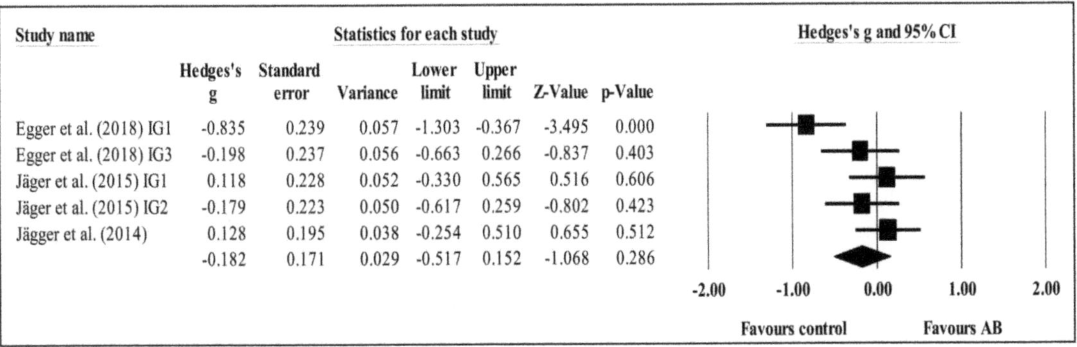

Study name	Hedges's g	Standard error	Variance	Lower limit	Upper limit	Z-Value	p-Value
Egger et al. (2018) IG1	-0.835	0.239	0.057	-1.303	-0.367	-3.495	0.000
Egger et al. (2018) IG3	-0.198	0.237	0.056	-0.663	0.266	-0.837	0.403
Jäger et al. (2015) IG1	0.118	0.228	0.052	-0.330	0.565	0.516	0.606
Jäger et al. (2015) IG2	-0.179	0.223	0.050	-0.617	0.259	-0.802	0.423
Jägger et al. (2014)	0.128	0.195	0.038	-0.254	0.510	0.655	0.512
	-0.182	0.171	0.029	-0.517	0.152	-1.068	0.286

Figure 5. Forest plot of changes in shifting in school-age students participating in active breaks (AB) compared to controls. Values shown are effect sizes (Hedges's g) with 95% confidence intervals (CI). The size of the plotted squares reflects the statistical weight of each study. The black diamond reflects the overall result. This trend is not statistically significant. IG: intervention group.

3.4.6. Effects on Sustained Attention/Vigilance

ANT, d2 and PVT tests were employed in each of the three studies that measured sustained attention/vigilance [47–49]. No acute effects were found [49]. On the other hand, positive chronic effects after a four-week intervention were found [48], but it presented a high risk at ROBINS-I. Meta-analyses could not be run since there were less than three studies per analysis.

3.4.7. Effects on Other Outcomes

Only one study [47] measured orienting, in 9 to 12-year-old students, before and after a 9-week intervention of daily MVPA cognitive ABs. No chronic effects were found. In addition, one study, of 8 to 11-year-old students, reported a global outcome of attention by a compendium of different executive function tasks [50]. This study found favorable acute effects after 10-15 min of AB involving aerobic MPA. Due to the reduced number of studies reporting data for these outcomes, a meta-analysis was precluded.

4. Discussion

4.1. Discussion of Evidence

The present systematic review with meta-analysis scrutinized and synthesized the literature related to the effects of ABs on students' (of any age) attention when compared to control conditions. The results do not point to any clear acute or chronic effects of ABs on students' overall attention, although some positive effects were found in terms of accuracy, concentration, inhibition, sustained attention, and (especially) selective attention. The meta-analysis revealed no statistical differences between AB and control groups regarding inhibition (Figure 2), concentration (Figure 3), or shifting (Figure 5). The trends for inhibition and concentration favored AB groups, and the trend for shifting favored the control groups. Nevertheless, all three meta-analyses included zero in their confidence intervals; therefore, no solid conclusions can be drawn. However, for selective attention, there was a significant difference between the AB and control groups (Figure 4) in favor of the former. These results will be discussed. In addition, overall, ABs did not compromise students' attention.

As a first approximation to the problem, we suggest that the small number of positive effects [6,23,35,42,44–46,48,50] in the different included outcomes could be attributed to the fact that performing any type of exercise provokes neurophysiological changes in the brain [52]. Nevertheless, there is much heterogeneity and a wide variety of ABs protocols encountered (i.e., durations ranging from 4 to 20 min; intensities of exercise ranging from moderate and vigorous; the inclusion of various types of PA such as aerobic, anaerobic, and muscular resistance; and the use of specific cognitive tasks to assess attention such as d2, ANT, and the flanker task). These differences do not provide clear evidence and have sparked controversy due to the non-existence of general guidelines for applying and implementing ABs.

Regarding the acute exercise paradigm, positive effects were observed only for accuracy [23], concentration [44], inhibition [42], and selective attention [6,35,45]. This suggests that cognitive activities performed after exercise lasting 4 to 20 min could produce overall benefits to students' attention. In addition, regarding the studies including chronic AB interventions, positive effects were observed after 10 weeks in terms of concentration and selective attention [46]. Likewise, positive chronic effects on sustained attention were found after a four-week intervention [48].

Despite the lack of support from the meta-analyses, the positive findings found regarding selective attention are in line with the study of Donnelly et al. (2016) [53], who showed that routinely practicing PA in schools enhances cognitive performance. In fact, chronic exercise positively influences different attention processes in children [54]. In all cases, an argument could be made for the importance of studying additional moderators since these could influence the effects of exercise [20]. However, the small number of studies found per outcome did not allow us to make robust distinctions about the effects according to the moderators [20].

Despite the lack of clear moderators explaining the relationship between exercise and cognitive function, the results suggest that the intensity of the exercise used in ABs plays a fundamental role in the literature exploring the specific effects of PA on cognition. In fact, the current research suggests that some attentional outcomes improved after ABs at MVPA intensities (40 to 80 of VO_2max). However, many studies did not monitor the intensities of ABs, nor did they measure the magnitudes of the changes in some physiological mechanisms (e.g., brain-derived neurotrophic factor, catecholamines, increased cerebral blood flow) [20,55] to predict their possible effects on behavior and cognitive performance.

Nevertheless, in most studies, ABs were carried out in the classroom and never under laboratory conditions. For this reason, measuring exercise intensities with a large sample is a truly complex matter. In fact, objective instruments (either a heart rate monitor [19,35,41–43,45,51] or accelerometry [6,24,47,49]) were used to calculate the loads of ABs. However, researchers have also relied on subjective measures controlled by the teachers—in some cases, the measures were simply not registered [23,44,46,48,50]. In all cases, a potential and valid proposal might be the use of the subjective perception of effort (e.g., Egger et al. 2018 [41], Schmidt et al. 2016 [19]). This approach, which helps calculate metabolic changes during exercise, could be an effective option to use in school children [56,57].

In light of the above discussion, another key factor that might moderate the effects of ABs is the person who applies the AB. In this sense, it would be appropriate for physical education teachers to be responsible for applying ABs in all interventions [1,27]. On the one hand, they have the capacity to guide research proposals since they have a deeper knowledge of training principles involved in any kind of PA. On the other hand, they could provide students and other teachers with techniques for controlling the intensity of the AB in each intervention, which is suggested as being a determinant of outcomes in the present work [26].

4.2. Study Limitations

The first limitation of this systematic review is the small number of studies found per outcome and associated effect of PA (i.e., acute or chronic) and the heterogeneity among these studies' designs. This leads to the second main limitation, which is that the ESs for each effect type cannot rely on a minimum of three studies in all cases. In addition, the heterogeneity was considerable for chronic effects in concentration and acute effects in shifting. Altogether, these limitations indicate that the results should be interpreted with caution.

4.3. Practical Implications

The outcomes of this study present implications for incorporating ABs into school lessons to improve students' attention. Through ABs students can reach higher levels of PA, which promotes a healthy lifestyle.

However, teachers are not usually adequately prepared to carry out ABs throughout the day during class. Thus, teachers should be trained on the correct implementation of ABs and the integration of movement on class days to ensure that ABs positively affect students' health and cognition. To achieve this, ABs should control physiological measures to objectively calculate exercise intensity. As a result, students might obtain more benefits from ABs if teachers are also trained in the use and interpretation of measures of PA intensity feasible for in-class use, such as the Borg scale [58–60].

Finally, as it relates to practical implications, the duration of ABs varied from 4 min to more than 20 min. However, previous research has shown that only exercise of more than 20 min had positive results on cognitive performance [20]. From an educational point of view, it could be thought that adding a 20-min break into current school timetables may compromise the learning time. Therefore, ABs of such a length may not be practical. Additionally, evidence on this matter is not clear, as a recent review found that the duration of PA was inversely related to attentional performance [12].

Notwithstanding, from an academic performance perspective, increasing the amount of school-based PA does not compromise academic achievement and can improve classroom behavior and academic achievement [14]. In addition, as seen in some of the included studies, ABs can include academic content [24,49] and, therefore, could also be included in the learning time. Although more conclusive evidence is needed on this topic, ABs could be included and adapted to different educational contexts and would be effective for improving students' health and cognitive outcomes as long as the required intensity and time are met.

5. Conclusions

There are no clear positive effects of ABs on students' attention. The heterogeneity in the designs and measurements of the studies and the small number of studies carried out in school environments are the main reasons for the lack of conclusive results. Notwithstanding, it seems that including PA in school time through ABs does not compromise students' attention, and it could positively affect selective attention.

The intensity and duration of the PA seem to play a key role in cognitive effects. Therefore, efforts should be made to help teachers understand how to motivate their students to reach the correct intensity levels when carrying out an AB.

Even though research on ABs started around a decade ago, clear evidence is still lacking regarding their effects on attention. The results presented here highlight that this topic is still of significant relevance.

Author Contributions: Conceptualization, S.G.-V. and F.M.C.; methodology, Á.I.-P., F.T.G.-F., H.S. and F.M.C.; software and formal analysis, R.R.-C.; investigation, Á.I.-P. and F.T.G.-F.; resources, H.S., S.G.-V. and F.M.C.; data curation, Á.I.-P., F.T.G.-F. and H.S.; writing—original draft preparation, Á.I.-P., F.T.G.-F. and F.M.C.; writing—review and editing, Á.I.-P., A.F.S., F.T.G.-F., R.R.-C. and F.M.C.; supervision, A.F.S., S.G.-V. and F.M.C.; funding acquisition, Á.I.-P., H.S., F.T.G.-F., S.G.-V. and F.M.C. All authors have read and agreed to the published version of the manuscript.

Funding: This research is funded by the Spanish Ministry of Universities, grant number EST19/00498; and by Consejería de Educación, Cultura y Deportes de Castilla-La Mancha cofinanciados por el Fondo Europeo de Desarrollo Regional (FEDER), grant number: SBPLY/19/180501/000147. No other specific sources of funding were used to assist in the preparation of this article.

Acknowledgments: Álvaro Infantes-Paniagua gratefully acknowledges that his research is supported by the Spanish Ministry of Education, Culture and Sports (FPU16/00082). Francisco Tomás González Fernández gratefully acknowledges that he is supported by Beca de Movilidad Erasmus+ para recibir formación (PDI) Erasmus 2020–2021. Hugo Sarmento gratefully acknowledges the support of a Spanish government subproject 'Integration ways between qualitative and quantitative data, multiple case development, and synthesis review as main axis for an innovative future in physical activity and sports research (PGC2018-098742-B-C31) (Ministerio de Economía y Competitividad, Programa Estatal de Generación de Conocimiento y Fortalecimiento Científico y Tecnológico del Sistema I+D+i)', which is part of the coordinated project 'New approach of research in physical activity and sport from mixed methods perspective (NARPAS_MM) (SPGC201800 × 098742CV0)'. Filipe Manuel Clemente gratefully acknowledges that this work is also funded by Fundação para a Ciência e Tecnologia/Ministério da Ciência, Tecnologia e Ensino Superior through national funds and when applicable co-funded EU funds under the project UIDB/50008/2020.

Conflicts of Interest: The authors declare no conflict of interest.

Appendix A

Table A1. Excluded full texts with reasons.

Excluded References	Reason for Exclusion
Adsiz et al. (2012) [61]	No AB
Amicone et al. (2018) [62]	No AB
Bartholomew et al. (2018) [63]	No attention: Time on task
Ben-Zeev et al. (2020) [64]	No AB: PE lesson
Blasche et al. (2018) [65]	No attention
Budde et al. (2008) [28]	No proper CG: PE lesson
Chou et al. (2020) [66]	No AB: PE lesson
Chrismas et al. (2019) [67]	No AB
Contreras et al. (2020) [68]	Not written in English. No control group
Egger et al. (2019) [69]	No proper CG: highly demanding cognitive lesson
Fenesi et al. (2018) [70]	No attention: self-reported mind wandering question
Flippin et al. (2020) [71]	No AB
Gonzalez et al. (2020) [72]	Not written in English
Grieco et al. (2016) [73]	No attention: Time on task
Howie et al. (2015) [74]	No attention
Howie et al. (2014) [75]	No attention: Time on task
Kubesch et al. (2009) [76]	No AB: PE lesson
Mahar (2011) [77]	Review
Mavilidi et al. (2020) [78]	No attention: Time on task
Mazzoli et al. (2019) [79]	No AB
McGowan et al. (2020) [80]	Laboratory
Merriman et al. (2020) [81]	Report
Miklós et al. (2020) [82]	Laboratory
Napoli et al. (2005) [83]	No AB
Niedermeier et al. (2020) [84]	No pre-posttest on objective measure of attention
Ochoa et al. (2020) [85]	Not written in English
Owen et al. (2018) [86]	No AB
Pesce et al. (2013) [87]	No AB
Ruiz-Ariza et al. (2021) [88]	Not written in English
Sánchez-López et al. (2015) [89]	Protocol
Sugahara et al. (2018) [90]	No AB
Tan et al. (2016) [4]	No attention: Time on task
Vazou, et al. (2020) [91]	No attention: Observation Protocol
Watson et al. (2017) [92]	No attention: Classroom behavior
Watson et al. (2019) [93]	No attention: Time on task
Webster et al. (2015) [94]	

AB: active break, CG: control group, PE: physical education.

Table A2. Qualitative assessment.

Study	Outcome	D1a	D1b	D2	D3	D4	D5	DS	Overall
Altenburg et al. (2016) [35]	All	L	-	L	L	SC	SC	-	SC
Buchele et al. (2018) * [48]	All	C	L	S	L	NI	M	M	C
Egger et al. (2018) [41]	All	SC	-	SC	L	L	SC	-	SC
Hill et al. (2010) [50]	All	L	L	SC	L	L	SC	SC	SC
Jäger et al. (2014) [42]	All	SC	-	L	L	L	SC	-	SC
Jäger et al. (2015) [43]	All	SC	-	SC	L	L	H	-	H
Janssen et al. (2014) [6]	All	SC	-	SC	H	L	H	H	H
Ma et al. (2015) [23]	All	SC	-	SC	H	L	SC	SC	H
Niemann et al. (2013) [44]	All	L	-	SC	H	L	H	-	H
Ordóñez et al. (2019) [46]	All	L	L	SC	H	L	SC		H
Schmidt et al. (2016) [19]	All	SC	-	L	L	L	SC	-	SC
Schmidt et al. (2019) [24]	All	SC	L	SC	L	L	SC	-	SC
Tine et al. (2012) [45]	Selective attention	SC	-	SC	L	L	H	-	H
	Accuracy	SC	-	SC	H	L	H	-	H
van den Berg et al. (2016) [51]	All	L	L	SC	L	L	SC	L	SC
van den Berg et al. (2019) [47]	All but focused attention	L	L	SC	L	L	SC	-	SC
	Focused attention (concentration)	L	L	SC	L	H	SC	-	H
Wilson et al. (2016) [49]	All	L	L	L	L	L	SC	SC	SC

D1a: Randomization process; D1b: The timing of identification or recruitment of participants in a cluster-randomized trial; D2: Deviations from the intended interventions; D3: Missing outcome data; D4: Measurement of the outcome; D5: Selection of the reported result; DS: Period and carryover effects. C: Critical; H: High risk; L: Low risk; M: Moderate; NI: No information; S: Serious; SC: Some concerns. * Assessment from ROBINS-I: D1a: Bias due to confounding; D1b: Bias in selection of participants into the study; D2: Bias in classification of interventions; D3: Bias due to deviations from intended interventions; D4: Bias due to missing data; D5: Bias in measurement of outcomes; S: Bias in selection of the reported result.

Table A3. Protocols of interventions.

Attentional Outcome	Test/Task	Calculation
Accuracy	d2	Errors (%)
Concentration	FACES	Total correct responses–Errors (Commission)
	d2	Total correct responses–Commission errors
Global attention	Compendium of tasks (paced serial addition, size ordering, listening span, digit-span backwards, and digit-symbol encoding)	Overall score
Inhibition	ANT (flanker)	Incongruent–Congruent
	Flanker	Incongruent–Congruent
	Stroop	Incongruent–Congruent
Orienting	ANT (flanker)	Center cue–Spatial cue
Selective attention	d2	Total number of responses–Errors
	FACES	Total of right responses
	TEA-Ch	Time of pair identification–Time of motor performance
Shifting	Flanker	SCORE (Mixed block)–SCORE (Standard block)
	ANT (flanker)	SCORE no cue – SCORE center cue
Sustained attention/Vigilance	d2	Fluctuation rate
	PVT	Mean response time to a repeating visual stimulus

ANT: attentional network test; PVT: psychomotor vigilance task; TEA-Ch: Test of Selective Attention in Children.

References

1. Daly-Smith, A.J.; Zwolinsky, S.; McKenna, J.; Tomporowski, P.D.; Defeyter, M.A.; Manley, A. Systematic review of acute physically active learning and classroom movement breaks on children's physical activity, cognition, academic performance and classroom behaviour: Understanding critical design features. *BMJ Open Sport Exerc. Med.* **2018**, *4*, e000341. [CrossRef] [PubMed]
2. Masini, A.; Marini, S.; Gori, D.; Leoni, E.; Rochira, A.; Dallolio, L. Evaluation of school-based interventions of active breaks in primary schools: A systematic review and meta-analysis. *J. Sci. Med. Sport* **2020**, *23*, 377–384. [CrossRef] [PubMed]
3. Watson, A.; Timperio, A.; Brown, H.; Best, K.; Hesketh, K.D. Effect of classroom-based physical activity interventions on academic and physical activity outcomes: A systematic review and meta-analysis. *Int. J. Behav. Nutr. Phys. Act.* **2017**, *14*, 114. [CrossRef] [PubMed]
4. Goh, T.L.; Hannon, J.; Webster, C.; Podlog, L.; Newton, M. Effects of a TAKE 10! Classroom-Based Physical Activity Intervention on Third- to Fifth-Grade Children's On-task Behavior. *J. Phys. Act. Health* **2016**, *13*, 712–718. [CrossRef] [PubMed]
5. van den Berg, V.; Salimi, R.; de Groot, R.; Jolles, J.; Chinapaw, M.; Singh, A. "It's a Battle . . . You Want to Do It, but How Will You Get It Done?": Teachers' and Principals' Perceptions of Implementing Additional Physical activity in School for Academic Performance. *Int. J. Environ. Res. Public Health* **2017**, *14*, 1160. [CrossRef]
6. Janssen, M.; Chinapaw, M.J.M.; Rauh, S.P.; Toussaint, H.M.; van Mechelen, W.; Verhagen, E.A.L.M. A short physical activity break from cognitive tasks increases selective attention in primary school children aged 10-11. *Ment. Health Phys. Act.* **2014**, *7*, 129–134. [CrossRef]
7. Muñoz-Parreño, J.A.; Belando-Pedreño, N.; Torres-Luque, G.; Valero-Valenzuela, A. Improvements in Physical Activity Levels after the Implementation of an Active-Break-Model-Based Program in a Primary School. *Sustainability* **2020**, *12*, 3592. [CrossRef]
8. Vazou, S.; Saint-Maurice, P.; Skrade, M.; Welk, G. Effect of Integrated Physical Activities with Mathematics on Objectively Assessed Physical Activity. *Children* **2018**, *5*, 140. [CrossRef]
9. Heward, W.L. Ten Faulty Notions About Teaching and Learning That Hinder the Effectiveness of Special Education. *J. Spec. Educ.* **2003**, *36*, 186–205. [CrossRef]
10. Guthold, R.; Stevens, G.A.; Riley, L.M.; Bull, F.C. Global trends in insufficient physical activity among adolescents: A pooled analysis of 298 population-based surveys with 1·6 million participants. *Lancet Child Adolesc. Health* **2020**, *4*, 23–35. [CrossRef]
11. Álvarez-Bueno, C.; Pesce, C.; Cavero-Redondo, I.; Sánchez-López, M.; Martínez-Hortelano, J.A.; Martínez-Vizcaíno, V. The Effect of Physical Activity Interventions on Children's Cognition and Metacognition: A Systematic Review and Meta-Analysis. *J. Am. Acad. Child Adolesc. Psychiatry* **2017**, *56*, 729–738. [CrossRef]
12. Haverkamp, B.F.; Wiersma, R.; Vertessen, K.; van Ewijk, H.; Oosterlaan, J.; Hartman, E. Effects of physical activity interventions on cognitive outcomes and academic performance in adolescents and young adults: A meta-analysis. *J. Sports Sci.* **2020**, *38*, 2637–2660. [CrossRef] [PubMed]
13. de Greeff, J.W.; Bosker, R.J.; Oosterlaan, J.; Visscher, C.; Hartman, E. Effects of physical activity on executive functions, attention and academic performance in preadolescent children: A meta-analysis. *J. Sci. Med. Sport* **2018**, *21*, 501–507. [CrossRef]
14. Álvarez-Bueno, C.; Pesce, C.; Cavero-Redondo, I.; Sánchez-López, M.; Garrido-Miguel, M.; Martínez-Vizcaíno, V. Academic Achievement and Physical Activity: A Meta-analysis. *Pediatrics* **2017**, *140*, e20171498. [CrossRef] [PubMed]
15. Stadler, M.A. Role of attention in implicit learning. *J. Exp. Psychol. Learn. Mem. Cognit.* **1995**, *21*, 674–685. [CrossRef]
16. Steinmayr, R.; Ziegler, M.; Träuble, B. Do intelligence and sustained attention interact in predicting academic achievement? *Learn. Individ. Differ.* **2010**, *20*, 14–18. [CrossRef]
17. Janssen, M.; Toussaint, H.M.; van Mechelen, W.; Verhagen, E.A. Effects of acute bouts of physical activity on children's attention: A systematic review of the literature. *SpringerPlus* **2014**, *3*, 410. [CrossRef] [PubMed]
18. Kremer, J.M.; Moran, A.; Walker, G.; Craig, C. *Key Concepts in Sport Psychology*; Sage: Newcastle upon Tyne, UK, 2012. [CrossRef]
19. Schmidt, M.; Benzing, V.; Kamer, M. Classroom-Based Physical Activity Breaks and Children's Attention: Cognitive Engagement Works! *Front. Psychol.* **2016**, *7*. [CrossRef] [PubMed]
20. Chang, Y.K.; Labban, J.D.; Gapin, J.I.; Etnier, J.L. The effects of acute exercise on cognitive performance: A meta-analysis. *Brain Res.* **2012**, *1453*, 87–101. [CrossRef]
21. Etnier, J.L.; Nowell, P.M.; Landers, D.M.; Sibley, B.A. A meta-regression to examine the relationship between aerobic fitness and cognitive performance. *Brain Res. Rev.* **2006**, *52*, 119–130. [CrossRef] [PubMed]
22. Basso, J.C.; Suzuki, W.A. The Effects of Acute Exercise on Mood, Cognition, Neurophysiology, and Neurochemical Pathways: A Review. *Brain Plast.* **2017**, *2*, 127–152. [CrossRef] [PubMed]
23. Ma, J.K.; Le Mare, L.; Gurd, B.J. Four minutes of in-class high-intensity interval activity improves selective attention in 9-to 11-year olds. *Appl. Physiol. Nutr. Metab.* **2015**, *40*, 238–244. [CrossRef] [PubMed]
24. Schmidt, M.; Benzing, V.; Wallman-Jones, A.; Mavilidi, M.-F.; Lubans, D.R.; Paas, F. Embodied learning in the classroom: Effects on primary school children's attention and foreign language vocabulary learning. *Psychol. Sport Exerc.* **2019**, *43*, 45–54. [CrossRef]
25. van den Berg, V.; Saliasi, E.; Jolles, J.; de Groot, R.H.M.; Chinapaw, M.J.M.; Singh, A.S. Exercise of Varying Durations: No Acute Effects on Cognitive Performance in Adolescents. *Front. Neurosci.* **2018**, *12*. [CrossRef] [PubMed]
26. McMullen, J.; Kulinna, P.; Cothran, D. Chapter 5 Physical Activity Opportunities During the School Day: Classroom Teachers' Perceptions of Using Activity Breaks in the Classroom. *J. Teach. Phys. Educ.* **2014**, *33*, 511–527. [CrossRef]

27. González, F.T.; González-Víllora, S. Bases metodológicas desde la neuroeducación para el diseño-aplicación de los descansos activos (Methodological bases from neuroeducation for the design-application of active breaks). In *Neuroeducación: Ayudando a Aprender Desde Las Evidencias Científicas*; González-Víllora, S., Bodoque-Osma, A.R., Eds.; Morata: Madrid, Spain, 2021; pp. 105–128.

28. Budde, H.; Voelcker-Rehage, C.; Pietraßyk-Kendziorra, S.; Ribeiro, P.; Tidow, G. Acute coordinative exercise improves attentional performance in adolescents. *Neurosci. Lett.* **2008**, *441*, 219–223. [CrossRef]

29. Higgins, J.P.T.; Green, S. (Eds.) *Cochrane Handbook for Systematic Reviews of Interventions 2.6 [Updated September 2006]*; John Wiley & Sons, Ltd: Chichester, UK, 2006.

30. Moher, D.; Liberati, A.; Tetzlaff, J.; Altman, D.G. Preferred Reporting Items for Systematic Reviews and Meta-Analyses: The PRISMA Statement. *PLoS Med.* **2009**, *6*, e1000097. [CrossRef]

31. Sterne, J.A.C.; Savović, J.; Page, M.J.; Elbers, R.G.; Blencowe, N.S.; Boutron, I.; Cates, C.J.; Cheng, H.-Y.; Corbett, M.S.; Eldridge, S.M.; et al. RoB 2: A revised tool for assessing risk of bias in randomised trials. *BMJ* **2019**, l4898. [CrossRef]

32. Eldridge, S.; Campbell, M.K.; Campbell, M.J.; Drahota, A.K.; Giraudeau, B.; Reeves, B.C.; Siegfried, N.; Higgins, J.P. Revised Cochrane Risk of Bias Tool for Randomized Trials (RoB 2): Additional Considerations for Cluster-Randomized Trials (RoB 2 CRT); 2020. Available online: https://sites.google.com/site/riskofbiastool/welcome/rob-2-0-tool/rob-2-for-cluster-randomized-trials (accessed on 1 January 2021).

33. Higgins, J.P.T.; Savović, E.; Page, M.J.; Sterne, J.A.C. Revised Cochrane Risk of Bias Tool for Randomized Trials (RoB 2): Additional Considerations for Crossover Trials; 2020. Available online: https://sites.google.com/site/riskofbiastool/welcome/rob-2-0-tool/rob-2-for-crossover-trials (accessed on 1 January 2021).

34. Sterne, J.A.; Hernán, M.A.; Reeves, B.C.; Savović, J.; Berkman, N.D.; Viswanathan, M.; Henry, D.; Altman, D.G.; Ansari, M.T.; Boutron, I.; et al. ROBINS-I: A tool for assessing risk of bias in non-randomised studies of interventions. *BMJ* **2016**, i4919. [CrossRef]

35. Altenburg, T.M.; Chinapaw, M.J.M.; Singh, A.S. Effects of one versus two bouts of moderate intensity physical activity on selective attention during a school morning in Dutch primary schoolchildren: A randomized controlled trial. *J. Sci. Med. Sport* **2016**, *19*, 820–824. [CrossRef]

36. Hopkins, W.G.; Marshall, S.W.; Batterham, A.M.; Hanin, J. Progressive Statistics for Studies in Sports Medicine and Exercise Science. *Med. Sci. Sport. Exerc.* **2009**, *41*, 3–12. [CrossRef] [PubMed]

37. Higgins, J.P.T.; Thompson, S.G. Quantifying heterogeneity in a meta-analysis. *Stat. Med.* **2002**, *21*, 1539–1558. [CrossRef] [PubMed]

38. Egger, M.; Smith, G.D.; Schneider, M.; Minder, C. Bias in meta-analysis detected by a simple, graphical test. *BMJ* **1997**, *315*, 629–634. [CrossRef] [PubMed]

39. Duval, S.; Tweedie, R. Trim and Fill: A Simple Funnel-Plot-Based Method of Testing and Adjusting for Publication Bias in Meta-Analysis. *Biometrics* **2000**, *56*, 455–463. [CrossRef] [PubMed]

40. Shi, L.; Lin, L. The trim-and-fill method for publication bias: Practical guidelines and recommendations based on a large database of meta-analyses. *Medicine (Baltimore)* **2019**, *98*, e15987. [CrossRef]

41. Egger, F.; Conzelmann, A.; Schmidt, M. The effect of acute cognitively engaging physical activity breaks on children's executive functions: Too much of a good thing? *Psychol. Sport Exerc.* **2018**, *36*, 178–186. [CrossRef]

42. Jäger, K.; Schmidt, M.; Conzelmann, A.; Roebers, C.M. Cognitive and physiological effects of an acute physical activity intervention in elementary school children. *Front. Psychol.* **2014**, *5*, 1–11. [CrossRef]

43. Jäger, K.; Schmidt, M.; Conzelmann, A.; Roebers, C.M. The effects of qualitatively different acute physical activity interventions in real-world settings on executive functions in preadolescent children. *Ment. Health Phys. Act.* **2015**, *9*, 1–9. [CrossRef]

44. Niemann, C.; Wegner, M.; Voelcker-Rehage, C.; Holzweg, M.; Arafat, A.M.; Budde, H. Influence of acute and chronic physical activity on cognitive performance and saliva testosterone in preadolescent school children. *Ment. Health Phys. Act.* **2013**, *6*, 197–204. [CrossRef]

45. Tine, M.T.; Butler, A.G. Acute aerobic exercise impacts selective attention: An exceptional boost in lower-income children. *Educ. Psychol.* **2012**, *32*, 821–834. [CrossRef]

46. Ordóñez, A.F.; Polo, B.; Lorenzo, A.; Zhang, S. Effects of a School Physical Activity Intervention in Pre-adolescents. *Apunt. Educ. Fís. Deport.* **2019**, *136*, 49–61. [CrossRef]

47. van den Berg, V.; Saliasi, E.; de Groot, R.H.M.; Chinapaw, M.J.M.; Singh, A.S. Improving cognitive performance of 9–12 years old children: Just dance? A randomized controlled trial. *Front. Psychol.* **2019**, *10*, 1–14. [CrossRef]

48. Buchele Harris, H.; Cortina, K.S.; Templin, T.; Colabianchi, N.; Chen, W. Impact of Coordinated-Bilateral Physical Activities on Attention and Concentration in School-Aged Children. *BioMed Res. Int.* **2018**, *2018*. [CrossRef]

49. Wilson, A.N.; Olds, T.; Lushington, K.; Petkov, J.; Dollman, J. The impact of 10-minute activity breaks outside the classroom on male students' on-task behaviour and sustained attention: A randomised crossover design. *Acta Paediatr.* **2016**, *105*, e181–e188. [CrossRef]

50. Hill, L.; Williams, J.H.G.; Aucott, L.; Milne, J.; Thomson, J.; Greig, J.; Munro, V.; Mon-Williams, M. Exercising Attention within the Classroom. *Dev. Med. Child Neurol.* **2010**, *52*, 929–934. [CrossRef] [PubMed]

51. van den Berg, V.; Saliasi, E.; de Groot, R.H.M.; Jolles, J.; Chinapaw, M.J.M.; Singh, A.S. Physical activity in the school setting: Cognitive performance is not affected by three different types of acute exercise. *Front. Psychol.* **2016**, *7*, 1–9. [CrossRef] [PubMed]

52. Park, S.; Etnier, J.L. Beneficial Effects of Acute Exercise on Executive Function in Adolescents. *J. Phys. Act. Health* **2019**, *16*, 423–429. [CrossRef] [PubMed]
53. Donnelly, J.E.; Hillman, C.H.; Castelli, D.; Etnier, J.L.; Lee, S.; Tomporowski, P.; Lambourne, K.; Szabo-Reed, A.N. Physical Activity, Fitness, Cognitive Function, and Academic Achievement in Children: A Systematic Review. *Med. Sci. Sports Exerc.* **2016**, *48*, 1197–1222. [CrossRef]
54. Xue, Y.; Yang, Y.; Huang, T. Effects of chronic exercise interventions on executive function among children and adolescents: A systematic review with meta-analysis. *Br. J. Sports Med.* **2019**, *53*, 1397–1404. [CrossRef]
55. Brisswalter, J.; Collardeau, M.; René, A. Effects of Acute Physical Exercise Characteristics on Cognitive Performance. *Sport. Med.* **2002**, *32*, 555–566. [CrossRef]
56. Groslambert, A. Validation of a Rating Scale of Perceived Exertion in Young Children. *Int. J. Sports Med.* **2001**, *22*, 116–119. [CrossRef]
57. Parfitt, G.; Shepherd, P.; Eston, R.G. Reliability of effort production using the children's CALER and BABE perceived exertion scales. *J. Exerc. Sci. Fit.* **2007**, *4*, 49–55.
58. Hortigüela Alcalá, D.; Hernando Garijo, A.; Pérez-Pueyo, Á.; Fernández-Río, J. Cooperative Learning and Students' Motivation, Social Interactions and Attitudes: Perspectives from Two Different Educational Stages. *Sustainability* **2019**, *11*, 7005. [CrossRef]
59. Yuretich, R.F.; Khan, S.A.; Leckie, R.M.; Clement, J.J. Active-Learning Methods to Improve Student Performance and Scientific Interest in a Large Introductory Oceanography Course. *J. Geosci. Educ.* **2001**, *49*, 111–119. [CrossRef]
60. Borg, G. *Borg's Perceived Exertion and Pain Scales*; Human Kinetics: Champaign, IL, USA, 1998.
61. Adsiz, E.; Dorak, F.; Ozsaker, M.; Vurgun, N. The influence of physical activity on attention in Turkish children. *HealthMED* **2012**, *6*, 1384–1389.
62. Amicone, G.; Petruccelli, I.; De Dominicis, S.; Gherardini, A.; Costantino, V.; Perucchini, P.; Bonaiuto, M. Green Breaks: The Restorative Effect of the School Environment's Green Areas on Children's Cognitive Performance. *Front. Psychol.* **2018**, *9*, 1579. [CrossRef] [PubMed]
63. Bartholomew, J.B.; Golaszewski, N.M.; Jowers, E.; Korinek, E.; Roberts, G.; Fall, A.; Vaughn, S. Active learning improves on-task behaviors in 4th grade children. *Prev. Med. (Baltim)* **2018**, *111*, 49–54. [CrossRef] [PubMed]
64. Ben-Zeev, T.; Hirsh, T.; Weiss, I.; Gornstein, M.; Okun, E. The Effects of High-intensity Functional Training (HIFT) on Spatial Learning, Visual Pattern Separation and Attention Span in Adolescents. *Front. Behav. Neurosci.* **2020**, *14*. [CrossRef]
65. Blasche, G.; Szabo, B.; Wagner-Menghin, M.; Ekmekcioglu, C.; Gollner, E. Comparison of rest-break interventions during a mentally demanding task. *Stress Health* **2018**, *34*, 629–638. [CrossRef] [PubMed]
66. Chou, C.-C.; Chen, K.-C.; Huang, M.-Y.; Tu, H.-Y.; Huang, C.-J. Can Movement Games Enhance Executive Function in Overweight Children? A Randomized Controlled Trial. *J. Teach. Phys. Educ.* **2020**, *39*, 527–535. [CrossRef]
67. Chrismas, B.C.R.; Taylor, L.; Cherif, A.; Sayegh, S.; Bailey, D.P. Breaking up prolonged sitting with moderate-intensity walking improves attention and executive function in Qatari females. *PLoS ONE* **2019**, *14*, e0219565. [CrossRef] [PubMed]
68. Contreras Jordan, O.R.; Leon, M.P.; Infantes-Paniagua, A.; Prieto-Ayuso, A. Effects of active breaks in the attention and concentration of Elementary School students. *Rev. Interuniv. Form. Profr.* **2020**, *95*, 145–160. [CrossRef]
69. Egger, F.; Benzing, V.; Conzelmann, A.; Schmidt, M. Boost your brain, while having a break! The effects of long-term cognitively engaging physical activity breaks on children's executive functions and academic achievement. *PLoS ONE* **2019**, *14*. [CrossRef] [PubMed]
70. Fenesi, B.; Lucibello, K.; Kim, J.A.; Heisz, J.J. Sweat So You Don't Forget: Exercise Breaks During a University Lecture Increase On-Task Attention and Learning. *J. Appl. Res. Mem. Cognit.* **2018**, *7*, 261–269. [CrossRef]
71. Flippin, M.; Clapham, E.D.; Tutwiler, M.S. Effects of using a variety of kinesthetic classroom equipment on elementary students' on-task behaviour: A pilot study. *Learn. Environ. Res.* **2020**. [CrossRef]
72. Gonzalez Fernandez, F.T.; Baena Morales, S.; Vila Blanch, M.; Garcia-Taibo, O. Chronic effects in cognition of actives-breaks. *Sport. Tech. J. Sch. Sport Phys. Educ. Psychomot.* **2020**, *6*, 488–502. [CrossRef]
73. Grieco, L.A.; Jowers, E.M.; Errisuriz, V.L.; Bartholomew, J.B. Physically active vs. sedentary academic lessons: A dose response study for elementary student time on task. *Prev. Med. (Baltim)* **2016**, *89*, 98–103. [CrossRef] [PubMed]
74. Howie, E.K.; Schatz, J.; Pate, R.R. Acute Effects of Classroom Exercise Breaks on Executive Function and Math Performance: A Dose–Response Study. *Res. Q. Exerc. Sport* **2015**, *86*, 217–224. [CrossRef]
75. Howie, E.K.; Beets, M.W.; Pate, R.R. Acute classroom exercise breaks improve on-task behavior in 4th and 5th grade students: A dose-response. *Ment. Health Phys. Act.* **2014**, *7*, 65–71. [CrossRef]
76. Kubesch, S.; Walk, L.; Spitzer, M.; Kammer, T.; Lainburg, A.; Heim, R.; Hille, K. A 30-Minute Physical Education Program Improves Students' Executive Attention. *Mind Brain Educ.* **2009**, *3*, 235–242. [CrossRef]
77. Mahar, M.T. Impact of short bouts of physical activity on attention-to-task in elementary school children. *Prev. Med. (Baltim)* **2011**, *52*, S60–S64. [CrossRef]
78. Mavilidi, M.F.; Drew, R.; Morgan, P.J.; Lubans, D.R.; Schmidt, M.; Riley, N. Effects of different types of classroom physical activity breaks on children's on-task behaviour, academic achievement and cognition. *Acta Paediatr.* **2020**, *109*, 158–165. [CrossRef]
79. Mazzoli, E.; Teo, W.-P.; Salmon, J.; Pesce, C.; He, J.; Ben-Soussan, T.D.; Barnett, L.M. Associations of Class-Time Sitting, Stepping and Sit-to-Stand Transitions with Cognitive Functions and Brain Activity in Children. *Int. J. Environ. Res. Public Health* **2019**, *16*, 1482. [CrossRef]

80. McGowan, A.L.; Ferguson, D.P.; Gerde, H.K.; Pfeiffer, K.A.; Pontifex, M.B. Preschoolers exhibit greater on-task behavior following physically active lessons on the approximate number system. *Scand. J. Med. Sci. Sports* **2020**, *30*, 1777–1786. [CrossRef]
81. Merriman, W.; González-Toro, C.M.; Cherubini, J. Physical Activity in the Classroom. *Kappa Delta Pi Rec.* **2020**, *56*, 164–169. [CrossRef]
82. Miklós, M.; Komáromy, D.; Futó, J.; Balázs, J. Acute Physical Activity, Executive Function, and Attention Performance in Children with Attention-Deficit Hyperactivity Disorder and Typically Developing Children: An Experimental Study. *Int. J. Environ. Res. Public Health* **2020**, *17*, 4071. [CrossRef] [PubMed]
83. Napoli, M.; Krech, P.R.; Holley, L.C. Mindfulness Training for Elementary School Students. *J. Appl. Sch. Psychol.* **2005**, *21*, 99–125. [CrossRef]
84. Niedermeier, M.; Weiss, E.M.; Steidl-Müller, L.; Burtscher, M.; Kopp, M. Acute Effects of a Short Bout of Physical Activity on Cognitive Function in Sport Students. *Int. J. Environ. Res. Public Health* **2020**, *17*, 3678. [CrossRef] [PubMed]
85. Ochoa Diaz, C.E.; Machado Maliza, M.E.; Guarneri Chacha, K.A. Active break strategy to improve students' attention to teaching activities. *Rev. Conrado* **2020**, *16*, 285–290.
86. Owen, K.B.; Parker, P.D.; Astell-Burt, T.; Lonsdale, C. Effects of physical activity and breaks on mathematics engagement in adolescents. *J. Sci. Med. Sport* **2018**, *21*, 63–68. [CrossRef]
87. Pesce, C.; Crova, C.; Marchetti, R.; Struzzolino, I.; Masci, I.; Vannozzi, G.; Forte, R. Searching for cognitively optimal challenge point in physical activity for children with typical and atypical motor development. *Ment. Health Phys. Act.* **2013**, *6*, 172–180. [CrossRef]
88. Ruiz-Ariza, A.; Lopez-Serrano, S.; Mezcua-Hidalgo, A.; Martinez-Lopez, E.J.; Abu-Helaiel, K. Acute effect of physically active rests on cognitive variables and creativity in Secondary Education. *Retos Nuevas Tend. Educ. Fís. Deport. Recreacion* **2021**, 635–642.
89. Sánchez-López, M.; Pardo-Guijarro, M.J.; del Campo, D.G.-D.; Silva, P.; Martínez-Andrés, M.; Gulías-González, R.; Díez-Fernández, A.; Franquelo-Morales, P.; Martínez-Vizcaíno, V. Physical activity intervention (Movi-Kids) on improving academic achievement and adiposity in preschoolers with or without attention deficit hyperactivity disorder: Study protocol for a randomized controlled trial. *Trials* **2015**, *16*, 456. [CrossRef] [PubMed]
90. Sugahara, S.; Dellaportas, S. Bringing active learning into the accounting classroom. *Meditari Account. Res.* **2018**, *26*, 576–597. [CrossRef]
91. Vazou, S.; Long, K.; Lakes, K.D.; Whalen, N.L. "Walkabouts" Integrated Physical Activities from Preschool to Second Grade: Feasibility and Effect on Classroom Engagement. *Child Youth Care Forum* **2020**. [CrossRef]
92. Watson, A.; Timperio, A.; Brown, H.; Hesketh, K.D. A primary school active break programme (ACTI-BREAK): Study protocol for a pilot cluster randomised controlled trial. *Trials* **2017**, *18*, 433. [CrossRef]
93. Watson, A.J.L.; Timperio, A.; Brown, H.; Hesketh, K.D. A pilot primary school active break program (ACTI-BREAK): Effects on academic and physical activity outcomes for students in Years 3 and 4. *J. Sci. Med. Sport* **2019**, *22*, 438–443. [CrossRef]
94. Webster, E.K.; Wadsworth, D.D.; Robinson, L.E. Preschoolers' Time On-Task and Physical Activity During a Classroom Activity Break. *Pediatr. Exerc. Sci.* **2015**, *27*, 160–167. [CrossRef]

Article

Physical Activity and Inhibitory Control: The Mediating Role of Sleep Quality and Sleep Efficiency

Lin Li [1,2,†], Qian Yu [3,4,†], Wenrui Zhao [1,2], Fabian Herold [5,6], Boris Cheval [7,8], Zhaowei Kong [9], Jinming Li [3,4], Notger Mueller [5,6], Arthur F. Kramer [10,11], Jie Cui [1,2], Huawei Pan [1,2], Zhuxuan Zhan [1,2], Minqiang Hui [1,2] and Liye Zou [3,4,*]

[1] Key Laboratory Ministry of Education of Adolescent Health Assessment and Exercise Intervention, East China Normal University, Shanghai 200241, China; lilin.xtt@163.com (L.L.); zhaowenrui108@163.com (W.Z.); cuijie1615@163.com (J.C.); 13758114178@163.com (H.P.); zzx951105@163.com (Z.Z.); m13213595838@163.com (M.H.)

[2] School of Physical Education and Health, East China Normal University, Shanghai 200241, China

[3] Institute of KEEP Collaborative Innovation, Shenzhen University, Shenzhen 518060, China; yuqianmiss@163.com (Q.Y.); jinmingli1999@gmail.com (J.L.)

[4] Exercise Psychophysiology Laboratory, School of Psychology, Shenzhen University, Shenzhen 518060, China

[5] Department of Neurology, Medical Faculty, Otto von Guericke University, Leipziger Str. 44, 39120 Magdeburg, Germany; fabian.herold@st.ovgu.de (F.H.); notger.mueller@dzne.de (N.M.)

[6] Research Group Neuroprotection, German Center for Neurodegenerative Diseases (DZNE), Leipziger Str. 44, 39120 Magdeburg, Germany

[7] Swiss Center for Affective Sciences, University of Geneva, 1205 Geneva, Switzerland; Boris.Cheval@unige.ch

[8] Laboratory for the Study of Emotion Elicitation and Expression (E3Lab), Department of Psychology, FPSE, University of Geneva, 1205 Geneva, Switzerland

[9] Faculty of Education, University of Macau, Macao 999078, China; zwkong@um.edu.mo

[10] Center for Cognitive and Brain Health, Department of Psychology, Northeastern University, Boston, MA 02115, USA; a.kramer@northeastern.edu

[11] Beckman Institute, University of Illinois at Urbana-Champaign, Champaign, IL 61801, USA

*   Correspondence: liyezou123@gmail.com

†   Equally contributed to this study.

Citation: Li, L.; Yu, Q.; Zhao, W.; Herold, F.; Cheval, B.; Kong, Z.; Li, J.; Mueller, N.; Kramer, A.F.; Cui, J.; et al. Physical Activity and Inhibitory Control: The Mediating Role of Sleep Quality and Sleep Efficiency. *Brain Sci.* **2021**, *11*, 664. https://doi.org/10.3390/brainsci11050664

Academic Editors: Filipe Manuel Clemente and Ana Filipa Silva

Received: 7 May 2021
Accepted: 17 May 2021
Published: 19 May 2021

Publisher's Note: MDPI stays neutral with regard to jurisdictional claims in published maps and institutional affiliations.

Abstract: Objectives: the current study aimed to investigate the relationship between physical activity (PA) level and inhibitory control performance and then to determine whether this association was mediated by multiple sleep parameters (i.e., subjective sleep quality, sleep duration, sleep efficiency, and sleep disturbance). Methods: 180 healthy university students (age: 20.15 ± 1.92 years) from the East China Normal University were recruited for the present study. PA level, sleep parameters, and inhibitory control performance were assessed using the International Physical Activity Questionnaire (IPAQ), the Pittsburgh Sleep Quality Index Scale (PSQI), and a Stroop test, respectively. The data were analyzed using structural equation modeling. Results: A higher level of PA was linked to better cognitive performance. Furthermore, higher subjective sleep quality and sleep efficiency were associated with better inhibitory control performance. The mediation analysis revealed that subjective sleep quality and sleep efficiency mediated the relationship between PA level and inhibitory control performance. Conclusion: our results are in accordance with the literature and buttress the idea that a healthy lifestyle that involves a relatively high level of regular PA and adequate sleep patterns is beneficial for cognition (e.g., inhibitory control performance). Furthermore, our study adds to the literature that sleep quality and sleep efficiency mediates the relationship between PA and inhibitory control performance, expanding our knowledge in the field of exercise cognition.

Keywords: physical activity; sleep; inhibitory performance; mediating effects

1. Introduction

Brain health and cognitive enhancement are important factors for successful studies, better career opportunities, and a higher quality of life and thus has received widespread attention from different disciplines including medicine and education [1]. Emerging evidence demonstrates that regular physical activity (PA) can enhance cognitive performance, especially in the domain of executive functioning (EF) [2–5]. Among the diverse subcomponents of EF, controlled inhibition (the capacity to suppress irrelevant information or prepotent responses in monitoring and updating information) has been reported to benefit more from PA than other aspects of EF (e.g., shifting or updating) [6]. Considering that the brain develops until the age of 30 years [7–10] and the important role of inhibitory control in mastering life successfully [11], there is an urgency to investigate how regular PA influences inhibitory control performance and how to maximize PA-induced cognitive benefits among youths. Of note, existing research mostly focuses on children and older individuals, while there is a lack of studies on young to middle-aged adults [1]. Additionally, although PA has been proven to maintain and improve neurocognitive function, much less is known about the pathway by which it exerts salutary effects on inhibitory control performance. Among the suggested potential mediators, sleep has been proposed as an ideal candidate to explain the PA–cognition links, although empirical evidence in younger adults is currently lacking [12].

The improved cognitive performance among individuals who engaged in regular PA may be explained, at least partly, by sleep health. Indeed, young adults regularly experience restricted sleep (70% of people sleeping less than 6 h at least one night a week) due to educational, vocational, and social responsibilities [13]. Mounting evidence shows that the increased PA contributes to sleep health (i.e., sleep duration, efficiency, and quality). For example [14,15], randomized controlled trials conducted in 1997 were the first to indicate the PA (i.e., moderate-intensity exercise, daily activity)-induced benefits in self-reported sleep quality. A review by Vanderlinden et al. indicated that moderate intensity exercise intervention (lasted 12 weeks to 6 months), with a frequency of three times per week, showed the highest improvements in sleep outcomes in older adults [16]. Even among patients with sleep disruptions, exercise intervention was also found to positively influence sleep outcomes [17].

The provided evidence supports the idea that better sleep was linked to improved cognitive performance [18–20]. For example, sufficient sleep quality has been shown to be essential for an optimal cognitive performance, especially in the domain of EF (i.e., inhibitory control) [21]. However, although the exact neurobiological mechanisms of the association between sleep and cognitive performance are not fully understood. Sleep might work through complex ultradian, circadian, and homeostatic regulations to positively influence the prefrontal circuits that are known to be important neural correlates for EF [22–24]. The improved prefrontal circuits could be well reflected by higher voltage and slower brain waves during non-rapid eye movement (NREM), coinciding with the deactivation of dorsolateral prefrontal cortex during rapid eye movement (REM) [25].

To the best of our knowledge, there are only two studies examining the potential mediating role of sleep health on the association between PA and cognitive function [26,27]. Wilckens et al. found that sleep efficiency may be one pathway by which PA benefits executive control, but this study only investigated two sleep health indicators (sleep efficiency and total sleep time as measured by accelerometer-based sleep assessment) [26]. Contrary to the outcomes of Wilckens et al., Falck et al. reported that PA and sleep quality interacted with cognitive function via independent mechanisms without mediation effects [27] among the elderly (above 55 years). To address the gap and inconsistency, the current study attempts to explore how PA influences young adults' inhibitory control via a wider range of sleep parameters [28]. We aimed (i) to investigate the relationships among regular PA level, inhibitory control and multiple sleep parameters (subjective sleep quality, sleep duration, sleep efficiency, and sleep disturbance) among university students and (ii) to examine how sleep parameters mediate the relationship between PA level and inhibitory control performance.

2. Methods

2.1. Participants

In total, 180 healthy university students (age: 20.15 ± 1.92 years; 109 females and 71 males) from the East China Normal University were recruited in the present study. The PA level of participants, their sleep status, and their inhibitory control performance were assessed via the International Physical Activity Questionnaire (IPAQ), the Pittsburgh Sleep Quality Index Scale (PSQI), and Stroop test, respectively. Moreover, all participants were screened for depression and anxiety on the Self-rating depression scale (SDS) and the Self-Rating Anxiety Scale (SAS). Exclusion criteria included (1) SDS score $\geq$ 53, (2) SAS score $\geq$ 50, (3) failure to complete all of the assessments, and (4) left-handedness. The whole procedure was approved by the ethic committee of the East China Normal University (No. HR 085-2018), and all participants signed informed consent before the onset of this experiment.

2.2. Demographic Information

A self-developed scale on demographic information was used to acquire participants' information such as gender, age, height, and weight. Height and weight were assessed by a Height Weight Meter. Body Mass Index (BMI) was calculated as weight (in kilograms) divided by the square of height (in meters).

2.3. International Physical Activity Questionnaire

The IPAQ was used to assess the PA level of each individual (content validity = 0.98, calibration validity = 0.41, and test–retest reliability = 0.78). The IPAQ totally consists of five parts: (1) job-related PA; (2) transportation PA; (3) housework, house maintenance, and caring for family; (4) recreation, sport, and leisure-time PA; and (5) time spent sitting. MET (Metabolic Equivalent of Task) minutes per week, as a continuous variable, represents the amount of energy expended carrying out physical activity.

2.4. Pittsburgh Sleep Quality Index Scale

The subjective level of sleep quality was assessed using the PSQI, which queries possible factors related to sleep disturbance over the past four weeks [29]. It includes 19 items within seven sleep-related domains: subjective sleep quality, sleep latency, sleep duration, habitual sleep efficiency, sleep disturbances, use of sleep medication, and daytime dysfunction [29]. The score of each domain is rated from 0 to 3 and seven sub-scores can be added to produce a total sleep quality score ranging from 0 to 21 [29], with higher total score indicating lower sleep quality [29].

2.5. Measurement of Inhibitory Control Performance

Inhibitory control performance was tested by a Stroop task (Figure 1) in which participants were asked to differentiate stimulation in the mixed congruent and incongruent conditions [30]. The stimuli used in this task were composed of four words (red, yellow, blue, and green in Chinese). In the congruent condition, the word was shown in matching colors; in the incongruent condition, the word did not match the color. The task lasted for 240 s, including two blocks with 32 trials (8 congruent and 24 incongruent trails) in total. In each trial, a fixed cross was presented first to attract participants' attention and then the color words were shown randomly, with each one lasting for 2000 ms. The inter-stimulus intervals were also random (2000, 4000, 6000, 8000, or 10,000 ms). The participants were asked to report the correct color of the word by pressing the corresponding button on a keyboard as quickly and accurately as possible, ignoring the interference by the word's semantic meaning. Subjects performed 20 practice trials before performing the experimental trial blocks. Accuracy and reaction time of each condition were also recorded. According to the operational definition of inhibitory performance in the Stroop test, the efficiency of inhibitory performance was characterized by the difference in reaction time.

The smaller the Stroop effect ($\Delta RT = RT_{incongruent} - RT_{congruent}$), the stronger the inhibitory performance.

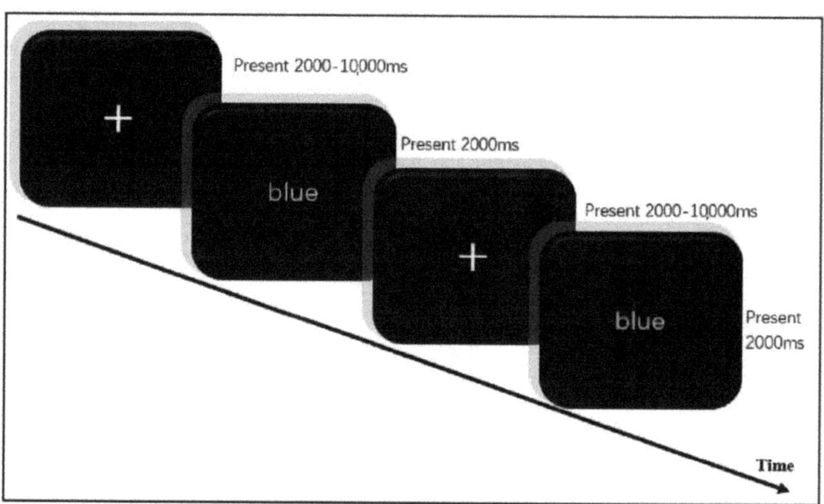

Figure 1. Stroop test.

2.6. Statistical Analysis

Statistical analyses were conducted using SPSS (Statistical Product and Service Solutions) software, version 23. Mean and standard deviation for all variables were calculated. Unless noted, all effects were described as significant at $p < 0.05$. Partial correlation analyses, with sex, age, and BMI as covariates, were used to examine the relationships between PA level (MET score), inhibitory control performance, and seven sleep parameters (subjective sleep quality, sleep latency, sleep duration, habitual sleep efficiency, sleep disturbances, use of sleep medication, and daytime dysfunction). Moreover, structural equation modeling (SEM) with the help of AMOS (Analysis of Moment Structure) was conducted to assess whether the sleep parameters mediated the association between PA and inhibitory control performance. The indices used to evaluate the overall fit of model included Chi-square (χ^2) test, normed Chi-square (χ^2/df), normed fit index (NFI), incremental fit index (IFI), and comparative fit index (CFI). Usually, a well-fitted model should meet the following requirements: (1) a nonsignificant Chi-square; (2) 0.90 or less NFI, IFI, and CFI values; and (3) 0.06 or less normed Chi-square value [31]. The paths from the mediation model were examined using direct and indirect effects of study variables (PA level, sleep parameters, and the Stroop effect); bootstrapping approach was used to test the statistical significance of mediation.

3. Results

The results of the demographic and anthropometric parameters, the IPAQ, the PSQI, and the performance on the Stroop test are shown in Table 1.

Table 1. Demographic information on the variables.

	Males (n = 71)	Females (n = 109)	Total (n = 180)
	Demographic and Anthropometric Parameters		
Age (years)	20.31 ± 1.45	19.68 ± 1.64	19.93 ± 1.59
BMI (kg/m^2)	21.66 ± 2.05	21.20 ± 2.55	21.37 ± 2.37
	International Physical Activity Questionnaire		
MET-min/w	4967.05 ± 1447.06	5171.22 ± 1556.39	5090.69 ± 1513.41
	Pittsburgh Sleep Quality Index		
Subjective sleep quality	0.68 ± 0.69	0.76 ± 0.79	0.73 ± 0.75
Sleep time	0.62 ± 0.70	0.54 ± 0.69	0.57 ± 0.69
Sleep efficiency	0.73 ± 0.72	0.68 ± 0.76	0.70 ± 0.74
Sleep disorders	0.80 ± 0.71	0.97 ± 0.55	0.91 ± 0.62
	Stroop Test		
Congruent accuracy (%)	95.00 ± 9.35	97.00 ± 5.80	96.00 ± 7.46
Incongruent accuracy (%)	95.00 ± 8.65	97.00 ± 4.70	96.00 ± 6.66
Congruent reaction time (ms)	899.98 ± 107.20	867.53 ± 135.39	875.47 ± 141.42
Incongruent reaction time (ms)	1000.18 ± 107.79	964.34 ± 148.10	973.07 ± 152.56
Stroop effect (ms)	100.21 ± 45.64	96.81 ± 47.13	98.15 ± 46.45

Note. BMI: body mass index; MET: metabolic equivalent of task.

3.1. Correlation Analysis

As shown in Table 2, the results indicate negative correlations of regular PA level with (1) subjective sleep quality (r = −0.42, p < 0.001), (2) habitual sleep efficiency (r = −0.52, p < 0.001), and (3) the Stroop effect (r = −0.39, p < 0.001) (with a higher score in the PSQI indicating worse sleep status and a smaller Stroop effect indicating better inhibitory control performance). Additionally, the scores of subjective sleep quality and sleep efficiency were found to positively correlate with the Stroop effect (r = 0.83, p < 0.001; r = 0.61, p < 0.001).

Table 2. Correlation analyses among the outcome measures.

	MET-min/w	S1	S2	S3	S4	S5	S6	S7	Acc.1	Acc.2	RT1	RT2	Stroop Effect
MET	-												
S1	−0.425 ***	-											
S2	−0.033	0.068	-										
S3	−0.006	0.082	0.107	-									
S4	−0.519 ***	0.652 ***	0.033	0.126	-								
S5	0.180 **	0.100	0.186 **	−0.106	−0.008	-							
S6	0.153 **	0.060	0.125	−0.210 **	0.081	0.331 ***	-						
S7	0.058	0.047	0.289 ***	−0.039	0.002	0.198 **	0.412 ***	-					
Acc.1	−0.018	0.057	0.064	0.074	−0.023	0.103	−0.019	0.136	-				
Acc.2	−0.056	−0.032	−0.016	0.116	−0.062	−0.002	−0.106	0.017	0.486 ***	-			
RT1	0.036	0.139	0.021	−0.053	0.043	0.047	0.116	0.033	−0.076	−0.107	-		
RT2	−0.101	0.419 ***	0.036	−0.006	0.252 ***	0.060	0.142	0.032	−0.027	−0.109	0.938 ***	-	
Stroop effect	−0.386 ***	0.831 ***	0.045	0.125	0.611 ***	0.045	0.098	0.003	0.125	−0.026	0.008	0.355 ***	-

Note. MET = metabolic equivalent of task; S1 = subjective sleep quality; S2 = sleep latency; S3 = sleep duration; S4 = habitual sleep efficiency; S5 = sleep disturbances; S6 = use of sleep medication; S7 = daytime dysfunction. Acc.1 = congruent accuracy; Acc.2 = incongruent accuracy; RT1 = congruent reaction time; RT2 = incongruent reaction time. ** means that p value is less than 0.01; *** means that p value is less than 0.001.

3.2. Multiple Mediation Model

Based on the outcomes of correlation analyses, only two sleep parameters (subjective sleep quality and sleep efficiency) were included in the mediation model. The structured equation model was used to determine the mediating role of sleep parameters on the relationship between PA and inhibitory control (Figure 2). In the initial model, RMSEA (0.07) and SRMR (0.06) were higher than 0.05, which meant that the initial model failed to present a good fit. Therefore, the model was modified as specified by modification indices (χ^2/df = 64.43, NFI = 0.83, IFI = 0.83, and CFI = 0.82). After removing the nonsignificant path between physical activity and inhibitory control performance, the model was re-analyzed and all indices within the new model show a good fit (χ^2/df = 0.06, NFI = 0.92, IFI = 0.92, and CFI = 0.93) (Table 3). In the final model, the mediating effect explained 76% of variance variation in the dependent variable, and subjective sleep quality and sleep efficiency explained 53.6% and 22.4% of variance, respectively. In order to reduce the interaction between the mediating variables, the residual correlation analysis between the two mediating variables was conducted (path coefficient = 0.55).

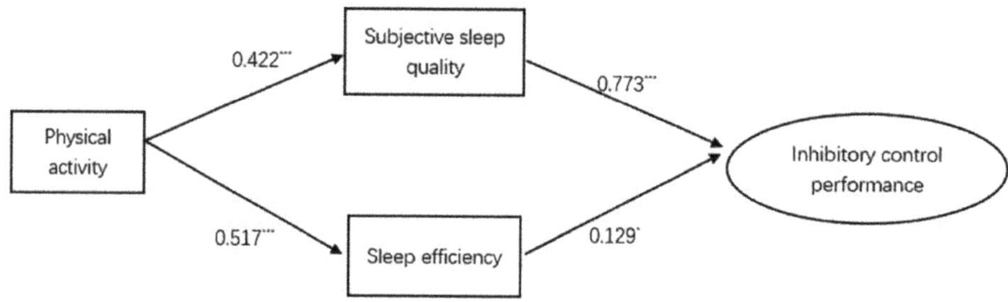

Figure 2. Multiple mediation model 2. Note. * means that *p* value is less than 0.05; *** means that *p* value is less than 0.001.

Table 3. Modification indices in multiple mediation model 2.

Model	χ^2	*p*	*f*	χ^2/DF	NFI	IFI	CFI	RMSEA	SRMR
Initial model	64.43	0.001	1	64.43	0.83	0.83	0.82	0.07	0.06
Final model	0.06	0.06	1	0.06	0.92	0.92	0.93	0.05	0.04
$\Delta\chi^2$	64.37								

The paths from the mediation model were examined using direct and indirect effects of the study variables (Table 4; Figure 2). The results showed the PA-induced indirect effect on inhibitory control performance while adjusting for sleep parameters among the university students. Thus, sleep parameters (subjective sleep quality and sleep efficiency) played a mediating role in the indirect relationships. Additionally, the nonsignificant direct effect of physical activity on inhibitory control performance is 0.001.

Table 4. The bootstrap analysis of multiple mediation model 2.

Route	Indirect Effects of Standardization	Average Indirect Effect	95% Confidence Interval		*p*
			Lower Limit	Upper Limit	
Subjective Sleep Quality	0.42 × 0.77 = 0.32	0.41	0.27	0.55	0.01
Sleep Efficiency	0.53 × 0.13 = 0.07	0.51	0.39	0.63	0.01

4. Discussion

Based on previous literature, the mediating role of sleep parameters on the relationship between PA and EF (i.e., inhibitory control) is expected, but this test has not been formally assessed in a young sample and with a wider range of sleep health indicators [12,32]. In the present study, we observed in our cohort of healthy younger adults (1) a negative correlation between PA and inhibitory control performance, (2) a negative correlation between PA and subjective sleep quality as well as sleep efficiency ((number of sleep hours/number of hours spent in bed) × 100), and (3) positive correlations between inhibitory control and sleep parameters (subjective sleep quality and sleep efficiency). Moreover, the mediation analysis reveals that subjective sleep quality and sleep efficiency play mediating roles in the PA–EF (inhibitory control).

In this study, positive correlations were observed between PA level and sleep indicators (subjective sleep quality and habitual sleep efficiency), which is in line with the findings of existing studies [26,33]. Over the last decades, it has been shown that, among younger adults, a higher PA level is positively associated with sleep health [34,35]. Notably, recent studies proposed that the PA–sleep relationship may be influenced by PA slots [36]. To be specific, some researchers emphasized that light PA in the daytime was correlated with better sleep status, whereas PA at night may reduce the sleep duration and sleep time window [37]. However, as we used the IPAQ to assess PA, the time of day in PA engagement was not used in the present study but such an assessment should be considered in future studies to deepen our understanding of the relationship between PA, sleep, and cognitive performance.

The results from our partial correlation analysis indicate that EF (inhibitory control) was positively correlated with both subjective sleep quality and sleep efficiency but not with sleep duration. Such results are consistent with previous researches in which sleep parameters were measured with an actigraphic device and polysomnography [38–40]. PA–sleep relationships may be partially explained by the following neurobiological mechanisms: (1) higher sleep efficiency and sleep quality enhanced the efficacy of slow-wave sleep in the restoration of prefrontal cortex function (which is responsible for inhibitory control performance) [41] and (2) prior sleep deprivation, sleep disorders, sleep fragmentation, and medication effects usually occur alongside longer sleep durations, which does not reflect real sleep health [42–44]. Non-restorative sleep caused by recurrent awakenings and stage shifts can even be observed in the population with normal sleep duration [45]. Due to this, sleep duration cannot be regarded as a reliable indicator for sleep health.

Only an indirect relationship (mediated by sleep parameters) rather than a direct relationship between PA and cognitive performance was observed among university students in the present study. This finding is partially supported Wilckens et al., who found that PA level was indirectly correlated with inhibitory control in both adolescent (aged 21–30) and older adults (aged 55–80) and that sleep efficiency but not total sleep time or sleep duration played a mediating role. In the present study, we observed that both sleep efficiency and subjective sleep quality played important roles in the PA–EF relationship. Based on these outcomes, it could be inferred that ease of falling asleep and staying asleep rather than any aspect of sleep affect the PA–cognition connections. Moreover, it is essential to highlight that the direct relationship between PA and inhibitory control performance reflects the overall relationship with cognition, which consists of sleep parameters and other mediators not tested here. Additionally, some of the potential mediators may suppress the association between PA and inhibitory control [46,47].

Besides the primary outcome carried out from the present study, some novel findings were summarized based on a comparison with previous research [28,29,48]. It is suggested that the mediating role of sleep efficiency in the PA–EF (inhibitory control) relationship is observed across different age groups [26]. PA can even attenuate the negative impact of low sleep efficiency on executive function, with the clearest effects observed using direct measurements of sleep and PA [49]. Even though sleep quality and sleep duration were also found to be positively associated with PA and cognition among young adults [33], an apparent functional weakening in such relationships appear in the aging population [27]. It seems that sleep measures such as sleep quality and sleep duration are more sensitive to aging-related neurobiology (e.g., neural atrophy, nocturnal hypoxia, neuroendocrine changes, and altered neuromodulation), which may reduce the potential to impact cognitive performance through strategies such as PA [50].

Finally, this study has a certain theoretical and practical significance, but some limitations must be acknowledged when interpreting our findings. First, the cross-sectional design does not allow us to make strong conclusions regarding the causality of the observed correlations. Secondly, this study did not account for the time of day when participants engaged in PA, which is known to influence sleep patterns and thus might influence cognitive performance, too. Thirdly, our sleep pattern assessment relies on the PSQI, which is an international recognized questionnaire with good psychometric properties but relies on subjective ratings rather than objective ratings. Given difference in subjective and objective ratings of sleep [48,51], this might have influenced our analysis. Thus, caution should be taken when comparing our results with findings of other studies quantifying sleep parameters by objective tools (e.g., actigraphy) although those studies point in the same direction (better sleep patterns are associated with better cognitive performance) [52]. In this regard, we recommend that further studies investigating the relationship between regular PA level, sleep patterns, and cognitive performance should consider assessing both subjective ratings and objective ratings of sleep. For studies with more stern requirements, polysomnography, the gold standard for sleep research, is recommended. Lastly, we focused only on one cognitive parameter (inhibitory control), preventing the generalizability of our results to other EF components and cognitive measures.

5. Conclusions

Sleep efficiency and subjective sleep quality were found to statistically mediate the significant relationship between physical activity and inhibitory control in a population of university students. Joint measurements including both objective and subjective assessment and participants in a wider age range should be considered in further studies to test the potential neurobehavioral mechanisms and moderators.

Author Contributions: Conceptualization, L.L., Q.Y. and L.Z.; methodology, Q.Y.; software, W.Z.; validation, J.C., H.P. and Z.Z.; formal analysis, W.Z.; investigation, M.H. and J.L.; resources, L.L.; data curation, W.Z.; writing—original draft preparation, Q.Y. and F.H.; writing—review and editing, N.M., B.C., Z.K., A.F.K. and L.Z.; visualization, L.Z.; supervision, L.Z.; project administration, L.L.; funding acquisition, L.L. All authors have read and agreed to the published version of the manuscript.

Funding: This research received no external funding.

Institutional Review Board Statement: The study was conducted according to the guidelines of the Declaration of Helsinki, and approved by the Ethics Committee of the East China Normal University (No. HR 085-2018; 27 April 2018).

Informed Consent Statement: Informed consent was obtained from all subjects involved in the study.

Data Availability Statement: The data presented in this study are available on request from the corresponding author.

Conflicts of Interest: The authors declare no conflict of interest.

References

1. Erickson, K.I.; Hillman, C.; Stillman, C.M.; Ballard, R.M.; Bloodgood, B.; Conroy, D.E.; Macko, R.; Marquez, D.X.; Petruzzello, S.J.; Powell, K.E. Physical Activity, Cognition, and Brain Outcomes: A Review of the 2018 Physical Activity Guidelines. *Med. Sci. Sports Exerc.* **2019**, *51*, 1242–1251. [CrossRef]
2. Garber, C.E.; Blissmer, B.; Deschenes, M.R.; Franklin, B.A.; Lamonte, M.J.; Lee, I.-M.; Nieman, D.C.; Swain, D.P. American College of Sports Medicine position stand. Quantity and quality of exercise for developing and maintaining cardiorespiratory, musculoskeletal, and neuromotor fitness in apparently healthy adults: Guidance for prescribing exercise. *Med. Sci. Sports Exerc.* **2011**, *43*, 1334–1359. [CrossRef]
3. Verburgh, L.; Königs, M.; Scherder, E.J.; Oosterlaan, J.J. Physical exercise and executive functions in preadolescent children, adolescents and young adults: A meta-analysis. *Br. J. Sports Med.* **2014**, *48*, 973–979. [CrossRef]
4. Li, J.W.; O'Connor, H.; O'Dwyer, N.; Orr, R. The effect of acute and chronic exercise on cognitive function and academic performance in adolescents: A systematic review. *J. Sci. Med. Sport* **2017**, *20*, 841–848. [CrossRef]

5. Xue, Y.; Yang, Y.; Huang, T. Effects of chronic exercise interventions on executive function among children and adolescents: A systematic review with meta-analysis. *Br. J. Sports Med.* **2019**, *53*, 1397–1404. [CrossRef]
6. Miyake, A.; Friedman, N.; Emerson, M.J.; Witzki, A.H.; Howerter, A.; Wager, T.D. The Unity and Diversity of Executive Functions and Their Contributions to Complex "Frontal Lobe" Tasks: A Latent Variable Analysis. *Cogn. Psychol.* **2000**, *41*, 49–100. [CrossRef]
7. Lenroot, R.K.; Giedd, J.N. Brain development in children and adolescents: Insights from anatomical magnetic resonance imaging. *Neurosci. Biobehav. Rev.* **2006**, *30*, 718–729. [CrossRef]
8. Whitford, T.J.; Rennie, C.J.; Grieve, S.M.; Clark, C.R.; Gordon, E.; Williams, L.M. Brain maturation in adolescence: Concurrent changes in neuroanatomy and neurophysiology. *Hum. Brain Mapp.* **2006**, *28*, 228–237. [CrossRef]
9. Lebel, C.; Walker, L.; Leemans, A.; Phillips, L.; Beaulieu, C. Microstructural maturation of the human brain from childhood to adulthood. *NeuroImage* **2008**, *40*, 1044–1055. [CrossRef]
10. Tamnes, C.K.; Herting, M.M.; Goddings, A.-L.; Meuwese, R.; Blakemore, S.-J.; Dahl, R.E.; Güroğlu, B.; Raznahan, A.; Sowell, E.R.; Crone, E.; et al. Development of the Cerebral Cortex across Adolescence: A Multisample Study of Inter-Related Longitudinal Changes in Cortical Volume, Surface Area, and Thickness. *J. Neurosci.* **2017**, *37*, 3402–3412. [CrossRef]
11. Diamond, A.; Lee, K. Interventions Shown to Aid Executive Function Development in Children 4 to 12 Years Old. *Science* **2011**, *333*, 959–964. [CrossRef]
12. Stillman, C.M.; Cohen, J.; Lehman, M.E.; Erickson, K.I. Mediators of Physical Activity on Neurocognitive Function: A Review at Multiple Levels of Analysis. *Front. Hum. Neurosci.* **2016**, *10*, 626. [CrossRef]
13. Wong, M.L.; Lau, E.Y.Y.; Wan, J.H.Y.; Cheung, S.F.; Hui, C.H.; Mok, D.S.Y. The interplay between sleep and mood in predicting academic functioning, physical health and psychological health: A longitudinal study. *J. Psychosom. Res.* **2013**, *74*, 271–277. [CrossRef]
14. King, A.C.; Oman, R.F.; Brassington, G.S.; Bliwise, D.L.; Haskell, W.L. Moderate-intensity exercise and self-rated quality of sleep in older adults. A randomized controlled trial. *JAMA* **1997**, *277*, 32–37. [CrossRef]
15. Singh, N.A.; Clements, K.M.; Fiatarone, M.A. Sleep, Sleep Deprivation, and Daytime Activities–A Randomized Controlled Trial of the Effect of Exercise on Sleep. *Sleep* **1997**, *20*, 95–101. [CrossRef] [PubMed]
16. Vanderlinden, J.; Boen, F.; Van Uffelen, J.G.Z. Effects of physical activity programs on sleep outcomes in older adults: A systematic review. *Int. J. Behav. Nutr. Phys. Act.* **2020**, *17*, 11. [CrossRef]
17. Lins-Filho, O.L.; Pedrosa, R.P.; Gomes, J.M.; Moraes, S.L.D.; Vasconcelos, B.C.E.; Lemos, C.A.A.; Pellizzer, E.P. Effect of exercise training on subjective parameters in patients with obstructive sleep apnea: A systematic review and meta-analysis. *Sleep Med.* **2020**, *69*, 1–7. [CrossRef] [PubMed]
18. Lo, J.C.; Groeger, J.; Cheng, G.H.; Dijk, D.-J.; Chee, M.W. Self-reported sleep duration and cognitive performance in older adults: A systematic review and meta-analysis. *Sleep Med.* **2016**, *17*, 87–98. [CrossRef]
19. Naismith, S.L.; Mowszowski, L.J. Sleep disturbance in mild cognitive impairment: A systematic review of recent findings. *Curr. Opin. Psychiatry* **2018**, *31*, 153–159. [CrossRef]
20. Smithies, T.D.; Toth, A.J.; Dunican, I.C.; Caldwell, J.A.; Kowal, M.; Campbell, M.J. The Effect of Sleep Restriction on Cognitive Performance in Elite Cognitive Performers: A Systematic Review. *Sleep* **2021**. [CrossRef]
21. Benitez, A.; Gunstad, J. Poor sleep quality diminishes cognitive functioning independent of depression and anxiety in healthy young adults. *Clin. Neuropsychol.* **2012**, *26*, 214–223. [CrossRef]
22. Horn, J.A. Human sleep, sleep loss and behavior. Implications for the prefrontal cortex and psychiatric behavior. *Br. J. Psychiatry* **1993**, *162*, 413–419. [CrossRef]
23. Harrison, Y.; Horne, J.A. Sleep loss impairs short and novel language tasks having a prefrontal focus. *J. Sleep Res.* **1998**, *7*, 95–100. [CrossRef]
24. Harrison, Y.; Horne, J.A.; Rothwell, A. Prefrontal Neuropsychological Effects of Sleep Deprivation in Young Adults—A Model for Healthy Aging? *Sleep* **2000**, *23*, 1067–1073. [CrossRef]
25. Muzur, A.; Pace-Schott, E.F.; Hobson, J. The prefrontal cortex in sleep. *Trends Cogn. Sci.* **2002**, *6*, 475–481. [CrossRef]
26. Wilckens, K.A.; Erickson, K.I.; Wheeler, M.E. Physical Activity and Cognition: A Mediating Role of Efficient Sleep. *Behav. Sleep Med.* **2018**, *16*, 569–586. [CrossRef]
27. Falck, R.S.; Best, J.R.; Davis, J.C.; Liu-Ambrose, T. The independent associations of physical activity and sleep with cognitive function in older adults. *J. Alzheimer's Dis.* **2018**, *63*, 1469–1484. [CrossRef] [PubMed]
28. Cheval, B.; Maltagliati, S.; Sieber, S.; Beran, D.; Chalabaev, A.; Sander, D.; Cullati, S.; Boisgontier, M.P. Why Are Individuals with Diabetes Less Active? The Mediating Role of Physical, Emotional, and Cognitive Factors. *Ann. Behav. Med.* **2021**. [CrossRef] [PubMed]
29. Buysse, D.J.; Reynolds, C.F., III; Monk, T.H.; Berman, S.R.; Kupfer, D.J. The Pittsburgh Sleep Quality Index: A new instrument for psychiatric practice and research. *Psychiatry Res.* **1989**, *28*, 193–213. [CrossRef]
30. Cui, J.; Zou, L.; Herold, F.; Yu, Q.; Jiao, C.; Zhang, Y.; Chi, X.; Müller, N.G.; Perrey, S.; Li, L.; et al. Does Cardiorespiratory Fitness Influence the Effect of Acute Aerobic Exercise on Executive Function? *Front. Hum. Neurosci.* **2020**, *14*, 569010. [CrossRef] [PubMed]
31. Hu, L.T.; Bentler, P.M. Cutoff criteria for fit indexes in covariance structure analysis: Conventional criteria versus new alternatives. *Struct. Equ. Model. Multidiscip. J.* **1999**, *6*, 1–55. [CrossRef]

32. Stillman, C.M.; Esteban-Cornejo, I.; Brown, B.; Bender, C.M.; Erickson, K.I. Effects of Exercise on Brain and Cognition Across Age Groups and Health States. *Trends Neurosci.* **2020**, *43*, 533–543. [CrossRef]

33. Youngstedt, S.D. Effects of exercise on sleep. *Clin. Sports Med.* **2005**, *24*, 355–365. [CrossRef] [PubMed]

34. Kalak, N.; Gerber, M.; Kirov, R.; Mikoteit, T.; Yordanova, J.; Pühse, U.; Holsboer-Trachsler, E.; Brand, S. Daily morning running for 3 weeks improved sleep and psychological functioning in healthy adolescents compared with controls. *J. Adolesc. Health* **2012**, *51*, 615–622. [CrossRef] [PubMed]

35. Lang, C.; Brand, S.; Feldmeth, A.K.; Holsboer-Trachsler, E.; Pühse, U.; Gerber, M. Increased self-reported and objectively assessed physical activity predict sleep quality among adolescents. *Physiol. Behav.* **2013**, *120*, 46–53. [CrossRef] [PubMed]

36. Kline, C.E. The bidirectional relationship between exercise and sleep: Implications for exercise adherence and sleep improvement. *Am. J. Lifestyle Med.* **2014**, *8*, 375–379. [CrossRef] [PubMed]

37. Wendt, A.; da Silva, I.C.M.; Gonçalves, H.; Menezes, A.; Barros, F.; Wehrmeister, F.C. Short-term effect of physical activity on sleep health: A population-based study using accelerometry. *J. Sport Health Sci.* **2020**, in press. [CrossRef]

38. Bastien, C.H.; Fortier-Brochu, E.; Rioux, I.; LeBlanc, M.; Daley, M.; Morin, C.M. Cognitive performance and sleep quality in the elderly suffering from chronic insomnia: Relationship between objective and subjective measures. *J. Psychosom. Res.* **2003**, *54*, 39–49. [CrossRef]

39. Foley, D.J.; Masaki, K.; White, L.; Larkin, E.K.; Monjan, A.; Redline, S. Sleep-disordered breathing and cognitive impairment in elderly Japanese-American men. *Sleep* **2003**, *26*, 596–599. [CrossRef]

40. Blackwell, T.; Yaffe, K.; Ancoli-Israel, S.; Schneider, J.L.; Cauley, J.A.; Hillier, T.A.; Fink, H.A.; Stone, K.L. Poor sleep is associated with impaired cognitive function in older women: The study of osteoporotic fractures. *J. Gerontol. Ser. A* **2006**, *61*, 405–410. [CrossRef]

41. Wilckens, K.A.; Erickson, K.I.; Wheeler, M.E. Age-Related Decline in Controlled Retrieval: The Role of the PFC and Sleep. *Neural Plast.* **2012**, *2012*, 624795. [CrossRef] [PubMed]

42. Harrison, Y.; Horne, J.A. Long-term extension to sleep—Are we really chronically sleep deprived? *Psychophysiology* **1996**, *33*, 22–30. [CrossRef] [PubMed]

43. Youngstedt, S.D.; Kripke, D.F. Long sleep and mortality: Rationale for sleep restriction. *Sleep Med. Rev.* **2004**, *8*, 159–174. [CrossRef]

44. Monk, T.H.; Buysse, D.J.; Begley, A.E.; Billy, B.D.; Fletcher, M.E. Effects of a Two-Hour Change in Bedtime on the Sleep of Healthy Seniors. *Chronobiol. Int.* **2009**, *26*, 526–543. [CrossRef] [PubMed]

45. Shrivastava, D.; Jung, S.; Saadat, M.; Sirohi, R.; Crewson, K. How to interpret the results of a sleep study. *J. Community Hosp. Intern. Med. Perspect.* **2014**, *4*, 24983. [CrossRef] [PubMed]

46. Preacher, K.J.; Hayes, A.F. Asymptotic and resampling strategies for assessing and comparing indirect effects in multiple mediator models. *Behav. Res. Methods* **2008**, *40*, 879–891. [CrossRef] [PubMed]

47. Zhao, X.; Lynch, J.G., Jr.; Chen, Q. Reconsidering Baron and Kenny: Myths and truths about mediation analysis. *J. Consum. Res.* **2010**, *37*, 197–206. [CrossRef]

48. Landry, G.J.; Best, J.R.; Liu-Ambrose, T. Measuring sleep quality in older adults: A comparison using subjective and objective methods. *Front. Aging Neurosci.* **2015**, *7*, 166. [CrossRef]

49. Lambiase, M.J.; Gabriel, K.P.; Kuller, L.H.; Matthews, K.A. Sleep and Executive Function in Older Women: The Moderating Effect of Physical Activity. *J. Gerontol. Ser. A* **2014**, *69*, 1170–1176. [CrossRef]

50. Scullin, M.K.; Bliwise, D.L. Sleep, cognition, and normal aging: Integrating a half century of multidisciplinary research. *Perspect. Psychol. Sci.* **2015**, *10*, 97–137. [CrossRef]

51. Hughes, J.M.; Song, Y.; Fung, C.H.; Dzierzewski, J.M.; Mitchell, M.N.; Jouldjian, S.; Josephson, K.R.; Alessi, C.A.; Martin, J.L. Measuring Sleep in Vulnerable Older Adults: A Comparison of Subjective and Objective Sleep Measures. *Clin. Gerontol.* **2018**, *41*, 145–157. [CrossRef] [PubMed]

52. Okano, K.; Kaczmarzyk, J.R.; Dave, N.; Gabrieli, J.D.; Grossman, J.C. Sleep quality, duration, and consistency are associated with better academic performance in college students. *NPJ Sci. Learn.* **2019**, *4*, 16. [CrossRef] [PubMed]

Study Protocol

Effects of Open-Skill Exercises on Cognition on Community Dwelling Older Adults: Protocol of a Randomized Controlled Trial

Wei Guo [1,2], Biye Wang [1,2], Małgorzata Smoter [3] and Jun Yan [1,*]

1   College of Physical Education, Yangzhou University, Yangzhou 225000, China; guowei@yzu.edu.cn (W.G.); wangbiye@yzu.edu.cn (B.W.)
2   Institute of Sports, Exercise and Brain, Yangzhou University, Yangzhou 225000, China
3   Gdańsk Academy of Physical Education and Sport, 80-001 Gdańsk, Poland; m.h.smoter@gmail.com
*   Correspondence: yanjun@yzu.edu.cn

Abstract: (1) Cognitive function may benefit from physical exercise in older adults. However, controversy remains over which mode of exercise is more beneficial. (2) The aim of the proposed study is to investigate the effect of open-skill exercise training on cognitive function in community dwelling older adults compared with closed-skill exercise, cognitive training, and active control. (3) One hundred and sixty participants, aged between 60 and 80 years old, will be recruited from community senior centers in Yangzhou, China and randomly assigned to one of four groups: open-skill exercise group, closed-skill exercise group, mobile game playing group, and active control group. All participants will join a 24-week program involving 50 min sessions three times a week. The primary outcome measure is visuospatial working memory. Secondary measures include subjective memory complaint, attention network, nonverbal reasoning ability, and physical activities. All participants will be measured before, mid-way, and immediately after intervention, and three months later. (4) If successful, this study is expected to provide evidence-based recommendations for older adults to select the most efficient and effective mode of exercise to improve cognitive function. Importantly, the three intervention groups provide an opportunity to separate the cognitive activity component from the physical activity component. Comparison of these components is expected to help elucidate possible mechanisms contributing to the additional cognitive benefit of open-skill exercises.

Keywords: open-skill exercises; closed-skill exercises; cognition; older adults

Citation: Guo, W.; Wang, B.; Smoter, M.; Yan, J. Effects of Open-Skill Exercises on Cognition on Community Dwelling Older Adults: Protocol of a Randomized Controlled Trial. *Brain Sci.* **2021**, *11*, 609. https://doi.org/10.3390/brainsci11050609

Academic Editors: Filipe Manuel Clemente and Ana Filipa Silva

Received: 30 March 2021
Accepted: 7 May 2021
Published: 10 May 2021

Publisher's Note: MDPI stays neutral with regard to jurisdictional claims in published maps and institutional affiliations.

1. Introduction

Data issued by the National Bureau of Statistics of China indicate that there were 249 million (represent 17.9% of the total) people aged over 60 and 167 million (represent 11.9% of the total) aged over 65 in China in 2019. The prevalence of dementia in elderly people over 65 years old is 3.2–9.9% [1], which means that there are 5.3–16.5 million dementia patients in China. Cognitive function declines with aging, especially in people with neurodegenerative diseases, such as dementia and Alzheimer's [2]. Aging-related cognitive decline and cognitive impairment impact the quality and expectancy of life in old people [3], and is of great public health and economic concern. Of note, certain types of cognitive variables remain intact, while others become impaired, with cognitive aging. Salthouse et al., (2010) demonstrated a nearly linear decline from early adulthood with respect to memory, spatial visualization, speed, and reasoning, whereas vocabulary knowledge could still increase until the age of 60 [4]. That is to say, "held" cognitive variables include vocabulary and general information in which the relevant acquisition occurred earlier in one's life, while the "non-held" variables tend to involve the manipulation and transformation of information. Thus, the proposed study will mainly focus on aspect of cognition that decline with aging.

Cognitive function reacts sensitively to environmental changes, and can be significantly improved after certain intervention. Several studies have shown that cognitive function may benefit from physical exercise in older adults [5–7]. Yet, Diamond et al., (2016) argued that aerobic exercise without any cognitive component produces little or no cognitive benefit. The authors predicted that, besides direct exercise training, successful approaches should incorporate cognitive, emotional, and social needs [8]. The authors also stressed that certain modes of physical exercise, including aerobic-exercise and resistance-training, are among the least effective ways of improving executive functions [9]. Therefore, physical training should extend beyond simple moving to moving with thought [10]. Fortunately, various studies have started to focus on identifying cognitive benefit from combinations of cognitive training and physical exercise. Some studies examined these components separately, while others conducted the two components simultaneously. The separate approach usually involves cognitive trainings (e.g., focus on memory, attention, executive function, visuospatial ability), followed by physical exercise (e.g., aerobic exercise, resistance training), or vice versa [11–16]. The simultaneous approach is usually conducted in dual-tasks, such as completing the cognitive task during physical exercise [17–20] or in exergames, which combine physical exercise with computer-simulated environments and interactive videogame features [7,21]. In general, that simultaneous approach tends to have a larger effect size than the separate approach on cognition [22,23]. However, concerns exist that the similarity between cognitive assessments and dual-tasks produce a learning effect that impacts the reliability of the results. Laboratory based intervention cannot easily be generalized to everyday life of the elderly.

Thus, the present study introduces a new approach with high ecological validity by combining physical exercise and cognitive training simultaneously, specifically open-skill exercises. Open-skill exercises are defined as exercises in which participants are required to react in a dynamically changing, unpredictable, and externally environment (e.g., table tennis, badminton). In contrast, closed-skill exercises are conducted in an environment that is relatively stable, predictable, and self-paced (e.g., running and swimming) [24]. Previous studies confirmed that older adults demonstrated better executive function when participating in open-skill exercises compared to closed-skill exercises [25–29]. Children who played open-skill exercises also exhibited better cognitive functions and academic achievement than those who played closed-skill exercises [28,30,31]. Open-skill exercises might be advantageous because participants must mentally compare the present situation with past ones, and use this information to predict likely outcomes; consequently, participants must use their cognitive functions.

Our previous cross-sectional study on healthy older adults showed that both closed-skill and open-skill exercisers exhibited better visuospatial working memory compared to sedentary older adults. Specifically, open-skill exercises that demand higher cognitive processing exhibited selective benefits for the passive maintenance of working memory [32]. However, the cross-sectional design could only reveal a possible relationship, not a causal relationship. Additionally, the reason why open-skill exercise produces additional cognitive benefits to working memory remains unknown. Therefore, the present study protocol provides a first step to initiate an intervention study to answer the question that which type of exercise is better for older adults to decay cognitive decline. The protocol will further reveal the mechanism by setting a cognitive training group besides the open-skill and closed-skill exercise groups. Open-skill exercises can be viewed as a combination of physical exercise and cognitive training, closed-skill exercises and cognitive trainings can be viewed as pure physical exercise or cognitive training. This design may provide the opportunity to separate the physical activity component and cognitive activity component by comparing differences among the groups.

The primary objective is to investigate the effect of 24 weeks of open-skill exercise training on the cognitive function of heathy older adults in comparison to the closed-skill exercise group, cognitive training group, and control group. It is hypothesized that both training regimes will have a positive effect on cognitive function compared to the control, with the largest effect occurring in the open-skill exercise group.

2. Materials and Methods

2.1. Study Design

This study is a double-blind, 24-week randomized controlled trail (RCT) with three experimental intervention groups and an active control group. Participants will be randomly allocated to one of the four groups. The outcome measures will be performed at baseline, mid-way through intervention, immediately after intervention, and at a 3-month post-intervention follow-up. All tests and measures will be conducted by trained professionals who are blinded to the group assignment. The study design is presented in Figure 1. The schedule of enrolment, interventions, and assessments is presented in Figure 2.

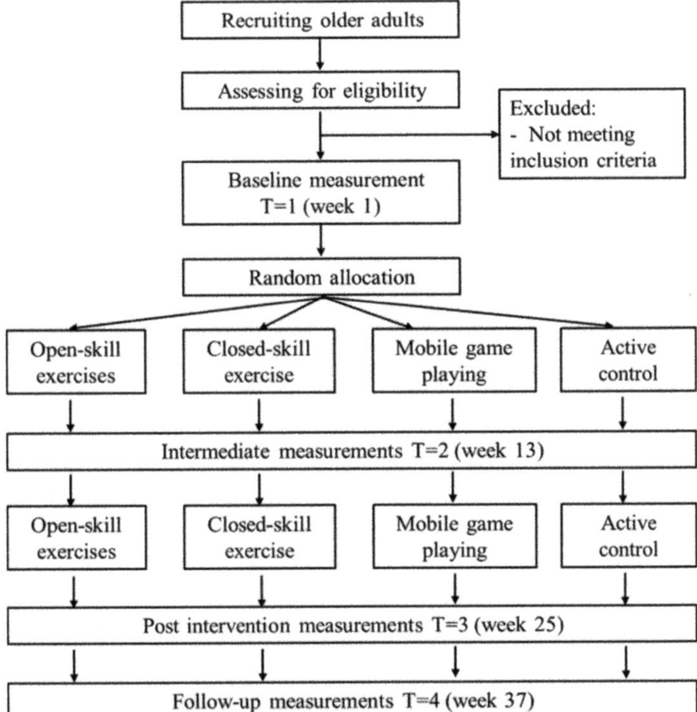

Figure 1. Flowchart of the study design.

The study protocol has been approved by the Ethics Committee of Yangzhou University (YXYLL-2020-106). The methods for consent were registered at Chinese Clinical Trail Registry (ChiCTR2000038733), and the data will be uploaded at the site when finishing the trial. The informed consent form will be obtained from all participants.

	STUDY PERIOD					
	Enrolment	Allocation	Post-allocation			Close-out
TIMEPOINT	$-t_1$	0	t_1	t_2	t_3	t_4
ENROLMENT:						
Eligibility screen	X					
Informed consent	X					
Allocation		X				
INTERVENTIONS:						
Open-skill exercise			←———————→			
Closed-skill exercise			←———————→			
Mobile game playing			←———————→			
Active control			←———————→			
ASSESSMENTS:						
Background characteristics	X	X				
Primary outcomes VWMT, VSMT, VMTT			X	X	X	X
Secondary outcomes MAC-Q, ANT, MRT, IPAQ, walking steps			X		X	X

Figure 2. Schedule of enrolment, interventions, and assessments of the trial. t_1: baseline (week 1), t_2: intermediate of intervention (week 13), t_3: immediately post-intervention (week 25), t_4: twelve weeks after intervention (week 37), VWMT: visuospatial working memory task, VSMT: visuospatial short-term memory task, VMTT: visuospatial mental rotation task, MAC-Q: Memory Complaint Questionnaire, ANT: attention network test, MRT: matrix reasoning test, IPAQ: International Physical Activity Questionnaire.

2.2. Participant Recruitment and Selection

Participants aged between 60 and 80 years will be recruited from the community senior centers in Yangzhou, China. Both male and female participants will be used. The older adults who are interested in participating in the research will be subject to a short interview to confirm their eligibility in the research. Potential participants will be screened using the Mini-Mental Status Examination (MMSE) [33] and exercise-related questionnaire [34].

Eligible participants are additionally required to satisfy the following criteria: (1) 60–80 years of age, (2) apparently healthy, free of cardiovascular disease, musculoskeletal problems, and psychiatric and neurological disorders, (3) normal body weight (body mass index is less than 25.0 and more than 18.5), (4) strong right handedness, (5) corrected visual acuity of at least 20/40, (6) a sedentary lifestyle, exercise at irregular base, and were assessed at the inactivity or low activity level, (7) voluntary participation in the study and willing to give written informed consent.

Participants will be excluded for any of the following: (1) objective cognitive impairment as measured by a MMSE score less than 25, (2) had experience playing table tennis in the past, (3) participated in similar research before, (4) any conflict with the objectives of this study.

2.3. Sample Size

The required sample size of the study was assessed by a priori power analysis (performed by G-Power). To attain a power of 0.8 (1-βerror probability), with a significant level at 0.05, 136 participants will be needed in the study.

Because the subjects are retired, older adults and due to the voluntary nature of enrollment, the attrition rate of the study is anticipated to be relatively low. When also considering other factors (such as health status, environment, and other emergency situations), a general attrition rate of 15% will be calculated in the recruitment plan task. Twenty-four additional participants will be recruited to compensate for the predicted drop-out rate, with 40 participants in each intervention condition.

To maximize compliance, the attendance and exercise performance of participants in the intervention will be monitored routinely throughout the trial.

2.4. Randomization

After signing the written informed consent, eligible participants will be randomly assigned to the active control group, and one of the three experimental intervention groups by the principal researchers. The randomization procedure will be run by an independent statistician using a computerized randomization program. The participants do not know the real purpose of the study, they are just told that they are participating in a project that aims to enrich the life of community dwelling older adults and that they are randomly allocated to different groups of the project according to a computerized randomization program. Participants and two principal researchers know the group assignments. Those who perform the intervention and outcome measures will be blinded to the allocation of participants.

2.5. Interventions and Control Condition

Participants in each intervention group will be told to enroll in a 24-week program and complete 50 min sessions three times a week, and to keep their regular daily life in normal living condition. All interventions will be conducted in the community center under the supervision of personnel who will be trained by the principle researchers to ensure the standardized administration of the intervention protocols.

The intervention conditions include: (1) open-skill exercise condition, which involves a combination of physical and cognitive training, (2) closed-skill exercise condition, which involves physical exercise training alone, (3) mobile game playing condition, which is considered as cognitive training alone, (4) active control condition.

2.5.1. Open-Skill Exercise Condition

The open-skill exercise group will participate in a specially designed table tennis training program. The exercise intervention will be performed in a sequence of increasing complexity under the instruction of trainers. The trainers will be chosen from college students majoring in table tennis training.

The procedure of each single intervention session includes a warm-up, training in the basic skills of table tennis, playing table tennis with each other, special tasks designed to increase the cognitive load, and a cool down period. The aim of basic skills training is to improve general skills. It includes the following component: serving, backhand blocking, forehand attacking, left side blocking, and right side attacking (including footwork). After the coaches have demonstrated the basic skills, participants will practice with each other (or with the coaches). Training for each skill begins with simple movement, and then progresses to more complex skills, according to their degree of mastery.

To add more cognitive load to physical exercises during the process of learning the basic skills, tasks must be planned and a program of movement will be designed. This approach will be structured to achieve a particular type of training that is expected to relate to executive function. Table-tennis balls will be successively thrown to participant by a serving machine on one side of the ping-pong table. Participants stand on the other side of the table, and must anticipant the ball trajectory to intercept it. Balls are of two different colors; half are white and half are yellow. Three different tasks were designed to increase cognitive load. Participants must intercept the white ball but ignore the yellow ball in task 1, which requires inhibition function during movement. Task 2 requires participants to hit the white ball to the left side and the yellow ball to the right side across the ping-pong table. which needs task-switch ability. Task 3 was designed to train the working memory ability by asking participants to hit the ball to the left side if the previous ball is white and to the right if previous ball is yellow.

All participants will be required to move continuously during the training to maintain an average individual target heart rate reserve (HRR) of 40–59%.

2.5.2. Closed-Skill Exercise Condition

The closed-skill exercise group will use a cycle ergometer or motor-driven treadmill with a moderate intensity (40–59% HRR). Each training session will include a warm-up, cycle ergometer riding or brisk walking/jogging for 30 min, and cool down period. Intensity and duration will be progressive during the training period. The Borg Rating of Perceived Exertion (RPE) will be used every 5 min during exercise to monitor the subjective feeling of exertion.

2.5.3. Mobile Game Playing Condition

The cognitive training group will play a table tennis mobile game on the mobile phone without body movement. The mobile game is called Table Tennis Touch, and is a free sport action mobile multiplayer game featuring stunning graphics, intuitive swipe controls, high speed gameplay and multiple game modes developed by the British Indie Studio Yakuto. It offers a fantastic simulation of real table tennis playing settings. Players can serve, spin, and smash their way to winning. The older adults will be trained the first time, and will subsequently play under the supervision of the trainers.

2.5.4. Active Control Condition

To control for the Hawthorne effect, which means participants may change their behavior just by being aware that they are being watched, the control group will also do some stretching and toning of the same duration and frequency as the intervention group. Social engagement will be similar among all groups.

2.6. Outcome Measures

Demographic variables will be collected during the evaluation, and include age, gender, weight, height, mental status, visual acuity, resting heart rate and blood pressure. The outcomes include objective and subjective measurements of cognitive and physical activities. All outcomes will be measured prior and immediately after the intervention, and at 3 months after the intervention. After 12 weeks, there will be an intermediate measurement, consisting of the primary outcome measures.

2.6.1. Primary Outcome

The primary outcome measure of this study is objective visuospatial working memory. It is measured by the visuospatial working memory task (VWMT) developed by Guo et al., (2016), which includes both the passive storage and active manipulation of visuospatial information [32]. Participants will be asked to hold a 4×4 matrix with four solid black squares as a probe stimulus in memory for 3 s, and mentally rotate the matrix 90° to the right or the left at the same time. The task is used to decide whether the testing stimulus is consistent with the mentally rotated probe stimulus 90° either to the left or right.

Passive storage and active manipulation will be measured separately to reveal the mechanism driving differences on visuospatial working memory. The passive storage of visuospatial information will be tested by visuospatial short-term memory task (VSMT). The experimental paradigm of VSMT is similar to that of VWMT, except that subjects are instructed to retain the memory stimulus during the retention interval. Participants must determine whether the testing stimulus is identical to the memory stimulus. Active manipulation of visuospatial information will be tested by the visuospatial mental rotation task (VMTT). A pair of matrices will be presented on the screen for 6 sec. Participants must compare the two matrices to determine whether the one on the right side corresponds to a 90° rotation of the one on the left side.

2.6.2. Secondary Outcomes

The subjective memory complaint (SMC) occurs long before the decline of working memory during aging [35]. It will also be measured in this study. SMC will be rated using the Memory Complaint Questionnaire (MAC-Q). The six-item MAC-Q requires participants to compare current memory function to memory function at earlier ages in daily scenarios. For each question, there are five possible answers, ranging from "much better now" (scored 5) to "much worse now" (scored 1). Serious SMC is defined as a total score below 15 [36]. MAC-Q demonstrates good internal consistency together with satisfactory test-retest reliability [37].

The attention network test (ANT) will be used to assess the alerting, orienting, and executive networks of the attentional system independently in a single test [38]. The test is a combination of the Flanker paradigm and a cueing task, with four cue types (no, central, double, and spatial) and three flanker types (congruent, incongruent, and neutral). The cue stimuli are asterisks, and are always valid to provide temporal information about the coming target stimuli. The target stimuli are five horizontal arrows that serve as flankers. Participants will be instructed to respond to the central arrow as fast and as accurately as possible [39,40].

A matrix reasoning test (MRT) will be used to measures abstract nonverbal reasoning ability. This test includes 35 items, in which the individual is required to look at an incomplete matrix and select the missing portion from five response options. This test is a subtest of the Wechsler Abbreviated Scale of Intelligence, and is relatively language free [41]. The score is generated based on the number of items completed correctly.

The Taiwan version of the International Physical Activity Questionnaire (IPAQ) will be used as a subjective measurement to estimate the total amount of physical activity in the preceding week.

The Mi Band wristband will be used to measure objective physical activity. Participants will be asked to wear the wristband all day from one week before the intervention, during the whole intervention, during the 3-month post-intervention. The wristband inserts a 3-axis acceleration sensor ADXL362 that measures the number of walking steps and distance travelled in everyday life. The wristband will also monitor the heart rate during intervention.

2.7. Data Collection, Management and Statistical Analysis

The outcome measures will be carried out conducted in the community center by trained research assistants who are not involved in the intervention. All outcome data will be collected and stored electronically. A data management team will be established to monitor the process of data collection, record and analysis once a week during the project. The data of the participants who withdraw will be treated as having no change from baseline after dropping out. Analysis of variance (ANOVA) or the χ^2-test will be used to assess differences across groups with respect to age, percentage of females, BMI, scores on MMSE, visual acuity, resting heart rate, and blood pressure. If group differences are observed at baseline, these variables will be included as covariates in further analysis. Repeated-measures analysis of variance will be used for the primary and sedentary outcome measures. Dependent variables include the accuracy and reaction times on VWMT, VSMT and VMTT, the scores on alerting, orienting, and executive networks of ANT, Matrix reasoning test, MAC-Q and IPAQ, and the number of steps and distance. Independent variables include group (open-skill exercise, closed-skill exercise, cognitive training and control group) as the between subject factor, and the test time points (pre-test, intermediate-test, post-test, and 3-month follow-up test) as the within subject factor. To reveal the relationship between subjective memory complaint and objective working memory ability, Pearson correlation will be adopted to compute the correlation coefficient between the accuracy of VWMT and the MAC-Q score at different test time points.

3. Discussion

Previous studies that evaluated how open/closed-skill exercises affected the cognitive functions of older people mainly used a cross-sectional design [25,32]. There has been only one longitudinal study using a 6-month randomized controlled trial [27]. The longitudinal study showed that older adults who participated in open-skill exercises had better executive function compared to those that participated in closed-skill exercises. However, the mechanism underlying the additional advantage of open-skill exercises remains subject to debate. One major strength of the proposed study protocol is its design with four groups. In addition to the open-skill, closed-skill, and control group, there will be a cognitive training group. The inclusion of a cognitive training group provides the opportunity to separate the physical activity component and cognitive activity component by comparing differences among the groups. This approach might reveal the possible mechanisms driving the additional cognitive benefits of open-skill exercises. The closed-skill group and cognitive training group will only have the physical activity and cognitive activity component, respectively, while the open-skill exercise group will involve both physical and cognitive components. The cognitive training group will use a table tennis mobile game that differs previous computerized cognitive training [42,43], as it will utilize the same cognitive component as the open-skill exercise group.

Several studies have explored various methods to combine cognitive and physical training as a tentative approach to improve cognitive functions. On one hand, according to the cardiovascular fitness hypothesis, participant in chronic exercises may lead to changes in cardiovascular system, this change can increase cerebral blood flow and the up-regulation of neurotransmitters [44]. On the other hand, cognitive training can guide and integrate new neurons and synapses into existing neural networks [45]. Therefore, the combination of the two interventions may induce superimposed cognitive benefits. The combination of cognitive and physical training typically being completed separately or simultaneously. It has been demonstrated that simultaneous combination has a larger effect on cognitive function compared to their being conducted separately. The simultaneous combination is usually conducted as dual-tasks, such as pedaling on a stationary bike while learning the memory strategies and practicing the memory drills presented by a trainer on the screen in front of them [46]. These combinations are not popular by the elderly population due to the difficulty of implementation. Thus, open-skill exercises that are played in a dynamically changing, unpredictable, and external environment, and require a cognitive component, represent better choice. Similar to table tennis used in this proposed study, Moreau et al., (2015) designed training that was loosely based on freestyle wrestling. This approach favored new motor coordination via demonstrations and active trial-and-error problem-solving with a partner. After 8 weeks training, the designed sport group showed the largest gains in all cognitive measures. [47]. Open-skill exercises can be viewed as high ecologically validated approaches that combine cognitive and physical training simultaneously.

This proposed project differs to previous studies by measuring visuospatial working memory as a primary outcome, rather than general cognitive function. This is key because visuospatial working memory ability declines more severely after the age of 60. Additionally, a previous cross-sectional study of older adults only revealed a possible relationship between exercise mode and visuospatial working memory. In comparison, the proposed longitudinal project will explore the reasons why open-skill exercises generate more benefits on visuospatial working memory. The subjective memory complaint (SMC) will also be measured alongside the objective measures of working memory. SMC is defined as the reporting of memory problems without pathological results on neuropsychological tests. SMC is associated with later cognitive decline and a higher incidence of dementia [48]. A randomized controlled study reported that 12 weeks of exercise interventions induced beneficial effects on memory complaints (tested by MAC-Q), but not on objective memory performance (tested by the Prose Test and Rey test) [49]. In addition to SMC, the proposed study will investigate the association between MAC and visuospatial working memory, which have been subject to controversy in previous studies.

A limitation of the study is that outcome measures mainly include cognitive domains. However, the general aim of exercise is to provide benefits on the overall aspects of well-being, including both cognitive functions and physical performance. Apart from the included measures of physical activities, other measures of cardiovascular fitness and physical performance such as mobility, balance, strength and endurance will also be added to evaluate the physical functions in future studies.

4. Conclusions

The overall goal of this study is to compare the effects of open-skill exercises on cognition in older adults. If successful, the results of the proposed study provide important evidence-based recommendations for older adults to select the most efficient and effective exercise mode to improve cognitive function. In particular, this study will reveal the mechanism driving the additional cognitive benefits of open-skill exercises compared to closed-skill exercises.

Author Contributions: Conceptualization, W.G. and J.Y.; methodology, B.W. and W.G.; validation, B.W., M.S. and J.Y.; resources, M.S.; writing—original draft preparation, B.W. and W.G.; writing—review and editing, M.S. and J.Y.; supervision, J.Y.; project administration, J.Y.; funding acquisition, W.G. All authors have read and agreed to the published version of the manuscript.

Funding: This project was funded by the Natural Science Foundation of Jiangsu Province, China, grant number BK20180926. The funder had no role in study design, decision to publish, or preparation of the manuscript, and will not be involved in the collection, analysis or interpretation of data.

Institutional Review Board Statement: The study was conducted according to the guidelines of the Declaration of Helsinki, and approved by the Ethics Committee of Yangzhou University (YXYLL-2020-106). The methods for consent were registered at Chinese Clinical Trail Registry (ChiCTR2000038733).

Informed Consent Statement: Informed consent will be obtained from all subjects involved in the study.

Data Availability Statement: The data are not publicly available, as they have not yet been collected. The data will be uploaded at the Chinese Clinical Trail Registry site when finishing the study.

Acknowledgments: We are grateful for cooperation with Yangzhou University and Wrocław University School of Physical Education joint laboratory on sport and brain science.

Conflicts of Interest: The authors declare no conflict of interest.

References

1. Jia, J.; Wang, F.; Wei, C.; Zhou, A.; Jia, X.; Li, F.; Tang, M.; Chu, L.; Zhou, Y.; Zhou, C. The prevalence of dementia in urban and rural areas of China. *Alzheimer's Dement.* **2014**, *10*, 1–9. [CrossRef]
2. Bobkova, N.V.; Poltavtseva, R.A.; Leonov, S.V.; Sukhikh, G.T. Neuroregeneration: Regulation in Neurodegenerative Diseases and Aging. *Biochemistry* **2020**, *85* (Suppl. 1), S108–S130. [CrossRef]
3. Wilson, R.S.; Boyle, P.A.; Segawa, E.; Yu, L.; Begeny, C.T.; Anagnos, S.E.; Bennett, D.A. The influence of cognitive decline on well-being in old age. *Psychol. Aging* **2013**, *28*, 304–313. [CrossRef] [PubMed]
4. Salthouse, T.A. Selective review of cognitive aging. *J. Int. Neuropsychol. Soc.* **2010**, *16*, 754–760. [CrossRef]
5. Erickson, K.I.; Voss, M.W.; Prakash, R.S.; Basak, C.; Szabo, A.N.; Chaddock, L.; Kim, J.S.; Heo, S.; Alves, H.; White, S.M. Exercise training increases size of hippocampus and improves memory. *Proc. Natl. Acad. Sci. USA* **2011**, *108*, 3017–3022. [CrossRef]
6. Fernandes, R.M.; Correa, M.G.; Dos, S.M.A.R.; Almeida, A.P.C.P.S.; Fagundes, N.C.F.; Maia, L.C.; Lima, R.R. The Effects of Moderate Physical Exercise on Adult Cognition: A Systematic Review. *Front. Physiol.* **2018**, *9*, 667. [CrossRef]
7. Leyland, L.A.; Spencer, B.; Beale, N.; Jones, T.; van Reekum, C.M.; Piacentini, M.F. The effect of cycling on cognitive function and well-being in older adults. *PLoS ONE* **2019**, *14*, e0211779. [CrossRef]

8. Diamond, A.; Ling, D.S. Conclusions about Interventions, Programs, and Approaches for Improving Executive Functions that appear Justified and those that, despite much hype, do not. *Dev. Cogn. Neurosci.* **2015**, *18*, 34–48. [CrossRef] [PubMed]

9. Diamond, A.; Ling, D.S. Aerobic-Exercise and resistance-training interventions have been among the least effective ways to improve executive functions of any method tried thus far. *Dev. Cogn. Neurosci.* **2019**, *37*, 100572. [CrossRef] [PubMed]

10. Diamond, A. Effects of Physical Exercise on Executive Functions: Going beyond Simply Moving to Moving with Thought. *Ann. Sports Med. Res.* **2015**, *2*, 1011.

11. Bae, S.; Lee, S.; Lee, S.; Jung, S.; Makino, K.; Harada, K.; Harada, K.; Shinkai, Y.; Chiba, I.; Shimada, H. The effect of a multicomponent intervention to promote community activity on cognitive function in older adults with mild cognitive impairment: A randomized controlled trial. *Complement. Med.* **2019**, *42*, 164–169. [CrossRef]

12. Shah, T.; Verdile, G.; Sohrabi, H.; Campbell, A.; Putland, E.; Cheetham, C.; Dhaliwal, S.; Weinborn, M.; Maruff, P.; Darby, D. A combination of physical activity and computerized brain training improves verbal memory and increases cerebral glucose metabolism in the elderly. *Transl. Psychiatry* **2014**, *4*, e487. [CrossRef]

13. Alesi, M.; Giordano, G.; Giaccone, M.; Basile, M.; Bianco, A. Effects of the Enriched Sports Activities-Program on Executive Functions in Italian Children. *J. Funct. Morphol. Kinesiol.* **2020**, *5*, 26.

14. Linde, K.; Alfermann, D. Single versus combined cognitive and physical activity effects on fluid cognitive abilities of healthy older adults: A 4-month randomized controlled trial with follow-up. *J. Aging Phys. Act.* **2014**, *22*, 302–313. [CrossRef]

15. Rahe, J.; Petrelli, A.; Kaesberg, S.; Fink, G.R.; Kessler, J.; Kalbe, E. Effects of cognitive training with additional physical activity compared to pure cognitive training in healthy older adults. *Clin. Interv. Aging* **2015**, *10*, 297–310. [CrossRef]

16. Li, R.; Zhu, X.; Yin, S.; Niu, Y.; Zheng, Z.; Huang, X.; Wang, B.; Li, J. Multimodal intervention in older adults improves resting-state functional connectivity between the medial prefrontal cortex and medial temporal lobe. *Front. Aging Neurosci.* **2014**, *6*, 39. [CrossRef] [PubMed]

17. Theill, N.; Schumacher, V.; Adelsberger, R.; Martin, M.; Jancke, L. Effects of simultaneously performed cognitive and physical training in older adults. *BMC Neurosci.* **2013**, *14*, 103. [CrossRef]

18. León, J.; Ureña, A.; Bolaños, M.J.; Bilbao, A.; Oña, A. A combination of physical and cognitive exercise improves reaction time in persons 61-84 years old. *J. Aging. Phys. Act.* **2015**, *23*, 72–77. [CrossRef]

19. Nishiguchi, S.; Yamada, M.; Tanigawa, T.; Sekiyama, K.; Kawagoe, T.; Suzuki, M.; Yoshikawa, S.; Abe, N.; Otsuka, Y.; Nakai, R. A 12-Week Physical and Cognitive Exercise Program Can Improve Cognitive Function and Neural Efficiency in Community-Dwelling Older Adults: A Randomized Controlled Trial. *J. Am. Geriatr. Soc.* **2015**, *63*, 1355–1363. [CrossRef]

20. Ji, Z.G.; Feng, T.; Wang, H.B. The Effects of 12-Week Physical Exercise Tapping High-level Cognitive Functions. *Adv. Cogn. Psychol.* **2020**, *16*, 59–66. [CrossRef]

21. Karssemeijer, E.G.A.; Aaronson, J.A.; Bossers, W.J.R.; Donders, R.; Olde Rikkert, M.G.M.; Kessels, R.P.C. The quest for synergy between physical exercise and cognitive stimulation via exergaming in people with dementia: A randomized controlled trial. *Alzheimer's Res.* **2019**, *11*, 3. [CrossRef]

22. Zhu, X.; Yin, S.; Lang, M.; He, R.; Li, J. The more the better? A meta-analysis on effects of combined cognitive and physical intervention on cognition in healthy older adults. *Ageing Res. Rev.* **2016**, *31*, 67–79. [CrossRef]

23. Guo, W.; Zang, M.; Klich, S.; Kawczynski, A.; Smoter, M.; Wang, B. Effect of Combined Physical and Cognitive Interventions on Executive Functions in OLDER Adults: A Meta-Analysis of Outcomes. *Int. J. Environ. Res. Public Health* **2020**, *17*, 6166. [CrossRef] [PubMed]

24. McMorris, T. *Acquisition and Performance of Sports Skills*; John Wiley & Sons: Hoboken, NJ, USA, 2014; pp. 223–224.

25. Dai, C.T.; Chang, Y.K.; Huang, C.J.; Hung, T.M. Exercise mode and executive function in older adults: An ERP study of task-switching. *Brain Cogn.* **2013**, *83*, 153–162. [CrossRef]

26. Huang, C.J.; Lin, P.C.; Hung, C.L.; Chang, Y.; Hung, T. Type of physical exercise and inhibitory function in older adults: An event-related potential study. *Psychol. Sport Exerc.* **2014**, *15*, 205–211. [CrossRef]

27. Tsai, C.L.; Pan, C.Y.; Chen, F.C.; Tseng, Y.T. Open- and Closed-Skill Exercise Interventions Produce Different Neurocognitive Effects on Executive Functions in the Elderly: A 6-Month Randomized, Controlled Trial. *Front. Aging Neurosci.* **2017**, *9*, 294. [CrossRef]

28. Gu, Q.; Zou, L.; Loprinzi, P.D.; Quan, M.; Huang, T. Effects of Open Versus Closed Skill Exercise on Cognitive Function: A Systematic Review. *Front. Psychol.* **2019**, *10*, 1707. [CrossRef]

29. Zhu, H.; Chen, A.; Guo, W.; Zhu, F.; Wang, B. Which Type of Exercise Is More Beneficial for Cognitive Function? A Meta-Analysis of the Effects of Open-Skill Exercise versus Closed-Skill Exercise among Children, Adults, and Elderly Populations. *Appl. Sci.* **2020**, *10*, 2737. [CrossRef]

30. Becker, D.R.; Mcclelland, M.M.; John, G.G.; Gunter, K.B.; Megan, M. Open-Skilled Sport, Sport Intensity, Executive Function, and Academic Achievement in Grade School Children. *Early Educ. Dev.* **2018**, *29*, 939–955. [CrossRef]

31. Formenti, D.; Trecroci, A.; Duca, M.; Cavaggioni, L.; D'Angelo, F.; Passi, A.; Longo, S.; Alberti, G. Differences in inhibitory control and motor fitness in children practicing open and closed skill sports. *Sci. Rep.* **2021**, *11*, 4033. [CrossRef]

32. Guo, W.; Wang, B.; Lu, Y.; Zhu, Q.; Shi, Z.; Ren, J. The relationship between different exercise modes and visuospatial working memory in older adults: A cross-sectional study. *PeerJ* **2016**, *4*, e2254. [CrossRef]

33. O'Neill, D. The Mini-Mental Status Examination. *J. Am. Geriatr. Soc.* **1991**, *39*, 733. [CrossRef]

34. Liou, Y.M.; Jwo, C.J.; Yao, K.G.; Chiang, L.C.; Huang, L.H. Selection of appropriate Chinese terms to represent intensity and types of physical activity terms for use in the Taiwan version of IPAQ. *J. Nurs. Res.* **2008**, *16*, 252–263. [CrossRef]
35. Jorm, A.F.; Christensen, H.; Korten, A.E.; Jacomb, P.A.; Henderson, A.S. Memory complaints as a precursor of memory impairment in older people: A longitudinal analysis over 7–8 years. *Psychol. Med.* **2001**, *31*, 441–449. [CrossRef]
36. Babberich, E.D.T.; Gourdeau, C.; Pointel, S.; Lemarchant, B.; Beauchet, O.; Annweiler, C. Biology of Subjective Cognitive Complaint Amongst Geriatric Patients: Vitamin D Involvement. *Curr. Alzheimer Res.* **2015**, *12*, 173–178. [CrossRef]
37. Crook, T.H., 3rd; Feher, E.P.; Larrabee, G.J. Assessment of memory complaint in age-associated memory impairment: The MAC-Q. *Int. Psychogeriatr.* **1992**, *4*, 165–176. [CrossRef]
38. Fan, J.; McCandliss, B.D.; Sommer, T.; Raz, A.; Posner, M.I. Testing the efficiency and independence of attentional networks. *J. Cogn. Neurosci.* **2002**, *14*, 340–347. [CrossRef]
39. Wang, B.; Guo, W.; Zhou, C. Selective enhancement of attentional networks in college table tennis athletes: A preliminary investigation. *PeerJ* **2016**, *4*, e2762. [CrossRef]
40. Wang, B.; Guo, W. Exercise mode and attentional networks in older adults: A cross-sectional study. *PeerJ* **2020**, *8*, e8364. [CrossRef]
41. Irby, S.M.; Floyd, R.G. Test review: Wechsler abbreviated scale of intelligence, second edition. *Can. J. Sch. Psychol.* **2013**, *28*, 295–299. [CrossRef]
42. Lampit, A.; Hallock, H.; Valenzuela, M. Computerized cognitive training in cognitively healthy older adults: A systematic review and meta-analysis of effect modifiers. *PLoS Med.* **2014**, *11*, e1001756. [CrossRef]
43. Harvey, P.D.; McGurk, S.R.; Mahncke, H.; Wykes, T. Controversies in Computerized Cognitive Training. *Biol. Psychiatry* **2018**, *3*, 907–915. [CrossRef] [PubMed]
44. Voss, M.W.; Nagamatsu, L.S.; Liu-Ambrose, T.; Kramer, A.F. Exercise, brain, and cognition across the life span. *J. Appl. Physiol.* **2011**, *111*, 1505–1513. [CrossRef]
45. Fissler, P.; Kuster, O.; Schlee, W.; Kolassa, I.T. Novelty interventions to enhance broad cognitive abilities and prevent dementia: Synergistic approaches for the facilitation of positive plastic change. *Prog. Brain Res.* **2013**, *207*, 403–434.
46. McEwen, S.C.; Siddarth, P.; Rahi, B.; Kim, Y.; Mui, W.; Wu, P.; Emerson, N.D.; Lee, J.; Greenberg, S.; Shelton, T.; et al. Simultaneous Aerobic Exercise and Memory Training Program in Older Adults with Subjective Memory Impairments. *J. Alzheimer's Dis.* **2018**, *62*, 795–806. [CrossRef] [PubMed]
47. Moreau, D.; Morrison, A.B.; Conway, A.R. An ecological approach to cognitive enhancement: Complex motor training. *Acta Psychol.* **2015**, *157*, 44–55. [CrossRef]
48. Waldorff, F.B.; Siersma, V.; Vogel, A.; Waldemar, G. Subjective memory complaints in general practice predicts future dementia: A 4-year follow-up study. *Int. J. Geriatr. Psychiatry* **2012**, *27*, 1180–1188. [CrossRef] [PubMed]
49. Iuliano, E.; Fiorilli, G.; Aquino, G.; Di Costanzo, A.; Calcagno, G.; Di Cagno, A. Twelve-Week Exercise Influences Memory Complaint but Not Memory Performance in Older Adults: A Randomized Controlled Study. *J. Aging Phys. Act.* **2017**, *25*, 612–620. [CrossRef]

brain
sciences

Article

The Effect of Acute Aerobic Exercise on Divergent and Convergent Thinking and Its Influence by Mood

Kohei Aga, Masato Inamura, Chong Chen *, Kosuke Hagiwara, Rikuto Yamashita, Masako Hirotsu, Tomoe Seki, Akiyo Takao, Yuko Fujii, Toshio Matsubara and Shin Nakagawa

Division of Neuropsychiatry, Department of Neuroscience, Yamaguchi University Graduate School of Medicine, Ube, Yamaguchi 755-8505, Japan; i001eb@yamaguchi-u.ac.jp (K.A.); g009eb@yamaguchi-u.ac.jp (M.I.); khagi@yamaguchi-u.ac.jp (K.H.); i096eb@yamaguchi-u.ac.jp (R.Y.); hirotsu@yamaguchi-u.ac.jp (M.H.); tseki@yamaguchi-u.ac.jp (T.S.); v053eb@yamaguchi-u.ac.jp (A.T.); yuko-f@yamaguchi-u.ac.jp (Y.F.); t-matsu@yamaguchi-u.ac.jp (T.M.); snakaga@yamaguchi-u.ac.jp (S.N.)
* Correspondence: cchen@yamaguchi-u.ac.jp

Citation: Aga, K.; Inamura, M.; Chen, C.; Hagiwara, K.; Yamashita, R.; Hirotsu, M.; Seki, T.; Takao, A.; Fujii, Y.; Matsubara, T.; et al. The Effect of Acute Aerobic Exercise on Divergent and Convergent Thinking and Its Influence by Mood. *Brain Sci.* **2021**, *11*, 546. https://doi.org/10.3390/brainsci11050546

Academic Editors: Filipe Manuel Clemente, Ana Filipa Silva and Stéphane Perrey

Received: 26 March 2021
Accepted: 26 April 2021
Published: 27 April 2021

Publisher's Note: MDPI stays neutral with regard to jurisdictional claims in published maps and institutional affiliations.

Abstract: Abundant evidence shows that various forms of physical exercise, even conducted briefly, may improve cognitive functions. However, the effect of physical exercise on creative thinking remains under-investigated, and the role of mood in this effect remains unclear. In the present study, we set out to investigate the effect of an acute bout of aerobic exercise on divergent and convergent thinking and whether this effect depends on the post-exercise mood. Forty healthy young adults were randomly assigned to receive a 15-min exercise or control intervention, before and after which they conducted an alternate use test measuring divergent thinking and an insight problem-solving task measuring convergent thinking. It was found that exercise enhanced divergent thinking in that it increased flexibility and fluency. Importantly, these effects were not mediated by the post-exercise mood in terms of pleasure and vigor. In contrast, the effect on convergent thinking depended on subjects' mood after exercise: subjects reporting high vigor tended to solve more insight problems that were unsolved previously, while those reporting low vigor became less capable of solving previously unsolved problems. These findings suggest that aerobic exercise may affect both divergent and convergent thinking, with the former being mood-independent and the latter mood-dependent. If these findings can be replicated with more rigorous studies, engaging in a bout of mood, particularly vigor-enhancing aerobic exercise, may be considered a useful strategy for gaining insights into previously unsolved problems.

Keywords: aerobic exercise; creativity; convergent thinking; divergent thinking; flexibility; insight problem-solving; cognitive functions; mood; vigor; pleasure

1. Introduction

A growing body of research has investigated the positive impact of a single bout of physical exercise on cognitive functions (for a meta-analysis, [1–5]; for a narrative review, [6–8]). For instance, a single bout of aerobic exercise may improve attention and executive functions [2,4,5], boost the speed of information processing [3], and enhance memory storage and retrieval [1]. Several neurobiological mechanisms have been proposed to explain these cognitive benefits, including increased production or release of lactate, cortisol, neurotrophins (e.g., BDNF, IGF-1), and neurotransmitters (e.g., serotonin, dopamine, endocannabinoids) as well as functional hemodynamic brain changes (particularly in the prefrontal cortex) following exercise [7,9–12]. Despite these fruitful findings, relatively little is known about the effects of physical exercise on creativity or creative thinking (for a brief literature review, see below).

It has been generally considered that creative thinking comprises two fundamental processes: divergent thinking, which involves stretching beyond existing ideas to create multiple, novel ones, and convergent thinking, which involves narrowing down and approaching a single correct, task-appropriate solution [13]. Performance on tests of divergent and convergent think has been reported to predict real-life creative potential, achievement, and creativity evaluated by others [14–17]. Since creative thinking is the key to invention and innovation and is indispensable for the progress of science, technology, business and management, education, art, and society as a whole [18], the search for strategies to enhance creative thinking may have great social significance.

We identified six papers [19–24] investigating the after-effects of acute physical exercise on creative thinking (i.e., creativity post-exercise) that included a control group or condition and with the full-text available (see Table S1, the literature search strategy is described in the footnote of Table S1). All of the studies evaluated divergent thinking with the original or adapted versions of the alternate uses test (AUT, [25,26]). In this test, subjects are asked to write down as many as possible uncommon, original uses of common objects, such as "bricks" and "tin cans". The number of generated uses (i.e., fluency), the number of different conceptual categories the alternative uses are from (i.e., flexibility), and the rareness of the generated uses (i.e., originality) are commonly used as indicators of creativity. In contrast, sometimes other indicators, such as the number of details in a given response (i.e., elaboration), have also been used [27,28]. Two of the studies [21,23] also investigated the impact of exercise on convergent thinking, one using the remote associates test (RAT, [29]) and the other its adapted version the compound remote associates test (CRA, [30]). In these tests, subjects must think of a single word associated with each of three seemingly unrelated words. For instance, for the three words, "time", "hair", and "stretch", the associate might be "long"; for the three words "cottage", "swiss", and "cake", the compound associate might be "cheese".

To summarize the findings, in brief, the effect of an acute bout of aerobic exercise on divergent and convergent thinking is inconsistent. Either no effect or an enhancing or impairing effect on some measures has been reported. Based on the identified studies, one trend might be speculated that exercise at too low intensity may not reliably improve divergent and convergent thinking [21–23]. For instance, cycling [21] or walking [23] at very light-to-light intensity (estimated based on age-estimated maximal heart rate or HR_{max}, [31,32]) failed to affect fluency, flexibility, originality, or elaboration of divergent thinking; nor did they affect the CRA or RAT performance. In contrast, a 4-min walk on a treadmill or outdoors (considered very light in intensity) increased the originality of divergent thinking compared to sitting, without affecting the CRA performance [22]. On the other hand, exercise at too high intensity, for instance, cycling with maximal effort [21], may reduce the flexibility of divergent thinking and impair the RAT performance (the latter in subjects without exercise habit).

Insights from the neurobiological literature suggest that an important factor that may provide a parsimonious explanation to the above inconsistent results is mood. Since exercise influences mood and cognition through similar neurobiological mechanisms, for instance, via neurotransmitters, such as serotonin, dopamine, and endocannabinoids [7], one may expect an association between the creativity and mood effects of exercise. Thus, it is possible that two previous studies failed to find the creativity-enhancing effect because their exercise interventions using cycling [21] or walking [23] at very light to light intensity failed to change subjects' moods. It is also possible that cycling with maximal effort impaired divergent and convergent thinking [21] because it caused exhaustion and central fatigue. Unfortunately, these studies did not examine this potential association, although an earlier study [19] reported that the creativity effect of approximately 20 min of aerobic workout or dance was independent of the mood effect. More research is required to confirm the association between the creativity and mood effects of exercise.

Furthermore, regarding the validity of the tests of divergent and convergent thinking that have been employed, the AUT has been found to reflect cognitive flexibility and suggested to be a validated measure of divergent thinking [33]. However, the RAT measuring convergent thinking has been criticized for having low validity due to its high dependence on verbal abilities (e.g., [34]). Thus, other more appropriate measures of convergent thinking or creative problem-solving are required for further research and confirmation of the effect of the exercise.

In the present study, we aimed to address these unsolved issues and test our speculations with a randomized controlled trial. We employed a 15-min automated exercise intervention program expected to be primarily moderate in intensity, the workload of which was greater than normal walking or cycling but lighter than cycling with maximal effort (as employed by Colzato et al. [21]). We predicted that exercise at this intensity might improve the participants' mood without causing central fatigue. Importantly, immediately following the exercise and control intervention, we measured subjects' mood at the moment using a visual analog scale in terms of pleasure, relaxation, and vigor. We predicted that the creativity effect of exercise might (at least partially) depend on its mood effect. Finally, we used two sets of creative problem-solving questions that require logical and visuospatial rather than verbal abilities to evaluate convergent thinking.

2. Materials and Methods

2.1. Participants

The study was approved by the Institutional Review Board of Yamaguchi University Hospital and preregistered on the University hospital Medical Information Network Clinical Trial Registry (UMIN-CTR, register ID: UMIN000041122). The study was a randomized controlled trial, with a between-subjects pre-test–post-test comparison design (see design and procedure below). Forty healthy subjects (all undergraduates, 11 females, 29 males, age: 22.98 ± 1.95 years) were recruited via posters placed on campus and department homepage and through word-of-mouth. This sample size was estimated with a priori power analysis based on data from Oppezzo and Schwartz [22]. With inputs of d = 0.999 (estimated from experiment 3, walk–sit vs. sit–sit), alpha = 0.05, and power of 0.8, 17 subjects per group were required, and we recruited 20 subjects per group.

The study was carried out following the latest version of the Declaration of Helsinki, and all subjects agreed to participate in this study and provided written informed consent after receiving a detailed explanation of the study. The inclusion criteria were being 20–29 years old at the time of the visit. The exclusion criteria were (1) having a history of diseases that greatly affect cardiopulmonary functions, such as chronic heart failure, (2) currently suffering from a mental illness or undergoing medical examination, (3) being a staff of our department, who receives the personnel evaluation directly by the principal investigator of this study, (4) being judged to be unsuitable for this study (e.g., bodyweight exceeding the applicable weight of the exercise bike). No participant was excluded owing to meeting any of the exclusion criteria.

2.2. Design and Procedure

Subjects were informed beforehand to (1) get enough sleep on the previous night, (2) refrain from drinking coffee and energy drinks, from smoking, and from engaging in intensive physical activities during the two hours before visiting the laboratory, and (3), contact the staff and reschedule the experiment if they were sick or did not feel well on the experimental day. On the experimental day, subjects first filled out a form answering their age, gender, history of smoking and alcohol drinking, education, and questions to confirm whether they adhered to the above instructions. Subjects also filled out the International physical activity questionnaire (IPAQ, [35]) indicating their level of physical activity during the past seven days and the positive and negative affect schedule (PANAS, [36]) indicating their current mood. They then completed bodyweight tests, height (for the calculation of BMI) and grip strength (Lafayette hydraulic hand dynamometer model J00105, the average of the two hands was used).

In this study, we used a between-subjects pre-test–post-test comparison design (Figure 1). Although a within-subjects crossover pre-test–post-test design may be more rigorous [8], the nature of the creative thinking tests precluded us from using this design. Specifically, for a within-subjects crossover pre-test–post-test design, participants must effectively perform the creative thinking tests four separate times. A critical characteristic of insight problem-solving tests for the evaluation of convergent thinking is that once subjects achieve insight into a problem, they will easily solve similar problems. As the available validated insight problems are limited at the moment, it is hard or almost impossible to appropriately schedule four assessments.

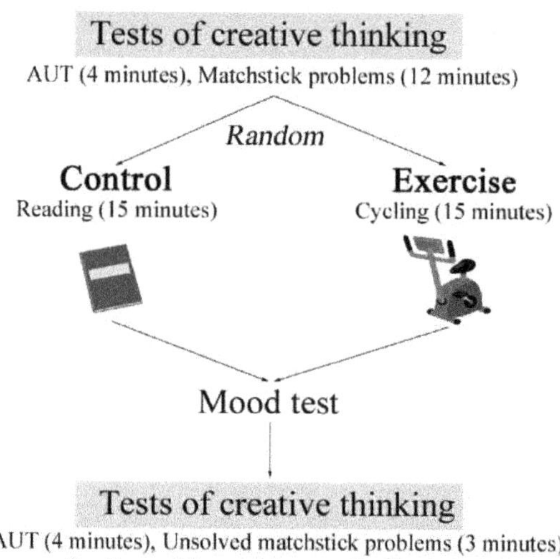

Figure 1. Schematic illustration of the study design.

Here, forty subjects were randomly assigned to receive a 15-min exercise or control intervention, before and after which they conducted tests of creative thinking that comprised a divergent thinking test and a convergent thinking test in a counterbalanced order. The statistical comparison revealed that subjects assigned to the two groups did not differ in their gender distribution, smoking and alcohol drinking habits, total metabolic equivalents (METs) of physical activity, BMI, grip strength, and positive and negative affect. The exercise group, however, had an older age than the control group (23.60 ± 2.14 vs. 22.35 ± 1.57 years, $t(38) = 2.110$, $p = 0.041$). Therefore, in our statistical analysis, we included age as a covariate.

Immediately following the intervention and before the second test of creative thinking, subjects were instructed to indicate their mood at the moment in terms of pleasure, relaxation, and vigor using a visual analog scale. This mood test was designed based on the valence-arousal two-dimensional affect grid [37]. The mood test generally took less than 30 s and immediately following the mood test, subjects conducted the second test of creative thinking. After the second test of creative thinking, subjects also filled out a set of questionnaires (measuring psychological stress, depressive and anxious symptoms, etc.) and cognitive tests (measuring decision-making and working memory), the results of which were not analyzed here.

2.3. Intervention

For the exercise intervention, we used the automated physical test program built in an exercise bike (Fukuda Denshi Wellbike BE-260). The program lasted 15 min and consisted of 10 min of physical test and 5 min of cooldown. Subjects were asked to sit quietly on the bike during the first minute to measure their resting state pulse rate and to start pedaling from the second minute at a pace of 50 rpm. The workload increased at 4 and 7 min after the start of the program according to the pulse rate of the subjects at the moment. For males, the workload increased to 12–25 N·m after 4 min and 15–35 N·m after 7 min, depending on the pulse rate of the subjects. For females, the workload increased to 8–15 N·m after 4 min and 11–25 N·m after 7 min, depending on the pulse rate of the subjects. Following the 10 min physical text, subjects performed a 5-min cooldown, pedaling at a self-selected pace.

Under this program, the heart rates of subjects are expected to rise to around 110–115 bpm after the first three minutes of pedaling (1–4 min), 123–135 bpm after the second three minutes of pedaling (4–7 min), and then maximally around 160 bpm after the last three minutes of pedaling (7–10 min). Since exercise intensity at <57% HR_{max} is considered very light, 57–63% HR_{max} light, 64–76% HR_{max} moderate, and 77–95% HR_{max} vigorous [31], the intensity of the current program is predicted to be very light to light during the first three minutes, light to moderate during the second three minutes, and moderate to vigorous during the third minute (assuming a mean age of 25 years and using the formula 220 minus mean age for the estimation of HR_{max}, [32]). As a whole, we expected that the intensity and workload of the exercise program fall between the two interventions employed investigated by Colzato et al. [21] (for non-athletes, mean HR for a 6-min normal cycling was 93.2 bpm, or 47% HR_{max}, and for a 6-min intensive cycling was 131.6 bpm, or 66% HR_{max}).

Based on health and ethical considerations, we also informed subjects before the program that if they experienced any progressively increasing chest pain, strong shortness of breath, strong feelings of fatigue, dizziness, vomiting, headache, stagger, or lower limb pain while exercising, they should inform the experimenter immediately. In these cases, the experimenter would immediately end the 10-min physical test. Furthermore, the physical test would also be terminated if the experimenter noticed any cyanosis, pallor of the face, cold sweat in the subject. As a result, four subjects reported strong feelings of fatigue during the 7–10 min of the physical test without any other subjective or objective symptoms. Therefore, the physical test was immediately terminated for these subjects, and all of them performed a cooldown for the rest of the intervention without reporting any other symptoms or request to quit the cooldown or withdraw from the experiment. Given there was no difference in the mean and maximal HR and self-reported mood after the intervention between these four subjects and the remaining 16 subjects in the exercise group, we included all 20 subjects in our final analysis.

For the control intervention, following Chang and Etnier [38], we asked subjects to read materials on the association between physical exercise and brain functions at a self-selected pace. The content of the materials was irrelevant to the creativity test and considered mood-neutral. Through pilot testing with volunteers, who had similar backgrounds to our subjects, we adjusted the amount of the materials to take roughly 15 min for most subjects to read. In the case of fast-readers, we prepared another set of materials on the same topic but with different contents.

During both interventions, subjects' heart rate was monitored with an Apple Watch Series 4 (Apple Inc., California, United States), which has been shown to have high accuracy [39]. Unfortunately, due to an initial setting problem, data of the first minute in the exercise group where subjects stayed seated quietly on the exercise bike were not recorded. That is, for the exercise group, we had only HR data after they started to pedal. Nevertheless, since the HR in the first minute is expected to be similar between the exercise and control group, we do not consider this recording problem a serious issue and, therefore, used HR across the 14 min as the data for the exercise group.

2.4. Divergent Thinking

We used the AUT for the measurement of divergent thinking. Subjects were presented three common objects and asked to write down as many as possible uncommon, original, and unique uses of those objects in 4 min on A4 size blank paper. We prepared two sets of objects, and for each subject, one set was randomly selected for pre-test (before the intervention) and the other post-test (after the intervention). For set A, "brick", "empty can", and "umbrella" were used; for set B, "pencil", "tissue box", and "newspaper" were used.

For the scoring of the AUT, we used fluency, flexibility, and originality as indicators of creativity [27,28]. Fluency was defined as the number of generated uses. Flexibility was the number of different conceptual categories the alternative uses are from. Originality refers to the rareness of generated uses and was defined here based on the conceptual category of the uses, rather than the uses, per se. Only if one single subject gave use(s) from a specific conceptual category, the category counted as original; if two or more subjects gave use(s) from the same conceptual category, the category did not count as original for both or all of them. For instance, given "brick", "road mark" belongs to one category, while "paperweight" and "stone weight" belong to another. If one subject gave "paperweight" and the other subject "stone weight", the response counted as original for neither of them. A primary coder scored all responses, and a secondary coder scored responses for a randomly selected object. The two coders reached an agreement of Cohen's $\kappa = 0.936$ for flexibility and Cohen's $\kappa = 0.706$ for originality, which is considered substantial or almost perfect [40].

2.5. Convergent Thinking

We used the matchstick arithmetic problems developed by Knoblich et al. [41] to evaluate insight problem-solving or convergent thinking in our pre-test. A matchstick arithmetic problem consists of a false equation written with Roman numerals, for example, II = III + I. Subjects were required to move a single stick to transform the equation into a correct one. In the above example, the correct response is to move one stick from "III" on the right side to "II" on the left side. Four different classes of problems can be identified based on the kind of move, including moving a matchstick from a numeral to another numeral (type A), moving a matchstick from an operator sign to another operator sign or numeral (type B), rotating the vertical matchstick of the plus sign to form an equal sign (type C), and sliding one of the matchsticks from the symbol X to form V (type D).

Based on data reported by Knoblich et al. [41] and our pilot testing, we selected six problems and set 12 min as the time limit for solving these problems in our pre-test. The six problems included two problems from type A and B each and one problem from type C and D each. We expected that a total of six problems was adequate such that subjects would not be overwhelmed while at the same time most of them would not be able to solve all the problems. Therefore, this allowed us to present the unsolved problems to the subjects again at post-test to investigate whether exercise could help them gain insights on these previously unsolved problems. In the post-test, subjects were given three minutes to solve the unsolved or incorrectly solved problems. Those who correctly solved all pre-test problems were presented three new problems to work on at the post-test.

Before solving these problems in the pre-test, all subjects first went through a training session to ensure that all were familiar with Roman numerals. They were instructed to study a correspondence table between modern numerals (1–15) and Roman numerals (I through XV) until they passed two tests: in the first test, they were presented all the fifteen modern numerals in random order and asked to write down the corresponding Roman numerals; in the second test, vice versa, they were presented with all the Roman numerals and asked to write down the corresponding modern numerals.

For convergent thinking at post-test, we also included another set of creative problem-solving puzzles that require visuospatial and logical reasoning [42]. Tago [42] is a selected collection of creative puzzles written by the Japanese educator Akira Tago, from his series *atama no taiso* or Mental Gymnastics first released in 1966. The puzzles in the series have been considered a valid measure of creativity in the Japanese culture and used to evaluate creativity in scientific studies by Japanese researchers [43]. Based on our pilot testing, we selected three puzzles for our test here. As an example, the first puzzle had the background stating that three friends were about to eat three pieces of isosceles triangle-shaped cakes when one more friend joined them; the question was, what is the minimum number of times required for cutting these three pieces of cakes into four equal parts, in order for four of them to eat. Subjects were encouraged not to give up and to do their best to solve all the puzzles. If subjects finished answering all the puzzles before the time limit (i.e., 9 min), they were given four more puzzles to solve, the results of which were not analyzed here.

Therefore, for convergent thinking, we created two measures for data analysis, one was a creative problem-solving (CPS) score consisted of matchstick arithmetic problems at pre-test and creative problem-solving puzzles at post-test; the other was a re-test score on the matchstick arithmetic problems (or matchstick re-test) obtained at post-test only.

After subjects finished all the above convergent thinking tests, they filled out a survey asking whether they have seen any of the tests presented. Two subjects (both from the exercise group) had seen exactly the same creative problem-solving puzzle previously (the second and third puzzle) and, therefore, were excluded from relevant data analysis (where n = 18 for the exercise group).

2.6. Statistical Analysis

The statistical analysis was conducted with IBM SPSS Statistics 26.0. The normality of the data was checked using the Shapiro–Wilk test. Student's t-tests or Mann–Whitney U tests were used for comparing between-group differences, while repeated measures ANOVAs were used for comparing between-group pre-test–post-test effects, with group (control vs. exercise) as between-group factor and time (pre-test vs. post-test) as the within-subjects factor. Due to an age difference between the two groups, age was included as a covariate in one-way ANCOVAs and repeated measures ANOVAs. Pearson's or Spearman correlation analysis was used to examining the association between mood and creative thinking measures. Median analysis was conducted with Mplus 8.0 (Muthén and Muthén, 2012). Effect size (Cohen's d) was calculated using G*Power Version 3.1.9.6 [44]. A significance level of $p < 0.05$ was used.

3. Results

3.1. Heart Rate and Mood Measures

As plotted in Figure 2, compared to the control group, subjects in the exercise group had significantly higher mean heart rate ($U = 4.000$, $p = 0.000$, $d = 3.72$) and maximal heart rate ($U = 0.000$, $p = 0.000$, $d = 5.47$) during the intervention, and higher heart rate in the last minute of the intervention ($t(38) = 6.766$, $p = 0.000$, $d = 2.14$). Subjects in the exercise group also reported higher feelings of pleasure ($t(38) = 3.707$, $p = 0.001$, $d = 1.17$) and vigor ($t(38) = 3.625$, $p = 0.001$, $d = 1.15$), but not relaxation ($t(38) = 0.015$, $p = 0.988$). All these between-group differences remained significant after incorporating age as a covariate (one-way ANCOVAs). Given that the mean heart rate was equivalent to 60.2% age-estimated HR_{max} [31,32], our exercise intervention was considered primarily light in intensity, despite the fact that the maximal heart rate here had reached 80.8% age-estimated HR_{max} (considered vigorous in intensity). Furthermore, since the two groups did not differ in their positive and negative affect at baseline, the above results suggest that exercise increased subjects' feelings of pleasure and vigor.

3.2. Divergent Thinking

The between-group pre-test–post-test comparisons of the AUT results are shown in Figure 3 and Table S2. There was a significant effect of group × time interaction on flexibility ($F_{(1,38)} = 5.158$, $p = 0.029$, $d = 0.37$) and a trend towards significance on fluency ($F_{(1,38)} = 3.588$, $p = 0.066$, $d = 0.31$). Similar results were obtained after, including age as a covariate ($F_{(1,37)} = 4.898$, $p = 0.033$ for flexibility, $F_{(1,37)} = 3.866$, $p = 0.057$ for fluency). As shown in Figure 3, exercise increased flexibility and fluency (with a trend) at post-test, without affecting originality.

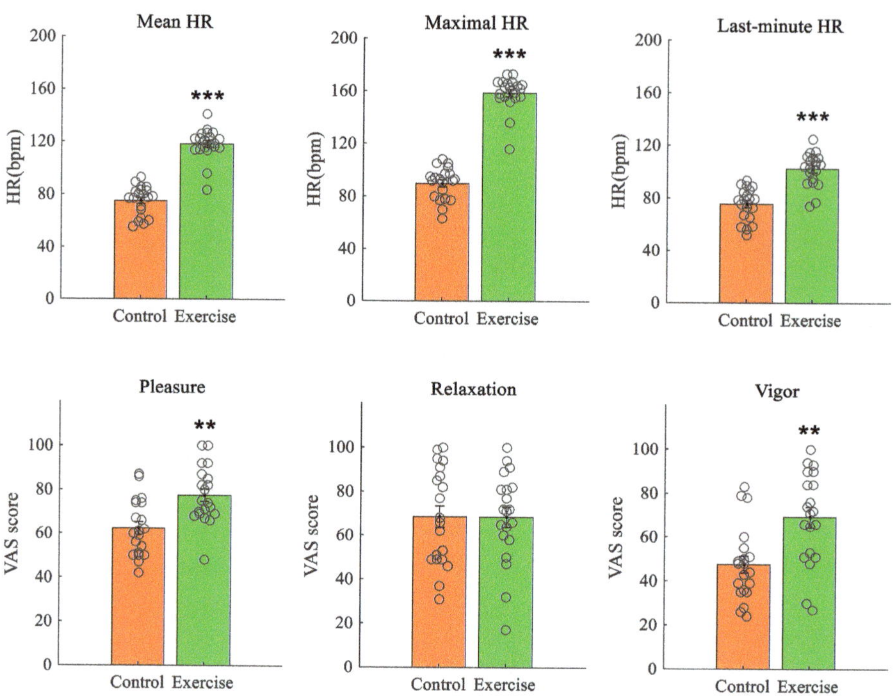

Figure 2. Intervention effect on heart rate (HR) and mood. **Upper** panels: HR; **lower** panels: mood. ** $p < 0.01$, *** $p < 0.001$ compared to control. Data shown as means $\pm$ SE, circles represent individual responses.

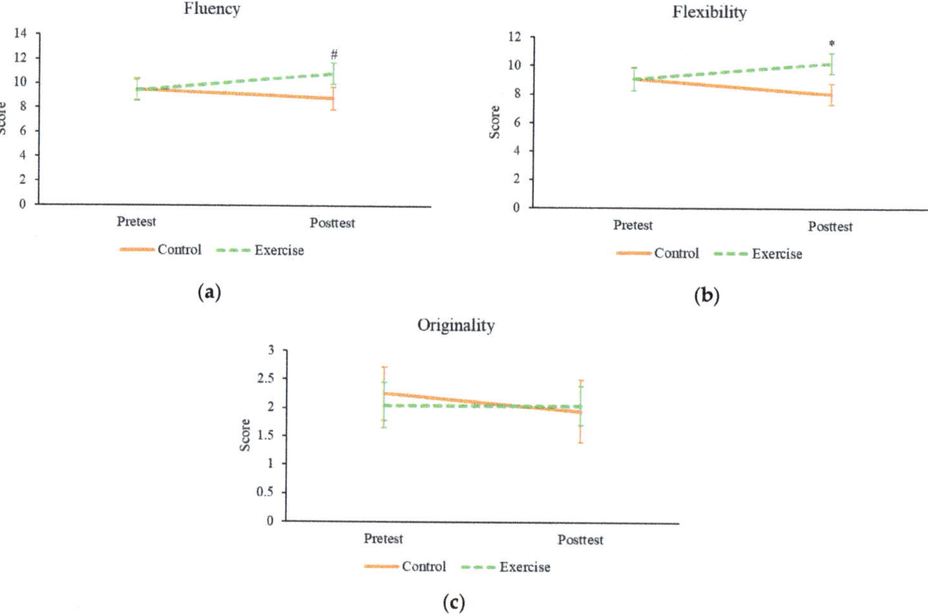

Figure 3. Exercise effect on divergent thinking (AUT). (**a**), fluency; (**b**), flexibility; (**c**), originality. * $p < 0.05$, indicating a significant effect of group $\times$ time interaction; #, $p = 0.066$ for the group $\times$ time interaction. Data shown as means $\pm$ SE.

3.3. Convergent Thinking

The comparisons of the convergent thinking scores are shown in Figure 4 and Table S2. No significant effect of group × time interaction was obtained for the CPS score, although a significant effect of time ($F_{(1,36)}$ = 57.134, p = 000, d = 1.26) indicating that subjects obtained significantly lower scores at post-test compared to pre-test was observed (Figure 4a). This difference, however, became nonsignificant after controlling age (Table S2). There was also no difference between the two groups in the score of the matchstick re-test (Figure 4b), which remained unchanged after controlling age.

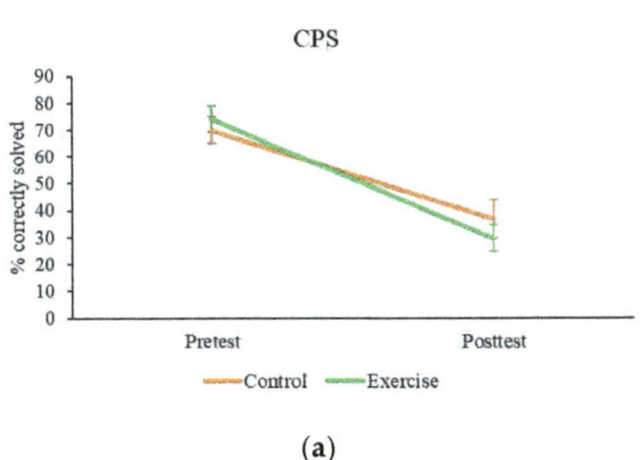

(a)

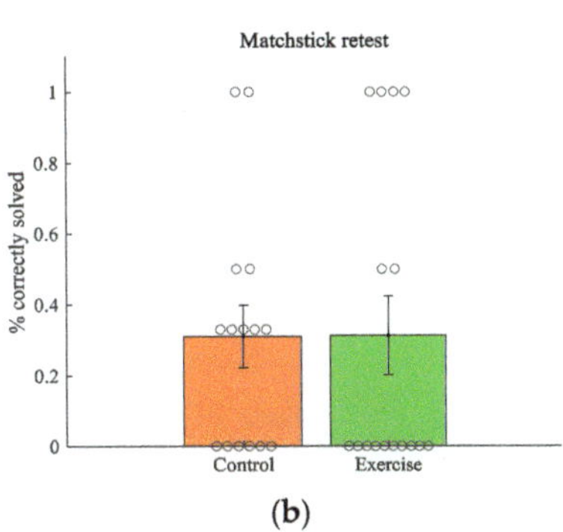

(b)

Figure 4. Exercise effect on convergent thinking. (**a**), CPS; (**b**), matchstick re-test. Data shown as means ± SE, circles represent individual responses.

3.4. Mood Effect

To explore whether mood regulated the exercise effect on divergent and convergent thinking, we conducted a correlation analysis for each group between creativity scores at post-test and subjects' self-reported mood following the intervention. As shown in Table S3 and Figure S1, neither pleasure nor vigor was associated with AUT measures in the control or exercise group (all p > 0.10), although relaxation was positively associated with AUT measures in the control group (r/rho = 0.511–0.572, all p < 0.05). Whereas exercise boosted fluency and flexibility and increased pleasure and vigor, there were no correlations between these mood and creativity measures.

Based on the literature review mentioned in the Introduction, we further tested the hypothesis that exercise improves fluency and flexibility through its effect on mood, namely pleasure and vigor, using mediation analysis. Specifically, we tested three models for each outcome variable (i.e., fluency and flexibility): In the first model, pleasure and vigor concurrently mediate the effect of exercise, while in the second and third models, pleasure or vigor alone mediates the effect of exercise. Contrary to our hypothesis, none of the models indicated a significant indirect mediation effect (all p > 0.10; Table S4).

For convergent thinking, notably, there was a positive association between matchstick re-test score and pleasure and vigor in the exercise group only (Table S3, Figure 5a). Subjects reporting high feelings of pleasure or vigor after exercise showed better performance on the matchstick re-test. These correlations (for pleasure, rho = 0.503, p = 0.047; for vigor, rho = 0.647, p = 0.007) are considered moderate to moderately high in magnitude [45]. When we divided the exercise group into low and high vigor groups based on a median split, compared to the control group, subjects in the high vigor group (Exercise: High Vigor) tended to have a higher score on the matchstick re-test (U = 34.00, p = 0.081, d = 0.83, Figure 5b), while those in the low vigor group (Exercise: Low Vigor) had a significantly lower score on the matchstick re-test (U = 24.00, p = 0.008, d = −1.13). That is, compared to the control group, subjects reporting high vigor after exercise tended to solve more matchstick problems that were unsolved before the intervention. On the other hand, subjects reporting low vigor became even less capable of solving the previously unsolved problems. However, there were no such differences when we conducted a similar analysis with pleasure.

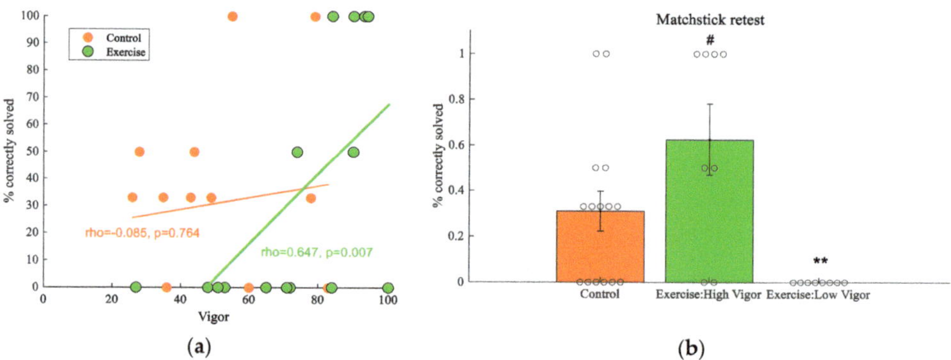

(**a**)   (**b**)

Figure 5. Exercise effect on matchstick re-test affected by vigor. (**a**) Scatterplot of the association (with a regression line) between post-intervention vigor and matchstick re-test score in each group; (**b**) comparison of matchstick re-test score between control and the two exercise groups (created based on a median split of vigor). **, $p < 0.01$; # $p = 0.081$, compared to control. Data shown as means $\pm$ SE, circles represent individual responses.

We also conducted another analysis to test whether subjects' self-reported vigor post-exercise was associated with their physical fitness and exercise habits, based on the assumption that subjects with a low level of physical fitness and habitual physical activity may be more easily exhausted. A comparison of the low to the high vigor group indicated that the low group tended to have lower grip strength ($t(14) = -1.948$, $p = 0.072$, $d = 0.97$) and conduct less physical activity in the past seven days (total METS, $t(10) = -1.895$, $p = 0.087$, $d = 0.95$).

4. Discussion

In the present study, we found an exercise program that enhanced subjects' self-reported mood (in terms of pleasure and vigor) improved the flexibility and fluency of divergent thinking. However, there was no correlation between flexibility and fluency and subjects' self-reported pleasure and vigor post-exercise, and the creativity effect of exercise was not mediated by post-exercise self-reported pleasure and vigor. In contrast, the exercise effect on convergent thinking depended on subjects' mood after exercise: subjects reporting high vigor tended to solve more insight problems unsolved previously, while those reporting low vigor became less capable of solving previously unsolved problems.

We used a built-in automatic physical test program for the exercise intervention, the intensity of which was considered falling between normal walking or cycling and cycling with maximal effort (as employed by Colzato et al. [21]). Whereas cycling with maximal effort impaired flexibility of divergent thinking in Colzato et al. [21], here our exercise intervention improved flexibility. This may partially support the hypothesis we speculated in the Introduction that exercises at low-to-moderate intensity might be better able to enhance divergent thinking than exercise at too low or too high an intensity ([21–23]). The hypothesis, however, still requires further confirmation by future studies with more homogenous and accurately prescribed intensities.

Our findings only partially confirm our prediction that the creativity effect of exercise may depend on its mood effect since the effect of exercise on convergent but not divergent thinking is mood-dependent. Our speculation that Colzato et al. [21] and Frith and Loprinzi [23] failed to find the divergent thinking-enhancing effect because they used exercise interventions that did not change subjects' mood is, therefore, not supported. Our prediction was considered possible because it was consistent with two lines of neurobiological findings: first, physical exercise improves mood at least partly through increasing neurotransmitters, including serotonin, dopamine, and endocannabinoids (e.g., [46–49]; for a review, see [7]); and second, these neurotransmitters play a critical role in cognition [50–52] and are also involved in the cognitive-enhancing effect of exercise [7]. Our results, however, showed that, while the effect of exercise on convergent thinking depends on mood, its effect on divergent thinking does not. This raises the possibility that convergent and divergent thinking might share distinct neurobiological mechanisms. Our findings are consistent with Steinberg et al. [19], who reported that the divergent thinking enhancing effect of exercise was independent of mood, although a Likert scale rather than visual analog scale [53] was used in that study. Future studies are required to test the possibilities that we suggested and uncover the underlying explanations of why the effect of exercise on convergent but not divergent thinking is mood-dependent.

For the first time, we showed that the effect of exercise on convergent thinking depends on post-exercise mood such that high vigor may enhance, while low vigor impairs creative problem-solving. This finding is in line with studies showing that exercise without causing changes in mood failed to improve convergent thinking using the RAT and CRA [21,23]. Previous research has indicated a critical role of serotonin, dopamine, and endocannabinoids in feelings of vigor [54,55] and behavioral indicators of vigor (e.g., performance or response vigor, [56,57]). Generally, exercise at an adequate intensity enhances vigor, while that at too high an intensity causes central fatigue and reduces vigor, likely through the interactions of serotonin and dopamine in the brain [58]. As such, it is possible that subjects reporting low vigor after exercise in our study may have been exhausted, which impaired their cognitive functions. Indeed, our analysis showed that these subjects tended to have lower grip strength, an important index of physical fitness [32], and conduct less physical activity in the past seven days. Considering that our exercise program reached vigorous intensity in the last three minutes, these subjects might have been more easily exhausted. This speculation is in line with Colzato et al.'s [21] findings that cycling with maximal effort (moderate intensity) impaired RAT performance in non-athletes but tended to improve RAT performance in athletes. It is also consistent with a recent report that jogging acutely enhanced visual attentional control, the effect of which was fully mediated by feelings of energy [59]. Future studies must further investigate these possibilities and clarify the association between post-exercise vigor and convergent thinking.

Our findings suggest that engaging in an acute bout of aerobic exercise enhances divergent thinking, and when it improves vigor, convergent thinking may also be enhanced. Therefore, aerobic exercise may be considered a simple but useful strategy for improving these cognitive functions. It may also have important therapeutic potential in psychiatric contexts since many psychiatric conditions, such as depression, have been associated with impaired divergent thinking and problem-solving abilities [60–64].

Several important limitations of our study should be noted. First, we used a built-in automatic physical test program for the exercise intervention and provided a post hoc estimation of the exercise intensity based on age-estimated maximal heart rate. A preferred way to account for the individual differences in aerobic capacity is to prescribe a homogenous intensity by referring to a percentage of the aerobic capacity reserve [8,31]. Neither did we have the participants rate their level of perceived exertion during the exercise intervention, which provides a subjective evaluation of the intensity [8]. A second limitation of the study is that we used a between-subjects pre-test–post-test design. As already stated in Materials and Methods, a within-subjects crossover pre-test–post-test design may be more rigorous [8]. However, the nature of the creative thinking tests hindered us from using this design. We hope more insight problems and convergent thinking tests will be developed and validated in the near future so that we will be able to use more rigorous designs. A third limitation is actually related to the insight problems we used to measure convergent thinking. As shown in Figure 4A, subjects on average performed poorly on the CPS at post-test. A closer look at subjects' responses indicated that 30 subjects (75%) failed to solve any puzzle or correctly solved merely one puzzle. Due to this high level of difficulty, a floor effect may have occurred that hindered us from detecting the specific, creativity-enhancing effect of exercise. Future research needs to use convergent thinking tests with an appropriate level of difficulty to confirm our findings.

Lastly, our sample size was relatively small, and although the data of AUT-originality and matchstick re-test were not normally distributed, we still used two-way ANOVA and one-way ANCOVA for our main analysis. These tests were considered rather robust, and few nonparametric tests existed for these situations [65]. Future studies with large sample sizes, more robust statistical tests and appropriate statistical powers are required to confirm the influence of mood on the creativity effects we reported here. We hope the current exploratory study may provide useful preliminary findings for more rigorous studies in the future.

5. Conclusions

In a randomized controlled trial with a between-subjects pre-test–post-test comparison design, we found that a 15-min exercise program improved divergent thinking in terms of fluency and flexibility, and the effect did not depend on subjects' self-reported mood after exercise. The exercise program, however, affected convergent thinking in a mood-dependent manner such that subjects reporting high vigor tended to solve more insight problems that were unsolved previously. In contrast, those reporting low vigor became less capable of solving previously unsolved problems. Although our sample size was relatively small, these findings suggest the possibility that engaging in a bout of mood, particularly vigor-enhancing aerobic exercise, may be considered a simple but useful strategy for gaining insights into previously unsolved problems. Future more rigorous studies are required to replicate our findings.

Supplementary Materials: The following are available online at https://www.mdpi.com/article/10.3390/brainsci11050546/s1, Table S1: Summary of previous studies on the after-effects of acute physical exercise on creative thinking; Table S2: A summary of main statistical results: repeated measures ANOVA; Table S3: Correlation between self-reported mood and divergent and convergent thinking at post-test; Table S4: Mediation models and results; Figure S1: Scatterplot (with a regression line) of correlation between mood and creative thinking measures at post-test.

Author Contributions: Conceptualization, C.C., S.N.; methodology, K.A., M.I., C.C., R.Y.; formal analysis, K.A., C.C., K.H., A.T.; investigation, K.A., M.I., C.C., K.H., R.Y., M.H.; resources, T.S., Y.F., T.M.; writing—original draft preparation, K.A., C.C.; writing—review and editing, all authors. All authors have read and agreed to the published version of the manuscript.

Funding: This research did not receive any specific grant from funding agencies in the public, commercial, or not-for-profit sectors.

Institutional Review Board Statement: The study was conducted according to the guidelines of the Declaration of Helsinki, and approved by the Institutional Review Board of Yamaguchi University Hospital (H2020-065).

Informed Consent Statement: Informed consent was obtained from all subjects involved in the study.

Data Availability Statement: The data that support the findings of this study are available from the corresponding author upon reasonable request.

Acknowledgments: We thank Huijie Lei (M. Psych.) for her assistance with mediation analysis.

Conflicts of Interest: The authors declare no conflict of interest.

References

1. Lambourne, K.; Tomporowski, P. The effect of exercise-induced arousal on cognitive task performance: A meta-regression analysis. *Brain Res.* **2010**, *1341*, 12–24. [CrossRef] [PubMed]
2. Chang, Y.K.; Labban, J.D.; Gapin, J.I.; Etnier, J.L. The effects of acute exercise on cognitive performance: A meta-analysis. *Brain Res.* **2012**, *1453*, 87–101. [CrossRef] [PubMed]
3. McMorris, T.; Hale, B.J. Differential effects of differing intensities of acute exercise on speed and accuracy of cognition: A meta-analytical investigation. *Brain Cogn.* **2012**, *80*, 338–351. [CrossRef] [PubMed]

4.   Ludyga, S.; Gerber, M.; Brand, S.; Holsboer-Trachsler, E.; Pühse, U. Acute effects of moderate aerobic exercise on specific aspects of executive function in different age and fitness groups: A meta-analysis. *Psychophysiology* **2016**, *53*, 1611–1626. [CrossRef] [PubMed]
5.   Liu, S.; Yu, Q.; Li, Z.; Cunha, P.M.; Zhang, Y.; Kong, Z.; Kong, Z.; Lin, W.; Chen, S.; Ca, Y. Effects of Acute and Chronic Exercises on Executive Function in Children and Adolescents: A Systemic Review and Meta-Analysis. *Front. Psychol.* **2020**, *11*, 3482. [CrossRef]
6.   Tomporowski, P.D. Effects of acute bouts of exercise on cognition. *Acta Psychol.* **2003**, *112*, 297–324. [CrossRef]
7.   Basso, J.C.; Suzuki, W.A. The effects of acute exercise on mood, cognition, neurophysiology, and neurochemical pathways: A review. *Brain Plast.* **2017**, *2*, 127–152. [CrossRef] [PubMed]
8.   Pontifex, M.B.; McGowan, A.L.; Chandler, M.C.; Gwizdala, K.L.; Parks, A.C.; Fenn, K.; Kamijo, K. A primer on investigating the after effects of acute bouts of physical activity on cognition. *Psychol. Sport Exerc.* **2019**, *40*, 1–22. [CrossRef]
9.   Cotman, C.W. Exercise builds brain health: Key roles of growth factor cascades and inflammation. *Trends Neurosci.* **2007**, *30*, 464–472. [CrossRef] [PubMed]
10.   Stimpson, N.J.; Davison, G.; Javadi, A.H. Joggin'the noggin: Towards a physiological understanding of exercise-induced cognitive benefits. *Neurosci. Biobehav. Rev.* **2018**, *88*, 177–186. [CrossRef]
11.   Herold, F.; Aye, N.; Lehmann, N.; Taubert, M.; Müller, N.G. The contribution of functional magnetic resonance imaging to the understanding of the effects of acute physical exercise on cognition. *Brain Sci.* **2020**, *10*, 175. [CrossRef] [PubMed]
12.   Herold, F.; Wiegel, P.; Scholkmann, F.; Müller, N.G. Applications of functional near-infrared spectroscopy (fNIRS) neuroimaging in exercise–cognition science: A systematic, methodology-focused review. *J. Clin. Med.* **2018**, *7*, 466. [CrossRef]
13.   Guilford, J.P. Creativity: Yesterday, today, and tomorrow. *J. Creat. Behav.* **1967**, *1*, 3–14. [CrossRef]
14.   Cropley, A.J. Defining and measuring creativity: Are creativity tests worth using? *Roeper Rev.* **2000**, *23*, 72–79. [CrossRef]
15.   Kim, K.H. Meta-analyses of the relationship of creative achievement to both IQ and divergent thinking test scores. *J. Creat. Behav.* **2008**, *42*, 106–130. [CrossRef]
16.   Runco, M.A.; Millar, G.; Acar, S.; Cramond, B. Torrance Tests of Creative Thinking as predictors of personal and public achievement: A fifty year follow up. *Creat. Res. J.* **2010**, *22*, 361–368. [CrossRef]
17.   Gralewski, J.; Karwowski, M. Are teachers' ratings of students' creativity related to students' divergent thinking? A meta-analysis. *Think. Ski. Creat.* **2019**, *33*, 100583. [CrossRef]
18.   Hennessey, B.A.; Amabile, T.M. Creativity. *Annu. Rev. Psychol.* **2010**, *61*, 569–598. [CrossRef]
19.   Steinberg, H.; Sykes, E.A.; Moss, T.; Lowery, S.; LeBoutillier, N.; Dewey, A. Exercise enhances creativity independently of mood. *Br. J. Sports Med.* **1997**, *31*, 240–245. [CrossRef] [PubMed]
20.   Netz, Y.; Tomer, R.; Axelrad, S.; Argov, E.; Inbar, O. The effect of a single aerobic training session on cognitive flexibility in late middle-aged adults. *Int. J. Sports Med.* **2007**, *28*, 82–87. [CrossRef] [PubMed]
21.   Colzato, L.S.; Szapora, A.; Pannekoek, J.N.; Hommel, B. The impact of physical exercise on convergent and divergent thinking. *Front. Hum. Neurosci.* **2013**, *7*, 824. [CrossRef] [PubMed]
22.   Oppezzo, M.; Schwartz, D.L. Give your ideas some legs: The positive effect of walking on creative thinking. *J. Exp. Psychol. Learn. Mem. Cogn.* **2014**, *40*, 1142. [CrossRef] [PubMed]
23.   Frith, E.; Loprinzi, P.D. Experimental effects of acute exercise and music listening on cognitive creativity. *Physiol. Behav.* **2018**, *191*, 21–28. [CrossRef]
24.   Román P Á., L.; Vallejo, A.P.; Aguayo, B.B. Acute aerobic exercise enhances students' creativity. *Creat. Res. J.* **2018**, *30*, 310–315. [CrossRef]
25.   Guilford, J.P. *Alternate Uses*; Form, A., Ed.; Sheridan Supply: Beverly Hills, CA, USA, 1960.
26.   Torrance, P.E. *Torrance Tests of Creative Thinking*; Personnel Press: Princeton, NJ, USA, 1966.
27.   Runco, M.A.; Acar, S. Divergent thinking as an indicator of creative potential. *Creat. Res. J.* **2012**, *24*, 66–75. [CrossRef]
28.   Reiter-Palmon, R.; Forthmann, B.; Barbot, B. Scoring divergent thinking tests: A review and systematic framework. *Psychol. Aesthet. Creat. Arts* **2019**, *13*, 144. [CrossRef]
29.   Mednick, M.T.; Mednick, S.A.; Mednick, E.V. Incubation of creative performance and specific associative priming. *J. Abnorm. Soc. Psychol.* **1964**, *69*, 84–88. [CrossRef]
30.   Bowden, E.M.; Jung-Beeman, M. Normative data for compound remote associate problems. *Behav. Res. Methods Instrum. Comput.* **2003**, *35*, 634–639. [CrossRef]
31.   American College of Sports Medicine; American College of Sports Medicine Position Stand. Quantity and quality of exercise for developing and maintaining cardiorespiratory, musculoskeletal, and neuromotor fitness in apparently healthy adults: Guidance for prescribing exercise. *Med. Sci. Sports Exerc.* **2011**, *43*, 1334–1359. [CrossRef]
32.   Plowman, S.A.; Smith, D.L. *Exercise Physiology for Health Fitness and Performance*; Lippincott Williams & Wilkins: Philadelphia, PA, USA, 2013.
33.   Benedek, M.; Konen, T.; Neubauer, A.C. Associative abilities underlying creativity. *Psychol. Aesthet. Creat. Arts* **2012**, *6*, 273–281. [CrossRef]
34.   Worthen, B.R.; Clark, P.M. Toward an improved measure of remote associational ability. *J. Educ. Meas.* **1971**, *8*, 113–123. [CrossRef]
35.   Craig, C.L.; Marshall, A.L.; Sjöström, M.; Bauman, A.E.; Booth, M.L.; Ainsworth, B.E.; Oja, P. International physical activity questionnaire: 12-country reliability and validity. *Med. Sci. Sports Exerc.* **2003**, *35*, 1381–1395. [CrossRef] [PubMed]

36. Watson, D.; Clark, L.A.; Tellegen, A. Development and validation of brief measures of positive and negative affect: The PANAS scales. *J. Personal. Soc. Psychol.* **1988**, *54*, 1063. [CrossRef]
37. Russell, J.A. Core affect and the psychological construction of emotion. *Psychol. Rev.* **2003**, *110*, 145. [CrossRef] [PubMed]
38. Chang, Y.K.; Etnier, J.L. Effects of an acute bout of localized resistance exercise on cognitive performance in middle-aged adults: A randomized controlled trial study. *Psychol. Sport Exerc.* **2009**, *10*, 19–24. [CrossRef]
39. Wang, R.; Blackburn, G.; Desai, M.; Phelan, D.; Gillinov, L.; Houghtaling, P.; Gillinov, M. Accuracy of wrist-worn heart rate monitors. *JAMA Cardiol.* **2017**, *2*, 104–106. [CrossRef] [PubMed]
40. Cohen, J. A coefficient of agreement for nominal scales. *Educ. Psychol. Meas.* **1960**, *20*, 37–46. [CrossRef]
41. Knoblich, G.; Ohlsson, S.; Haider, H.; Rhenius, D. Constraint relaxation and chunk decomposition in insight problem solving. *J. Exp. Psychol. Learn. Mem. Cogn.* **1999**, *25*, 1534. [CrossRef]
42. Tago, A. *Atama No Taiso Best [Mental Gymnastics: Best]*; Kobunsha: Tokyo, Japan, 2009; pp. 55–56, 75–76, 133–134.
43. Yokoyama, T.; Sato, K. Effects of positive mood induction on problem solving. *Annu. Rep. Fac. Educ. Gunma Univ. Cult. Sci. Ser.* **2005**, *54*, 233–247.
44. Faul, F.; Erdfelder, E.; Lang, A.-G.; Buchner, A. G*Power 3: A flexible statistical power analysis program for the social, behavioral, and biomedical sciences. *Behav. Res. Methods* **2007**, *39*, 175–191. [CrossRef]
45. Zhu, W. Sadly, the earth is still round ($p < 0.05$). *J. Sport Health Sci.* **2012**, *1*, 9–11.
46. Otsuka, T.; Nishii, A.; Amemiya, S.; Kubota, N.; Nishijima, T.; Kita, I. Effects of acute treadmill running at different intensities on activities of serotonin and corticotropin-releasing factor neurons, and anxiety-and depressive-like behaviors in rats. *Behav. Brain Res.* **2016**, *298*, 44–51. [CrossRef] [PubMed]
47. Chen, C.; Nakagawa, S.; Kitaichi, Y.; An, Y.; Omiya, Y.; Song, N.; Kusumi, I. The role of medial prefrontal corticosterone and dopamine in the antidepressant-like effect of exercise. *Psychoneuroendocrinology* **2016**, *69*, 1–9. [CrossRef] [PubMed]
48. Boecker, H.; Sprenger, T.; Spilker, M.E.; Henriksen, G.; Koppenhoefer, M.; Wagner, K.J.; Tolle, T.R. The runner's high: Opioidergic mechanisms in the human brain. *Cereb. Cortex* **2008**, *18*, 2523–2531. [CrossRef] [PubMed]
49. Fuss, J.; Steinle, J.; Bindila, L.; Auer, M.K.; Kirchherr, H.; Lutz, B.; Gass, P. A runner's high depends on cannabinoid receptors in mice. *Proc. Natl. Acad. Sci. USA* **2015**, *112*, 13105–13108. [CrossRef] [PubMed]
50. Bacqué-Cazenave, J.; Bharatiya, R.; Barrière, G.; Delbecque, J.P.; Bouguiyoud, N.; Di Giovanni, G.; De Deurwaerdère, P. Serotonin in animal cognition and behavior. *Int. J. Mol. Sci.* **2020**, *21*, 1649. [CrossRef] [PubMed]
51. Seamans, J.K.; Yang, C.R. The principal features and mechanisms of dopamine modulation in the prefrontal cortex. *Prog. Neurobiol.* **2004**, *74*, 1–58. [CrossRef]
52. Campolongo, P.; Trezza, V. The endocannabinoid system: A key modulator of emotions and cognition. *Front. Behav. Neurosci.* **2012**, *6*, 73. [CrossRef]
53. Reips, U.D.; Funke, F. Interval-level measurement with visual analogue scales in Internet-based research: VAS Generator. *Behav. Res. Methods* **2008**, *40*, 699–704. [CrossRef]
54. Ohmatsu, S.; Nakano, H.; Tominaga, T.; Terakawa, Y.; Murata, T.; Morioka, S. Activation of the serotonergic system by pedaling exercise changes anterior cingulate cortex activity and improves negative emotion. *Behav. Brain Res.* **2014**, *270*, 112–117. [CrossRef] [PubMed]
55. Brellenthin, A.G.; Crombie, K.M.; Hillard, C.J.; Koltyn, K.F. Endocannabinoid and mood responses to exercise in adults with varying activity levels. *Transl. J. Am. Coll. Sports Med.* **2017**, *2*, 138–145. [CrossRef] [PubMed]
56. Niv, Y.; Daw, N.D.; Joel, D.; Dayan, P. Tonic dopamine: Opportunity costs and the control of response vigor. *Psychopharmacology* **2007**, *191*, 507–520. [CrossRef] [PubMed]
57. Crockett, M.J.; Clark, L.; Apergis-Schoute, A.M.; Morein-Zamir, S.; Robbins, T.W. Serotonin modulates the effects of Pavlovian aversive predictions on response vigor. *Neuropsychopharmacology* **2012**, *37*, 2244–2252. [CrossRef] [PubMed]
58. McMorris, T.; Barwood, M.; Corbett, J. Central fatigue theory and endurance exercise: Toward an interoceptive model. *Neurosci. Biobehav. Rev.* **2018**, *93*, 93–107. [CrossRef] [PubMed]
59. Legrand, F.D.; Albinet, C.; Canivet, A.; Gierski, F.; Morrone, I.; Besche-Richard, C. Brief aerobic exercise immediately enhances visual attentional control and perceptual speed. Testing the mediating role of feelings of energy. *Acta Psychol.* **2018**, *191*, 25–31. [CrossRef] [PubMed]
60. Vosburg, S.K. The effects of positive and negative mood on divergent-thinking performance. *Creat. Res. J.* **1998**, *11*, 165–172. [CrossRef]
61. Mélendez, J.C.; Alfonso-Benlliure, V.; Mayordomo, T. Idle minds are the devil's tools? Coping, depressed mood and divergent thinking in older adults. *Aging Ment. Health* **2018**, *22*, 1606–1613. [CrossRef] [PubMed]
62. Rodrigue, A.L.; Perkins, D.R. Divergent thinking abilities across the schizophrenic spectrum and other psychological correlates. *Creat. Res. J.* **2012**, *24*, 163–168. [CrossRef]
63. Russo, M.; Mahon, K.; Burdick, K.E. Measuring cognitive function in MDD: Emerging assessment tools. *Depress. Anxiety* **2015**, *32*, 262–269. [CrossRef] [PubMed]
64. Kaser, M.; Zaman, R.; Sahakian, B.J. Cognition as a treatment target in depression. *Psychol. Med.* **2017**, *47*, 987–989. [CrossRef]
65. Zar, J.H. *Biostatistical Analysis*, 5th ed.; Prentice-Hall/Pearson: Upper Saddle River, NJ, USA, 2010.

MDPI
St. Alban-Anlage 66
4052 Basel
Switzerland
Tel. +41 61 683 77 34
Fax +41 61 302 89 18
www.mdpi.com

Brain Sciences Editorial Office
E-mail: brainsci@mdpi.com
www.mdpi.com/journal/brainsci